AF352486

MAESTRO OF SCIENCE

Omond McKillop Solandt and Government
Science in War and Hostile Peace, 1939–1956

Maestro of Science

Omond McKillop Solandt and Government Science in War and Hostile Peace, 1939–1956

JASON SEAN RIDLER

UNIVERSITY OF TORONTO PRESS
Toronto Buffalo London

© University of Toronto Press 2015
Toronto Buffalo London
www. utppublishing. com

ISBN 978-1-4426-4747-3

∞ Printed on acid-free paper

Library and Archives Canada Cataloguing in Publication

Ridler, Jason Sean, 1975–, author
Maestro of science : Omond McKillop Solandt and government science in war and hostile peace, 1939–1956 / Jason Sean Ridler.

Includes bibliographical references and index.
ISBN 978-1-4426-4747-3 (bound)

1. Solandt, O. M., 1909–. 2. Scientists – Canada – Biography. 3. Military research – Canada. 4. Science and state – Canada. 5. Scientists in government –
Canada. I. Title.

| Q143.S64R54 2015 | 509.2 | C2015-904403-0 |

All photographs in the illustration section are courtesy of the Thomas Fisher Rare Book Library, University of Toronto.

This book has been published with the help of a grant from the Canadian Federation for the Humanities and Social Sciences, through the Awards to Scholarly Publications Program, using funds provided by the Social Sciences and Humanities Research Council of Canada.

University of Toronto Press acknowledges the financial assistance to its publishing program of the Canada Council for the Arts and the Ontario Arts Council, an agency of the Government of Ontario.

Canada Council
for the Arts

Conseil des Arts
du Canada

Funded by the   Financé par le
Government   gouvernement
of Canada   du Canada

Canadä

To Omond McKillop Solandt (1909–1993)

"To study history means submitting to chaos and nevertheless retaining faith in order and meaning."

Herman Hesse, The Glass Bead Game

Contents

Illustrations follow page 112

Tables

Acknowledgments

While the historian's main tasks of research, reading, and writing are solitary affairs, no work of history is an author's alone. I have been grateful for the support of a diverse array of people in the creation of this work. I owe a tremendous debt to David Grenville. His years of research on Omond Solandt's life, including the many interviews he conducted with Solandt and his contemporaries, as well as his own insight into the man's accomplishments, were invaluable. For my own interviews and private papers, I thank Paul Hellyer, Cecil Law, George Lindsey, Archie Pennie, Michael Bliss, Robert Bothwell, Terry Copp, Don Phillipson, Kitty and George Fells, Kathy Solandt, and Vaire Solandt. Each provided perspectives on Solandt and his professional and personal life that enriched my understanding of the man and his work.

Outside of my own income, funding for my research and writing came solely from the following sources: the War Studies program, the History Department, and individual professors of the Royal Military College of Canada; a grant from the Canadian Operational Research Society; and my family. Specific thanks to Michael Hennessy, Brian McKercher, Doug Delaney, Jim Kenny, Sean Maloney, Bill Simms, Doug Paton, and James and Julie Ridler, for their support.

My research in Canada was greatly facilitated by Harold Averill, chief archivist for the Solandt fonds at University of Toronto Archive, and Steve Harris and his staff at the Directorate of History and Heritage, Ottawa. Thanks also to the staffs at the Queen's University Archive, Kingston, and the National Archive of Canada, Ottawa (now Library and Archives Canada), for their friendly assistance. In the United Kingdom, thanks to the staff of the National Archives, Richmond; the Imperial

War Museum, London; the University of East Anglia, Norwich; and the Wellcome Trust, London.

Special thanks to Sean Maloney for introducing me to Solandt's legacy and sharing his enthusiasm for my work, and to the late Lieutenant-Colonel (retired) John Marteinson for his support on this project during a difficult year. Personal thanks to my parents for their unfailing support through five hard years, and to Erin Hoffman, whose support at that time was invaluable. Finally, I thank my mentor Michael A. Hennessy, whose sage guidance, patience, and support not only channelled my own "aggressive enthusiasm" toward a successful end, but also proved invaluable to my intellectual development.

Abbreviations

1-CAORT	Canadian Army Operational Research Team
AACRG	Anti-Aircraft Command Research Group
AC	Army Council
ADC	Air Defence Command
ADRDC	Air Defence Research and Development Command
ADRDE	Air Defence Research and Development
AEC	Atomic Energy Canada
AECB	Atomic Energy Control Board
AFV	armoured fighting vehicle
AORG	Army Operational Research Group
AWRE	Atomic Weapons Research Establishment
BCATP	British Commonwealth Air Training Plan
BMJ	British Mission to Japan
BSP	Basic Security Plan
BW	Biological Weapons/Warfare
CACDS	Commonwealth Advisory Committee on Defence Science
CARDE	Canadian Armament Research and Development Establishment
CAS	chief of the Air Staff
CCOSC	chairman, Chiefs of Staff Committee
CDC	Cabinet Defence Committee
CDRB	chairman, Defence Research Board
CGS	chief of the General Staff
CN	Canadian National
CNS	chief of the Naval Staff
CO	carbon monoxide
COSC	Chiefs of Staff Committee

CPC	Combined Policy Committee
CPRSD	controller of Physical Research and Scientific Development
CW	chemical weapons/warfare
DATAR	Digital Automated Tracking and Resolving Computer System
DDP	Department of Defence Production
DDR	deputy director of research
DDSc	deputy director of science
DEW	Distant Early Warning Line
DND	Department of National Defence
DRB	Defence Research Board
DRCL	Defence Research Chemical Laboratory
DRKL	Defence Research Kingston Laboratory
DRML	Defence Research Medical Laboratory
DRNL	Defence Research Northern Laboratory
DRPC	Defence Research Policy Committee
DSA	deputy scientific advisor
DSDW	director of Staff Duty Weapons
DSI	Directorate of Scientific Intelligence
DSIR	Directorate of Scientific and Industrial Research
DTD	Directorate of Tank Design
DTRE	Defence Research Telecommunications Establishment
GEC	General Electric Company
GIES	Grosse Il Experimental Station
GSP	Global Strategy Paper
ICBM	Intercontinental Ballistic Missile
IDB	Industrial Defence Board
JIB	Joint Intelligence Bureau
JIC	Joint Intelligence Committee
MCC	Military Cooperation Committee
MIC	military industrial complex
MIT	Massachusetts Institute of Technology
MoS	Ministry of Supply
MPF	Manitoba Patriotic Fund
MPRC	Military Personnel Research Committee
MRC	Medical Research Council
MSF	Mobile Striking Force
NAE	National Aeronautical Establishment
NATO	North Atlantic Treaty Organization
NDRC	National Defence Research Council

NRC	National Research Council
NRE	Naval Research Establishment
OBE	Order of the British Empire
ONR	Office of Naval Research
OR	operational research
ORG	Operational Research Group
ORS	Operational Research Section
OSRD	Office of Scientific Research and Development
PEADQ	Panel on Economic Aspects of Defence Questions
PJBD	Permanent Joint Board on Defence
PNL	Pacific Naval Laboratory
QF	Quick Fire
RAF	Royal Air Force
RAMC	Royal Army Medical Corps
RCAMC	Royal Canadian Army Medical Corps
RDB	Research and Development Board
RW	radiological weapons/warfare
SA	scientific advisor
SA/AC	scientific advisor to the Army Council
SAC	Strategic Air Command
SACEUR	Supreme Allied Commander Europe
SACLANT	Supreme Allied Commander Atlantic
SAGE	Semi-Automatic Ground Environment
SEAC	South East Asia Command
SES	Suffield Experimental Station
TRE	Telecommunications Research Establishment
U of T	University of Toronto
USSBS	United States Strategic Bombing Survey
UTE	user trial establishments
WO	War Office

MAESTRO OF SCIENCE

Omond McKillop Solandt and Government
Science in War and Hostile Peace, 1939–1956

1 Introduction

He pioneered a new form of applied science most associated with physicists, but his background was in physiology. He helped redesign tanks for modern combat, but he'd never been a soldier or a professional engineer. He ran a British Army science organization during the Second World War, but was Canadian. Despite a distaste for war, he became an early expert on atomic warfare when he walked the battleground of Hiroshima and Nagasaki. And, for nearly a decade, he became a national "science tsar" during the tense days of the early Cold War, and was the only Canadian chief of staff to have visited Stalin's Russia.

If Dr Omond Solandt was anything, he was a polymath, one who thrived in challenging environments. Solandt was a witness, contributor, and leader within the changing military, scientific, and political landscape of Britain and Canada during the Second World War and Cold War. His career in government science was a direct result of Britain's wartime need for scientific capital and labour. It paralleled the growth of importance of state science during the war, but achieved its zenith during the early Cold War. Born in Winnipeg in 1909, Omond Solandt was one of the brightest young minds of his generation in Canada. The son of a respected Congregationalist minister father and university-educated mother, he excelled in his education and received top honours in medical studies at the University of Toronto under the mentorship of Dr Charles Best, co-discoverer of insulin. After completing his training in Britain at Cambridge under Sir Alan Drury, Solandt had prepared for a career in clinical research and teaching when the Second World War began.

In war, Solandt found his unique abilities in wide demand across a spectrum of tasks. He led the South-West London Blood Depot during the Battle of Britain, managing blood transfusion needs and research as part of Britain's civil defence. These successes caught the attention of the Medical Research Council (MRC) Secretary Sir Edward Mellanby, who selected Solandt to create and lead a research team and laboratory on tank problems that became known as the Physiological Laboratory at the Armoured Fighting Vehicle (AFV) School at Lulworth, Dorset, in late 1941. Solandt's leadership and pioneering work in operational research (OR) at Lulworth came to the notice of the Sir Charles Ellis, scientific advisor for the Army Council (SA/AC) in the War Office. Ellis selected Solandt to join the expanding Army Operational Research Group (AORG) and continue his tank research. However, by 1944, Solandt's administration and management skills had secured him the position of deputy and then superintendent of AORG. In this capacity, he abandoned his own research interests to work with senior government officials in the War Office, the Ministry of Supply, industry, and elsewhere, to support the AORG's work. Under his stewardship, AORG conducted research on diverse subjects such as lethality of weapons and Continental defence against German V1 and V2 rockets. In July 1945, Solandt was selected to be scientific advisor (SA) to Lord Louis Mountbatten's South East Asia Command (SEAC), but the dropping of the atomic bombs at Hiroshima and Nagasaki ended the war in the East. As a result of his reputation, ability, and medical training, Solandt was selected for the unique distinction of joining the British Mission to Japan (BMJ). The BMJ was sent by the War Office to investigate the effects of the atomic bombs on people and structures at Hiroshima and Nagasaki. He was the only Canadian to do so.

After the war, Ottawa found Solandt the only candidate suitable to lead the new Defence Research Board (DRB). For a decade, his leadership and convictions provided the organization with the strength to survive its birthing pains and become a contributor to Canada's national defence. While chairman, Solandt established the DRB as a significant research organization in such fields as OR, continental defence and atomic, biological, and chemical warfare. He earned a reputation as a tough and critical member of the Chiefs of Staff Committee (COSC), a confident speaker on the Cabinet Defence Committee (CDC), and respected scientific diplomat with Canada's American and British allies.

Themes and Contributions

The primary object here is to establish Solandt's unique contribution to government science as part of the growing body of literature on the history of state science in Britain and Canada in the twentieth century. The secondary objective is to use Solandt's career as a unique map of British and Canadian developments of defence and science policy during war and peace. Several essential questions are addressed. How and why did a Canadian physiologist, one with no interest in military affairs, rise to prominence in defence science circles beside contemporaries Vannevar Bush and Henry Tizard, among others? What historical insight is provided by examining Solandt's experiences amid the players and tension within government science? Indeed, what was the legacy of Omond Solandt?

The answers to these questions required traversing the historical fields of inquiry that Solandt's career touched, including medical, military, science, technology, industrial engineering, and national history in two countries. Such demands help explain why Solandt's life had yet to be given biographical treatment. Solandt's early development and family life are charted in chapter 2. Solandt found a keen supporter in Charles Best, who both mentored and championed Solandt and his education in Canada and Britain. This mentorship is the focus of chapter 3. Solandt's work at the Toronto General Hospital during the polio epidemic of 1937 demonstrated his ability to organize and maximize medical administration and research during a crisis, and is detailed in chapter 4.

From 1939 until 1945, Solandt maintained a steady trajectory of success in government science as he earned a reputation as a capable and reliable trouble-shooter and problem-solver. Chapter 5 examines his early management style and research efforts while director of the South West London Blood Depot. Chapter 6 establishes his pioneering work in OR at Lulworth and his early experiences in negotiating the relationships involved in government science. His conflict with Ministry of Supply on the MRC Sub Committee on Armoured Warfare introduced him to the frictions between science and government and the strategies required to negotiate between scientists, the civil service, industry, and the military. The chapter also explores how Solandt, along with others, became a champion of applying operational data from battlefields to government research, solidifying his place as a pioneer in operational research.

Solandt's rise from section leader, to deputy, and finally superintendent of AORG is examined in chapter 7. Solandt's tenure at AORG was one of expansion, diversification, and strain. His leadership style contrasted with that of his predecessor, South African physicist Brigadier

Basil Schonland. Both leadership styles are compared and analysed. The chapter also includes analysis of the conflict between Solandt and Schonland over the relationship between AORG and OR sections involved in the Normandy Campaign of 1944. Solandt's approach to administration and his view of the organizations' strengths and weakness is discussed, as is his working relationship with senior officials at the Ministry of Supply and War Office. Here, Solandt's view of Britain's application of science offers insight into the personalities, successes, and failings of Britain's approach to science in the Army. These insights also led, in part, to the creation of the DRB.

Solandt's experience conducting casualty research at Hiroshima and Nagasaki is the subject of chapter 8. His contribution to Britain's first post-war analysis of the realties of atomic warfare is charted, as is his own view of this deadly new weapon that would dominate his professional life for years to come.

Solandt's decade at the DRB is analysed through chapters 9 to 14. Chapter 9 covers the DRB's origin, Solandt's selection as director-general, and his relationships and confrontations with senior Canadian civil servants. Chapter 10 details Solandt's role as Canada's first scientific chief of staff, including the difficulties he faced and the successes he achieved in this unique role. Chapter 11 explores his efforts as chairman of the DRB with fellow military, industrial, and academic members of the board, with a focus on the McGill Fence project. Chapter 12 discusses Solandt's influence on major DRB projects undertaken during his chairmanship, including the Velvet Glove guided missile and chemical and biological warfare research. Chapter 13 establishes Solandt's reputation as a defence research diplomat with senior British and American allies. Chapter 14 chronicles Solandt's role as Canada's chief expert on atomic weapons and the DRB's role in providing guidance on the reality of nuclear warfare for the senior services (Army, Navy, and Air Force).

These chapters demonstrate that by 1956 Solandt had become a leader in government science management. This achievement was remarkable when you consider Solandt had never trained in managing government science nor ever worked for the NRC or any government science body before the Second World War. But Solandt had a natural gift for managing science, one based on his highly pragmatic intellect and built upon his limited experience managing medical and academic groups. These natural talents flourished within the opportunities generated by the Second World War and Cold War. Each management and leadership success built upon another as he learned from observation, experience, and his own utilitarian disposition how best to achieve scientific results at the behest of a government sponsor. His efforts were often met with opposition, from the military, other scientists, and powerful senior civil servants. Solandt did not always win in these confrontations, but he learned from every misstep so that his victories outnumbered his losses. Such an approach allowed him to rise from a researcher to a leader in government science.

Indeed, Solandt's story is a journey from aggressive researcher, to scientific manager, to seasoned government science leader and state-builder. While AORG was constructed as an ad hoc and temporary organization, Solandt and his cohorts intended the DRB to be an ongoing and thriving concern for the Canadian Armed Forces. At a time of massive demobilization and demilitarization, Solandt's accomplishments in creating, maintaining, and producing a first-rate defence research organization of over a thousand people across Canada and in Allied countries needs to be acknowledged. He was one of the first major post-war state-building civil servants, and, amazingly, did it in the difficult arena of military affairs while the rest of the senior services were being cut to the bone. The DRB was his crowning achievement in the realm of state science, and its destruction fifteen years after his retirement as chairman, one of his greatest heartaches (see Conclusion).

Limitations of Biography

While many histories have noted Solandt's achievements, none have delved deep into his education, wartime experience, or operation of the DRB. Even the official history of the DRB is quiet on the greater aspects of Solandt's leadership. None paid particular attention to the nature of Solandt's mind and dispositions as critical elements in his ascension in government science. None took stock of the full range of valuable contacts he had generated in science, industry, and elsewhere. It is for these reasons that a biographical form was chosen to explore Solandt and his arena of intellectual acumen and government science.

Biography has distinct limitations and capabilities as a historical form. The focus on one life provides necessary and useful boundaries for historical analysis and carries the risk of ascribing too many events to the actions of one person. A focus on Solandt's life means paying less attention to the origins of organizations he did not create but led, or the events that he participated in or witnessed. A degree of comprehensive focus is necessarily lost. In place of such attention, however, is a unique point of view, one that has not yet been braided into the historical dialogue on the Second World War or Cold War. This dialogue features friction between partners in defence research, case studies in learning by doing, and the initial and sustaining difficulties of governments attempting to harness science for direct ends. As such, Solandt's life provides new windows of clarity and perspective on the rise of British and Canadian government science, even if the picture seen through this window is, by necessity, limited.

Biographies also hold risks for analysis. If uncritical, they become hagiography. If too critical, they become a witch hunt. Acknowledgment of these two extremes is necessary and helpful. Michael Bliss often had to compare the famous Sir Frederick Banting, co-discover of insulin, to a "horse's ass, not always favourably." But in writing the biography of respected Canadian physician William Osler, Bliss found little cause to destroy the Osler legend.[1] I make the same attempt with the legacy of Omond Solandt. Solandt's successes were not made without resistance and friction from other colleagues, nor did he always succeed in his efforts. But his legacy does exist. Solandt's effect on government science policy was not absolute, but he had both a unique role and a unique influence. An appreciation of both facets, as well as their limitations, is integral to any understanding of how and why defence science policy emerged in Canada and, to a much lesser extent, Britain.

Primary Source Analysis

As the secondary literature on Solandt's life is weak, this work is based largely on print and oral primary sources, many declassified for the first time or uniquely available for my work. Both have enriched my analysis, and both have allowed aspects of Solandt's legacy to be written for the very first time.

In Britain, the MRC wartime collection at the National Archive, Kew, provided substantial documents on Solandt's work at Sutton and Lulworth, especially wartime research data and correspondence. Solandt's own research on CO poisoning in tanks was well represented. The minutes of the MRC Military Personnel Research Committee's Sub Committee on Armoured Fighting Vehicles were invaluable for assessing Solandt and his role in AFV policy. Solandt's often difficult relationship with senior MRC members is well documented, as is his early support and use of field teams for operational research. The War Office collection at the National Archive provided insight on Solandt's AORG days, particularly his relations with the War Office as superintendent. Solandt's Cold War relations with British science leaders through the Defence Research Policy Committee, the Commonwealth Advisory Committee on Defence Science, and the Ministry of Supply were essentially declassified and accessible. The Schonland papers at the Imperial War Museum, London, were essential for understanding the relationship between Schonland and Solandt. The Wellcome Trust Archive, London, and the Solly Zuckerman Archive, University of East Anglia, Cambridge, both held caches of supporting documents on Solandt's wartime research and relations.

In Canada, the highly classified nature of the DRB's work has generated more assumptions than arguments about its role and effectiveness. Recent changes have allowed these assumptions to be challenged. The Robert Lewis Raymont fonds were deposited at the Directorate of History and Heritage of the Department of National Defence in Ottawa in the 1990s. Raymont was executive staff officer to the Office of the Chairman Chief of Staff Committee (CCOSC), and his personal collection of documents included the private papers of Foulkes, complete collections of the COSC minutes, and other papers. In this work, Solandt's conduct as a chief of staff has been assessed largely through the examination of his role on the COSC from the Raymont collection. It is acknowledged that this important committee was not the end of discussion for national security policy in Canada, but a focal point for strategic discussion before major policy decisions were made by the minister of defence and the Cabinet Defence Committee (CDC). Solandt's role on the CDC, while not nearly as extensive as on the COSC, is also examined.

The declassification of many DRB documents at the National Archive of Canada (NAC) has been the second recent change. The minutes of all the DRB meetings from 1946 to 1954 are now available, though the whereabouts of the minutes from 1955 onward is unknown.[2] NAC also recently made available Solandt's personal correspondence and public addresses as chairman of the DRB, as well as collections of news clippings of the DRB during the Solandt era. While clearly a fragmentary collection, they add a necessary spine to the story of Solandt's DRB days. The DRB minutes collection also included most of the annual reports from Solandt during his chairmanship, budget estimates, and the minutes of the Selection Committee on Extra-Mural Grants. All of these documents have allowed a significant part of the DRB's history to be told.

Bolstering these caches of documents are the Solandt fonds at the University of Toronto Archive. Solandt organized the collection near the time of his death, with the aid of archivist Harold Averill. This personal collection of papers, letters, reports, and other documents span Solandt's entire life. They include documents regarding Solandt's early life, training, and education, his work at Lulworth and AORG, and a complete collection of files relating to his work at Hiroshima and Nagasaki. The collection is less robust regarding the DRB, but the voluminous personal correspondences of that era make up for the deficiencies in the collection at NAC, allowing a richer perspective of Solandt's relationships and efforts as chairman. The Queen's University Archive provided data on his parents. The Ronnie Sheppard collection at the Laurier Centre for Strategic and Disarmament Studies in Waterloo, ON, also provided a scattershot of useful material on Solandt's AORG days. Given the lack of any indexing system, however, the collection was of only moderate assistance.

I also conducted interviews with some of Solandt's colleagues. Cecil Law, a Second World War veteran who worked with a series of the DRB establishments, provided enthusiastic insight into Solandt's leadership style and personal approach to science management. Brian Austin, Schonland's biographer, shared his views on Solandt and Schonland's relationship and provided valuable feedback on their work at AORG's origins. Paul Hellyer, former minister of national defence, recounted his own experience with Solandt and Chute as a child, as well as insight into the personality of minister of national defence Ralph Campney, whom Solandt considered partly responsible for his leaving the DRB.[3]

The David Grenville Interview Collection

The unique and insightful resource for this work came from interviews conducted by David Grenville. Grenville and Solandt had become colleagues during the two-year royal commission investigating the 1982 *Ocean Ranger* tragedy. From 1985 to 1990, they worked together to write Solandt's biography.[4] Grenville conducted twenty-three ninety-minute interviews with Solandt, and twenty-six interviews of varied duration with those most closely associated with Solandt's life. These included Laurie Chute, a lifelong friend who was critical to Solandt's efforts at Lulworth; Patrick and Maggie Mollison, who had worked with Solandt at the Blood Depot; H.A. "Tony" Sargeaunt, who worked alongside and then under Solandt at Lulworth and AORG; E.A. Treadwell, an AORG engineer; George Lindsey, who worked under Solandt at AORG and at the DRB; Alec Fordyce, wartime staff officer and DRB comptroller; and Archie Pennie, who held critical posts at DRB, including secretary of the board. Interviews were also conducted with a number of Solandt's intimates and family, each providing a unique take of the man and his efforts.[5]

Unfortunately, Grenville was unable to finish the project. Instead, he, along with Lindsey and Law, commemorated Solandt's life and achievement by holding a symposium in his honour at Queen's University in 1994. The symposium papers became the collected work, *Perspectives of Science: The Legacy of Omond Solandt*, published by Queen's Quarterly Press, and edited by Law, Lindsey, and Grenville.[6] Many of the chapters were relevant to my work. In December 2006, Grenville graciously released all of the taped interviews to me for the purpose of writing my dissertation (on which this book is based), on one condition. I was to write summations of each tape so that they could be deposited at the Solandt archive. This task was completed in June 2007.

The Grenville interview collection is a rare and valuable resource. On the tapes, Solandt discussed his work, life, and perspective on a range of topics including nuclear weapons, operational research, and running scientific establishments in war and peace. There were also countless anecdotes regarding all the major players he had worked with in both the Second World War and Cold War, and Solandt was free with his opinion on all of them.

All such interviews were used cautiously. Memory is the most convincing of liars. Biases, errors in fact, and patching of past experience into a whole are all risks when relying on memory. Still, long-term memory of dramatic, unique, or unusual events has an uncanny staying power. According to one source, recent research shows that memory of dramatic events is generally more accurate than previously believed.[7] Solandt's career was certainly unique, dramatic, and unusual, with bursts of danger and risk, and his memory surprisingly accurate for his mid-seventies, though not perfect.[8] With these limitations in mind, the interviews have two unique values. First, they are evidence of things that were not recorded in print sources, including the private opinions of Solandt himself on a wide range of issues, subjects, and people. This access to otherwise unknowable data has illuminated blind spots in war and post-war government science history, though solely from one source. Wherever possible, interview data were also substantiated by other primary or secondary sources.

The second value of the tapes is the picture of Solandt that emerges from the interviews. Understanding the man who rose to his particular apex in state science is the essence of this work. Solandt did not keep a personal diary, and while he wrote many letters, most that remain are professional, not personal. The interviews enlighten our understanding of Solandt's personality, and thus his approach to state science. Both these values, illuminating undocumented events and establishing Solandt's personality, coalesce in this work, because Solandt the man cannot be separated from the events he lived and shaped.

Who Was Omond Solandt?

Solandt's life held peculiar contradictions. It was filled with dramatic events, yet his presence, while intense, was quiet. He was supremely confident in his abilities and research, had a deep capacity for work, and did not suffer fools gladly. But he was rarely arrogant and enjoyed working with and understanding people from a wide variety of backgrounds, from shop engineers to Nobel laureates to field marshals. Some found him cold and overbearing, others, a knowledgeable conversationalist on just about any topic. He was an utterly pragmatic man that some found boring, and others fascinating. Few could deny that his formidable intellect was at the forefront of his successes.

While his IQ is not known, there is little doubt, from interviews with contemporaries, family, and friends, that Omond Solandt was a genius. Science and psychology historian Michael Shermer has argued that popular appreciations of genius have focused on mystical, mysterious, and qualitative abilities of rare individuals (Mozart, for example), but an equally informative and more accurate definition of genius revolves around the quantitatively greater use of abilities we all have. "Thus one way to define a genius: *an individual with a quantitative difference in ability so great that he or she appears to be qualitatively different.*"[9] It is clear from all the evidence, from documents to interviews with his peers, that Solandt was just such a quantitative genius, though of a particular type. Far from being a specialist, Solandt had a multifaceted intellect for science that made him a master generalist, a critical component for any manager of a multi-science organization.[10]

Solandt knew he was an unemotional man, and this often led people to assume he was uncaring or unfeeling, and both were untrue. He just kept these feelings muted while doing his best to serve his country.[11] Chute witnessed Solandt angry only once, and it was a terrifying, cold, and focused anger.[12] He had a sharp and often dark sense of humour common to medical professionals, and led to jokes that mortified some of his junior colleagues, always worried he might make another dark gaffe. But for Solandt, control over his emotions was both a natural predilection and a professional means of conduct.

As someone who worked with dying and diseased patients, Solandt understood the need to keep emotions in check in the face of human suffering. He often noted that intuition, based on emotion, led to poor judgment in science. You did your patient no favours by relying on subjective states of mind instead of rational, objective analysis.[13] Maggie Mollison, who had worked for Solandt at the Blood Depot, felt that he retreated deeper into unemotional pragmatism after walking through the atomic battlefields of Japan. Solandt refuted this assertion. The war only solidified his unemotional outlook. After recovering dead and maimed victims of German air raids, then witnessing the devastation of modern war in Europe, he arrived in Japan "numb" in the face of human suffering. While his colleague Jacob Bronowski recalled Hiroshima and Nagasaki with sublime horror, Solandt concluded, as did the report he helped write, that while terrible, the atomic bomb needed to be considered rationally. Even here, at the cusp of a new age of science and warfare, Solandt remained cool and objective. Service would come before emotion. Anything else would be unprofessional, even dangerous.[14]

Such detachment and rationality allowed Solandt to be a steady hand on the levers of government science, especially during the Cold War. War scares in Europe, the outbreak of the Korean War, and increasing threat of chemical, biological, and nuclear weapons produced a climate of desperate, often hysterical, tension. Solandt's staunch pragmatism acted as a salve against wasted efforts or fear-driven research. His emotions were under his control, not vice versa. In comparison, Vannevar Bush, wartime director of the American Office of Scientific Research and Development (ORSD), had a vocal and combative personality that was largely insulated from career fallout because of his close relationship with President F.D. Roosevelt. With Roosevelt's death, Bush would pay for the bridges he had burned to make ORSD succeed.[15]

Solandt lacked Bush's attitude, but not his confidence. He was never shy with his opinions, nor unaware of his outsider status in any of the professional circles in which he found himself working or leading (a Canadian in British scientific circles, a physiologist among the armoured corps, a civilian chief of staff). But he made precious few enemies and won a great many admirers. His wide interests and genuine enthusiasm for people provided him the capacity to work well with differing groups of people involved in scientific enterprises: engineers, industrialists, soldiers, civil servants, and scientists among them – all of which help explain his success as a government scientist.

Solandt's rationality was occasionally seen as the cover for a cold disposition, but that is a gross simplification. While driven to succeed, he also maintained diverse friendships and interests. Arctic geologist G.S. Hattersley-Smith, who worked for the DRB in the Solandt era, found Solandt an endlessly fascinating conversationalist, with deep knowledge on a vast range of subjects.[16] Like Nick Adams, the protagonist of Ernest Hemingway's "Big Two-Hearted River," Solandt retreated from the world of man and found solace in the world of nature to rekindle his spirits. After the war, Solandt often spent his summers traversing Canada's waterways with a group of canoeing enthusiasts known as the Voyageurs, led by the legendary Scandinavian canoeist Sigurd Olson. These trips were a time for laughter, bonding, and adventure, as the group made the papers for attempting to retrace the path of the voyageurs of old. Solandt was often in charge of making the libations for the group, and many fondly recalled his great and awful experiments at making flavoured liquors, including a revolting mix of vodka and chocolate that he swore would be sublime. For Solandt, this escape into the natural world, far away from the grim realties with which he dealt daily, were healing times. As fellow Voyageur Denis Collican noted in the diary for their 400-mile journey in the summer of 1957, "The exertion, exercise, often hard work, and wet shoes; the solitude under a canoe on a portage, the companionship of a partner in the canoe and the community of the campsite make these good days. Omond, quoting his Mother, described them as days that 'take the wrinkles out of the soul.'"[17]

Adversaries and Supporters

Solandt did have detractors. E.A. Carmichael, secretary of the MRC's Military Personnel Research Committee during the war, originally dismissed Solandt as an aggressive but unfocused researcher who had no ability for government science policy and no business trying to learn it.[18] British physicist Patrick Johnson was Solandt's rival to replace Schonland to become scientific advisor to 21st Army Group in 1945. Johnson, who got the job, thought Solandt was a colossal bore.[19] Some AORG employees believed Solandt never lived up to Schonland's high reputation, and Nobel laureate Neville Mott, a physicist who briefly worked under Solandt during the war, never held his intellect or personality in high regard.[20] C.J. Mackenzie and several Canadian air marshals on the COSC tried to outmanoeuvre Solandt and mitigate his influence, perhaps concluding that the new, young, civilian member of an untested defence organization would be easily ignored or manipulated.[21] According to former DRB comptroller Alec Fordyce, Ned Steacie, NRC president during most of Solandt's tenure at the DRB, dismissed Solandt as a scientist and claimed he was merely a "charlatan."[22]

As proven in the following chapters, Solandt confronted each of these detractors by either exceeding expectation or establishing himself as an operator, and not a pushover, to contend with in government science. But Solandt had more admirers than detractors. Charles Best felt he was perhaps his greatest student.[23] Clinical researcher W.S. Feldberg believed Solandt's research work was entirely reliable, and one could trust that none could have been done it better.[24] Maggie and Patrick Mollison recalled Solandt's paternal leadership style inspiring each to do is best.[25] H.A. "Tony" Sargeaunt, a colleague from Lulworth who would succeed Solandt at AORG as superintendent, always felt Solandt was at his best on political committees, aggressively arguing his case and proving his points, backed by his reputation as an excellent researcher.[26] At the DRB, Lindsey believed Solandt's great attribute was the ability to work with large groups of competing interests that seemed in conflict and bring them to a consensus and way forward.[27] Archie Pennie, secretary for the board during Solandt's tenure, confirmed how hard Solandt fought for the board's interests against the senior services, and how well he was respected among the civil service mandarins in Canada. For minister of national defence Brooke Claxton, Solandt was a critical ally in Canada's defence policy arena.[28]

The Legacy of Omond Solandt

Solandt's legacy in government science rests in two areas: research and leadership. From the start of the Second World War until 1944, Solandt became a pioneer in the new field of operational research. His ability to view men and machines as part of one complex system, and his emphasis on the need for operational data as the bedrock of analysis, earned him a deserved reputation as a leader within this new applied science. He plied these skills to become an early expert on the effects of atomic weapons, contributing to the earliest known British analysis of the deadly new weapons. As important as those contributions were, Solandt's greatest achievements were in leading government science institutions.

Solandt's leadership style was like his presence: quiet, yet intense and forceful. Demanding excellence of his staff was coupled by acknowledgment of excellence. From the Sutton Depot to the DRB, Solandt introduced his best administrators, researchers, and technicians to all the "big shots" in government, industry, or academia. Credit was always given to the deserved. He had an effective and paternal way of making his staff feel as if they had failed him and the team if they were not doing their best.[29] While he took a keen interest in the work of all his organizations, Solandt did not micro-manage research. He considered himself a maestro of an orchestra. He got the best players in the game, provided guidance, and left them to find the tune so long as they played well.[30] For men of the calibre of Lindsey, Law, and Pennie, such freedom and responsibility was a boon.

Solandt often reflected that a government science administrator required one attribute above all else: "push," British slang for the ability to get things done within the complex and resistant machinery of government. For Solandt, that meant dealing with the "operators" or "big shots," those whose influence in government was required as catalysts for action. Solandt apprenticed in this field through the MRC subcommittees and later worked through Robert Watson-Watt, Cockcroft, and Tizard to achieve AORG's aims. In Canada, C.D. Howe, Claxton, and in the military sphere Charles Foulkes, were among the most powerful operators. Indeed, Solandt salvaged what might have been an adversarial relationship with Mackenzie, one of Howe's trusted friends and advisors, to maintain access to Howe's influence. Mackenzie was his "key" to Howe's door.[31]

Among the greatest attributes he brought to science organization and leadership was his diverse intellect and interest, fuelled by nearly bottomless energy. Some AORG scientists felt that Schonland, a physicist, was always more concerned with the radar side of AORG, while Solandt took a greater interest in the variety of tasks being done by other sections.[32] This personal interest and stake in their work was carried over to the DRB. Solandt held board meetings at every major DRB establishment and included tours of the facilities to introduce the board members to the working-level crews. Cecil Law well remembered Solandt's many trips to the Defence Research Northern Laboratory (DRNL) at Fort Churchill, Manitoba. Solandt not only toured the areas but talked to engineers and scientists and soldiers about their problems and needs, often providing answers or the means to get them. For Law, this ability to traverse across sciences and provide valuable judgment and guidance to members of the shop floor as well as the highest echelons of academia and industry was not shared by Solandt's successors.[33]

While many found his self-confidence and assuredness overbearing (children of family friends often referred to him as "Dr Perfect" because "his way" was always "the best" way), Solandt was largely free of arrogance, a character trait he found distasteful, especially in people with important positions. For example, Canadian diplomat Vincent Massey was one of the "vainest and most pompous men" Solandt had ever encountered in his life.[34] For Solandt, the respect of peers was earned by service and example, not station or a sense of entitlement. Chute felt that while Omond was almost always the smartest man in the room, he rarely looked down on anyone. Likewise, Solandt was rarely intimidated by anyone or anything and conducted himself with confidence in the presence of Nobel laureates, powerful civil servants, and highly decorated soldiers.[35]

The professional focus of this work largely omits in-depth analysis of Solandt's personal life beyond his early years, and it does not chronicle the vast array of important and influential work Solandt did after 1956 until his retirement in 1985. This work is not intended to be the final discussion on Solandt's life and influence, but a constructive beginning. Solandt always felt his most valuable role was as chairman of the DRB, but there remain large caches of documents and room for valuable analysis examining his role in industry and defence as a senior operator at Canadian National, De Havilland, Hawker Siddley, and the MITRE Board. His contribution to Canada's application of science in government also continued past 1956. He was the first president of the Science Council of Canada, as well as chairman of the Ontario-government-sponsored commissions (the "Solandt Commission") which investigated the location of transmission facilities in the Toronto area and their impact on the environment, and an expert consultant on the royal commission investigating the *Ocean Ranger* disaster of 1982.[36] Solandt's final years as a consultant, on topics ranging from drug-rehabilitation research to agrarian reform in arid countries, were also of note, and his expertise in managing complex organizations was held in high regard by international peers (see Conclusion). It is hoped that this work can act as a commencement point for research on the second epoch of Solandt's legacy.

The personal life of Omond Solandt is not the focus of this work, though its import is acknowledged. Sadly, Solandt kept no personal diary. While many letters contained discussions of family, it was usually just general courtesy. He used to send a family newsletter at Christmas filled with how great things were at home, yet most of his friends knew there were serious problems.[37] Solandt also faced many personal tragedies that may have also fostered a retreat from an emotionally driven life. He lost his father to a heart attack when he was twenty-seven. His brother Don, whom he had idolized as a child, suffered from health problems his entire life and, finally, a degenerative neurological problem that finally killed him in 1955. His wife Elizabeth found motherhood difficult, and so her children, Sigrid, Andrew, and Katherine paid an awful price while Solandt was working long hours away from home. Solandt had simply followed the family model he was raised on. Fathers and husbands worked hard, away from their family. Mothers ruled the house and raised the children. It was unfortunate this arrangement did not work for his immediate family as it had for his own idyllic childhood. Later in life, Elizabeth suffered from mental health issues and ended her own life in 1971. Andrew also suffered from his own mental concerns. While Solandt and his colleagues made attempts to help him, Andrew took his own life six years after his mother's death.[38] These were terrible tragedies for any father, but especially for one who had trained in medicine. As for many men of his ability and station, family often paid a price for his achievements and responsibilities in the form of their father's absence. Recalling these hardships, Solandt noted that one of the few regrets he had was not being a better family man.[39] But, as he had for his whole life, he kept his storms of emotion to himself.

Despite all these resources, unfortunate but necessary limits remain on this work. As noted, a lack of relevant secondary sources literature mandated I do most of my research from primary sources. There were no comprehensive general histories of Canada's defence production efforts in the Cold War and its relation to the DRB, for example. Nor has there been a sufficiently robust account of Canada and defence research in the Arctic. Chapter 11's discussion of the McGill Fence radar system attempts to bring both together as a case study of his influence, though some will no doubt wish more depth in this and other areas. Others would enjoy more details on the information exchanged and the access provided via Solandt's diplomatic efforts with senior American and British scientists, much of which remains classified. Still more would like to see more in-depth reviews of Solandt's work by his international peers, in particular the United States and Britain, than have been provided here. Given the lack of secondary literature, however, this work is a necessary spearhead to many of these critical subjects in defence research history, and while some of these subjects may not have been given as robust a treatment as I would have liked, they are the first scholarly attempt to use hitherto untapped primary sources in Canada and the United Kingdom and ground them in the historical context of their day. This work is a beginning, not an end. More work remains to be done by future scholars, and I hope this assessment of Solandt's life and achievements can act as a lodestar to even more work in the field.

Solandt never thought of himself as a historical figure. He always wished his life story could be included as part of the more important story of the development of science in Canada.[40] The history of government science in Britain and Canada could not be told without the legacy of Omond Solandt.

2 Idyllic Childhood: The Early Years of Omond McKillop Solandt, 1909–1927

Omond Solandt's early home life nurtured and shaped his interests and abilities. It was a home that respected service, education, and faith. Here Solandt witnessed and would inherit his father's work ethic and sense of duty, as well as his ability to apply himself to a variety of trades. His mother provided bottomless support and encouragement. His elder brother provided leadership and comradeship as both boys engaged their interests in nature, invention, and cars during their childhood in Manitoba.

When the family moved to Toronto in 1920, Solandt flourished in academics and technical training. While still following his brother's lead in such interests as radio telegraphy, flight, and medicine, Omond Solandt was establishing himself as his own man with his own star to chase. His experiences in high school provided invaluable training in the mechanical arts that served him well throughout his professional life. He turned many of these skills into job opportunities, including summer work as a radio operator for the Ontario forestry division. By the time he began university, he had a reputation as a gifted student in theory and practice.

Family History

Omond McKillop Solandt was born on 2 September 1909 at his family's home on 566 Maryland Street, Winnipeg, Manitoba. He was the second and last son of Reverend Donald McKillop Solandt and Edith MacLennan Young. The couple's first son, Donald Young Solandt, known as Don, was born on 11 March 1907 in Ottawa near Edith's family, but the family returned to Winnipeg where Reverend Solandt had accepted the job of assistant minister at the Knox Presbyterian Church in 1905. The Solandt family would call Winnipeg home until 1920.[1] Like his future American colleague in defence science, Vannevar Bush, Omond Solandt was singled out with a unique first name that was often mispronounced and misspelled throughout his life.[2] Coupled with an exotic last name, Omond inherited a distinct family heritage of hard work, achievement, and service.

Donald McKillop Solandt was born 31 May 1871,[3] the youngest of eleven children who were raised in the small farming community of Inverness in Quebec's Eastern Townships. Inverness was founded in 1829 by a group of Scottish settlers from the Isle of Arran/Ranza and led by Archibald McKillop, a former tax collector for the Duke of Hamilton. McKillop's youngest daughter, Jane, married Reverend André Solandt, a *"colporteur évangéliste"* who had come to Quebec from Strausbourg, via Geneva, to bring the Protestant faith to the Catholic province. Donald Solandt was their last child.[4]

As the youngest, Donald had to make his own way in life. He was raised largely by his eldest sister, whose future husband carried the surname Omond, and for whom Donald's second son would be named.[5] After working on the family farm, he attended the Inverness Academy, where he excelled at languages.[6] He spent three summers travelling to Europe with other young men on a cattle boat to sell stereoscopic views of Canada. On one of these return journeys, he survived a terrible shipwreck at Belle Isle, off the coast of Newfoundland. This may have been the famous 1889 sinking of the HMS *Lilly*, en route to meet a mail-ship from Montreal, but Omond Solandt could not verify this family story.[7] At the age of twenty Donald won a teacher's certificate and headed out west to instruct for three years at Aylmer East Academy, Carnduff, Saskatchewan, 1892–1995, far away from the Atlantic coast.

Teaching did not satisfy him. In December 1895, at the age of twenty-four, he returned to student life at Queen's University in Kingston, Ontario. Upon registration, he boldly listed his intended profession as "undecided" and set out to explore his varied interest and skills, from history to science to theology.[8] In all of them he excelled, receiving prizes and awards for his academic work and literary skills. He received a bachelor of arts (1900), a bachelor of divinity (1902), and a master of arts in philosophy and political science (1905). No mere bookworm, Donald Solandt also triumphed in sports, winning the university's all-round championship in field sports in 1900. The Canadian economic historian Professor Adam Shortt, and philosopher and Nietzsche scholar John Watson, two of the university's most respected professors, supervised Solandt's graduate work. Shortt would maintain a correspondence and friendship with Solandt throughout his life. It was during his time at Queen's that he began his career in the Congregationalist church. Solandt served on field missions in Maynooth (1900) and New Liskeard (1901), Ontario, and was ordained a Congregationalist minister in 1902. While completing his graduate work, he served as pastor of the First Congregationalist Church of Kingston.

Soon after, he met his future wife.[9] Edith MacLennan Young was born in Kingston, Ontario. Her father James, a bookbinder, had come to Canada from Dundee, Scotland, in 1857 and married Kate MacLennan, whose family were United Empire Loyalists. James worked for a variety of publishing ventures, including the Queen's Printer in Toronto and Quebec City, as well as the newly created Stationery Office in Ottawa. Edith and her sister Margaret both graduated from Lisgar Collegiate in Ottawa and then pursued degrees at Queen's in 1905, no small feat for women at the time. Here Edith met Donald Solandt. She would not finish her degree, for on 5 September 1905 she and Donald were married alongside Margaret and one of Donald's friends in a double ceremony.[10]

In 1905 Donald Solandt accepted a job as assistant minister to the Knox Church in Winnipeg, Manitoba, leaving Kingston behind.[11] In Winnipeg, Solandt initially served as assistant to Reverend Doctor DuVal at Knox Church until the doctor's retirement in 1916. Here, the Solandt brothers Don and Omond would call home until 1920.

Winnipeg Years, 1909–1920

Reverend Solandt was a quiet example of manhood for Omond, and this example was mostly experienced by Solandt's absence from home. According to family lore, Reverend DuVal concerned himself only with the important city members of Knox Church, so Donald Solandt was sent into the wilds to maintain contact with parishes and do ancillary work. Omond recalled that his father was completely immersed in his duties during this time, often away until the evening. He was by far the busiest member of the church. This high capacity for work was an example both brothers learned. Men worked hard, even if it meant long absences from home.[12]

Donald Solandt was too old for military service when the First World War began, but he still contributed. In 1915, the Manitoba provincial government approached him to become the managing secretary of the Manitoba Patriotic Fund (MPF). His work was so intense that by 1916 he had left Knox Church. It is unclear if he was ever offered DuVal's job as minister.[13] The MPF began in August 1915 and sought to raise support for soldiers' dependants in the province and served as a recruitment device. By 31 August, executive and financial committees were established, and funds were being pledged for $249,518.25. According to Solandt's report on the fund's war efforts, Manitoba was the premier province to "organize for the care of its soldiers' dependents."[14] The fund became more critical as the war marched on, and by 1916 the fund was incorporated. After a successful subscription run in Winnipeg, the fund turned to the regions outside the capital. In April 1916 Solandt was "associated" with J.H. Bowles to collect monies throughout the province. Bowles had worked through provincial patriotic societies raising subscriptions. Solandt's social skills, connections with the parishes outside the city, and reputation provided by his profession and conduct made him a solid choice to help secure donations. According to Solandt's report, "During the summer of 1916 practically every municipal council in the province was approached for a grant in addition to the tax levy, and almost all responded. Thus, in addition to the tax levy, during 1916, a sum of $261,037.86 was secured from points outside Winnipeg."[15]

Solandt was rarely home during the war. There were always alarms and excursions when people died. According to Omond, undertakers were his father's chief nemesis, taking advantage of poor, grieving widows and surviving family members by insisting on thousand-dollar funerals for their lost loved ones, assuming that the Patriotic Fund would always foot the bill. Solandt fought these men with a vehemence his youngest son well remembered. At the Solandt home, undertakers were considered swindlers who took advantage of the grieving by sucking the fund dry.[16]

By war's end, the Patriotic Fund had distributed more than six million dollars to the dependant families whose men had served overseas. In his review and assessment of the fund, Solandt praised the executive, female, and labour members who contributed to the fund's successes. The report's publication coincided with Solandt's transition into industrial relations.[17]

Not long after the influenza epidemic of 1919, the family endured the unrest and strife of the Winnipeg General Strike of August 1919. Omond recalled watching from the window as a mob attacked and derailed a streetcar.[18] In its aftermath, the Manitoba government created the Council of Industry. Its first chairman was the Reverend Charles Gordon, the famous Canadian pulp novelist who wrote such national bestsellers as *Skypilot* under the penname Ralph Connors.[19] Gordon had a reputation for objectivity and experience in arbitration. According to his autobiography, the council's job was to bring together "the best men representing all classes interested in industrial affairs to work together and prevent future strife like that of August 1919." Gordon had "secured an assistant whose salary I became responsible [*sic*]." This nameless assistant was Donald Solandt, who was chosen as Gordon's deputy on 10 June 1920.[20] According to Gordon, disenfranchised and dispirited soldiers were the main cause of labour trouble.[21] Solandt's war work made him an appropriate deputy. The council existed for four years. According to Gordon, their efforts were an unmitigated success whose record was "unparalleled by that of any organization of its kind in the world." Of the 107 cases registered, the council had not once failed in its duty.[22] "Solandt disagreed. After becoming deputy, he wrote Prof. Shortt a letter full of uncertainty about himself, and the council's job. Solandt was not enthusiastic for the work, nor was he enthralled with Gordon, whose stinginess was legendary. Gordon never carried any money on him and thus borrowed from everyone. He also never paid them back. Solandt's office was often filled with citizens looking to collect on Gordon's borrowing."[23] After conversing with Shortt on labour relations and his dissatisfaction with the council,[24] Solandt reported in late 1920 that "the Presbyterian Church, through different Presbyteries, is suggesting that I should be appointed Business Manager of the Presbyterian Publications in Toronto, and if that appointment is favoured Mrs Solandt and myself would rather prefer living near the friends in the East."[25] The appointment was favoured and Solandt left the council after less than three months on the job. He took his family to Toronto in the fall of 1920 to begin his work at Presbyterian Publications on 1 November.[26]

Childhood and Development

Omond recalled his Winnipeg childhood as idyllic, and in it are the seeds of his later intellectual and professional development. These interests were fostered by both Omond's mother and brother. Reverend Solandt's influence on his sons was important but not dominant.

He taught them practical skills, the basics of carpentry and woodworking, and hunting, to allow them to build their own toys and be self-reliant.[27] But the greatest lesson he provided was by example. He was what a man should be for his family and community: strong, with a deep capacity for work, serving needs of his office and people under his charge, even if it kept him far from home. He also showed that a man with a good mind could succeed at almost anything, if he was willing to work hard. Omond's ability to thrive in fields initially foreign to him, especially in adverse circumstances, paralleled the unconventional career and experiences of his father.

While most of the documentary evidence on the Solandt family revolves around Reverend Solandt, it is clear that Edith Solandt was the greatest parental influence on young Omond.[28] The home was her domain and the boys were her responsibility. She likely noticed very early on that she had two very gifted children. She made it her priority to see that their intellects and interests were nursed and fed so they could come to fruition, as they did. She showed a tireless effort in raising her two sons to be strong, smart, and able, encouraging them and providing the means for them to excel. Omond always recalled his mother as a practical person who supported their interests and often participated in them, making their childhood both engaging and character-building. She even helped Don construct a motorcycle from scrounged parts. More than anyone else in Omond's childhood, Edith provided the encouragement for his natural abilities to grow.[29]

Omond admitted that Don, his elder brother, was the leader of the two in those days. Don was very bright and able from an early age. According to his first wife, Edith was harder on him than Omond. He was the first to be disciplined and learn to be obedient in Edith's home, with no one to show him the ropes. Omond, though, could follow Don. This generated a small level of resentment, but Don nonetheless provided Omond an exceptional guide in various interests.[30]

One of Omond's earliest memories was of helping Don construct radios out of coils and old Quaker cereal boxes.[31] During the First World War, Don went to Kelvin High School, where their mechanical shop allowed him to make parts. By the time he was ten, Omond helped with simple tasks as Don made crystal receivers and then transmitters, scrounging spark plugs from a neighbour's Model T and returning them when they were caught. During the war they created a transmitter at low cost and low range. It was noisy, and hitting the key caused the houselights in the neighbourhood to dim. This made it hard for the

Solandt boys to hide the fact that they were playing with a transmitter: illegal to operate because of the war. Don stopped using it, but both boys would take their interest in machines to Toronto.[32] Don was also fascinated by vehicles of all stripes, and Omond followed suit. Cars, trains, planes, and boats were lifelong interests.[33]

Omond witnessed the value of teamwork with the family trips to their lot in Minaki, 115 miles from Winnipeg, where his father had built a cabin at Pistol Lake. After taking the train, the family hiked into the bush to arrive at the cottage.[34] In the early summer of 1920, the last in Manitoba, their father had bought an old cabin cruiser that had yet to be finished. All of the family worked to get the boat up and running, fixing it every day, leading to a confrontation with Omond's school's truant officer.[35] Donald Solandt was a good carpenter and worked on the boat's body. Young Don worked on the electrical system that used dry batteries, then the ignition and engine. Omond helped his mother in making cushions and painting. The family efforts to bring that boat to life stayed with Omond for seventy years.[36]

Under his mother's guidance, Omond, like Don, excelled at school and youthful activities. He recalled not only counting the days to summer vacation, but counting the days until school started, a great indicator of his joy in both freedom and education. His Winnipeg childhood was filled with winter sports, the stink of felt boots, sleighs, and snowball fights on the way to and from school. The family also took yearly trips, including ones to a rented cottage on Lake Melissa in North Dakota before they bought the lot on Pistol Lake. Omond even survived a severe storm when an iron gate, flung from its hinges, smacked him in the head. One of the best Canadian minds of a generation was thankfully not damaged by the freak storm.[37]

The family was, by Omond's account, poor, but they never knew it because they made the best of it. Yet, throughout his life he would often recall moments of fiscal want as a barrier to his interests, ranging from learning to fly, performing experiments with expensive physiological laboratory equipment, or fighting for government funds against C.D. Howe. The home was warmed by a wood furnace instead of coal, to save money, and the boys stacked wood and kept the stove warm for the house during those cold Manitoba mornings. They had no allowance. Instead, to teach them fiscal responsibility, their parents left money in a book in the library from which the boys could take. They would put a note in the book saying the purpose for the funds and how much they took, and it was just assumed they would know the family's needs and not take too much. By the time the boys were in their teens, they were working summers and after school to support their interests. The family struggled on a churchman's salary, but, through careful management, had enough for summer trips and a plot of land for a cottage.[38]

Reverend Solandt maintained a moderately religious household. Grace was said with every meal, and Don and Omond went to Sunday school and church events. But they were not very religious. Omond attributed this to his mother. Edith Solandt's observations of church leaders, the ones who kept her husband away from home, confirmed in her mind that one did not require the formalities of church to be a devout Christian. It was an attitude he would share.[39]

Finding His Own Path in Toronto

Reverend Solandt accepted the job at Presbyterian Publications in Toronto and would work in Canadian publishing for the rest of his life. The family lived at 297 Glen Road while the Solandt brothers pursued their public school education. Despite financial difficulties, the family lived in the affluent Toronto neighbourhood of Rosedale, with Reverend Solandt making approximately $3500 a year.[40] For the next seven years, Omond would attend Rosedale Public School (1920–2), Central Technical School (1922–6), and Jarvis Collegiate Institute (1926–7).[41]

Rosedale held few memorable moments for Solandt. He was keen to follow his brother Don to high school at Central Tech,[42] a "splendid" forward-thinking school that provided basic training in any scientific or technical field, including aeronautical engineering. When Omond worked for De Havilland in the 1960s, almost all the factory supervisors were Central Tech graduates.[43] Many of his interests in sports, music, and the arts were rooted in the Central Tech experience.[44] He worked longer hours by taking both matriculation courses as well as vocational subjects. These included architecture and mechanical drawing, woodworking, steel-turning, blacksmithing, electrical equipment, machinery operation and repair, and photography. Only boiler maintenance gave him trouble.[45]

At Central Tech, Solandt met future lifelong friend and colleague Laurie Chute. They had a lot in common. Chute's parents were missionaries and believed their son needed to be good with his hands as well his mind. The school was highly competitive, Chute recalled. But no one could touch Omond. Every year, he got top grades. Eventually, everyone stopped trying to beat him and competed among themselves.

Despite these achievements, Chute never saw Solandt as condescend-
ing to any students. He was too busy enjoying himself in his activities
with his friends. Never a gregarious personality, he was quiet but active
in group activities, more known for observing and listening than making
a racket – even in sports.[46]

Indeed, it was in physical activity that Omond Solandt decided to
differentiate himself from his brother. Don was a brilliant student with
wide interests, but he suffered from duodenal ulcers as well as other
health problems. He had no interest in athletics but enjoyed and had a
flair for writing and literature, later composing his own general-health
textbook for young adults.[47] As a teen, Omond was over six-foot tall
and almost two hundred pounds, with a stocky build that resembled a
wrestler. Solandt made football his sport, wading through opponents
with size and muscle and little skill.[48]

Solandt's Central Tech years were not confined to studies. He was
extremely social, even if quiet, by participating in athletics, outdoor
activities, attending musical performances, editing the school newspaper, the *Vulcan*, and maximizing his time in all of them. But if he
wanted the option of taking a medical degree at university, he would
require honours Latin, and that was not offered at Central Tech. So in
his last year he attended Jarvis Collegiate. Later in life, when asked if he
were a medical man or an engineer, he would think back to his Central
Tech days and say he was medical but with a strong practical experience in engineering. These early skills and interest in the mechanical
and engineering arts provided him the vocabulary, knowledge, and
practical know-how for working with machines during the Second
World War. He could "talk the talk" with engineers as much as doctors
or scientists.[49]

Jarvis Collegiate was an initial disappointment. Don was now in university and Chute was at Harbord Collegiate, so Omond was without
his friends. He was initially placed in a fifth-year class with all the culls:
students who had failed the previous year, or were Jewish. Solandt felt
the Jewish students had been unfairly misplaced, since many were brilliant. Instead of suffering through these conditions, Solandt contacted
Principal Warren at Central Tech. to see if he could change this situation.
Warren spoke to the administration at Jarvis, and Omond was moved to
a better class. Soon, Solandt distinguished himself with the best grades
of the year. Clearly, the young man did not lack confidence.[50]

Solandt filled his final year as he had the rest, with football, skat-
ing, dances, skiing, and dating girls for each event and learning from

them to do the activities that were new to him. Fun and learning were synonymous throughout his life.[51] But as university loomed, his varied interests focused on three general areas: science and technology, medicine, and nature. The Solandt brothers' interest in cars continued, as Don and Omond worked with friends John and Art McFarlane fixing old cars.[52]

Don had taken courses in radio telegraphy and flying in high school, and used these skills to get employment up north working on lake boats. Again, Omond followed Don's lead. On 23 April 1927 he earned a certificate of "Proficiency in Radiotelegraphy," Second Class, granted from the Ministry of Marine and Fisheries. Solandt was tested on a ½ Marconi Cabinet Ship Station and the Standard Marconi Emergency Aerial Equipment and examined on adjustment, operation, and care of the apparatus, on the transmission and sound reading at a speed of not fewer than twelve words per minute, and knowledge of the requisite regulations on the equipment and London Convention of 1912 regarding radio telegraphy. He also made a legal declaration that he would "preserve the secrecy of correspondences." This was his youthful initiation into top secret work.[53]

The Marconi School was unpleasant. Solandt studied there three nights a week, from October 1926 into the early spring of 1927. The curriculum was designed to train young men on old rotary spark gap Marconi equipment used solely on the lake boats in Ontario. The Marconi Company supplied them to the ships and trained the students at their school so that they graduated at boating season in spring. Solandt mastered the academic and theoretical side but struggled to receive his second-class operator licence. Never a fast typist, the required twelve words per minute was a challenge.[54]

In the summers of 1928 and 1929 Solandt was employed in the north of Ontario for the Forestry Branch. He worked with early vacuum tube radios, models that were built in the branch's shops. They often broke down, and their inventor would scream at Solandt because Solandt constantly called for help. So Solandt attempted to "diagnose" the problem by reviewing the "symptoms" he observed of the technology's failure, and solved the problem through experimentation. All the years working on cars and his Central Tech education proved valuable during his three summers as a radio operator in the north. Like in the intolerable situation at Jarvis, Solandt was never keen to endure problems or situations that could be fixed by applied knowledge, especially if he could get the answers himself.[55]

Before university, he also received two certificates from the Royal Life Saving Society after passing elementary tests in swimming and life-saving techniques.[56] He would have been tested in rescue and release drills, the Schafer method of resuscitation, and the promotion of warmth and circulation, performing the four methods of the rescue and three of release in water. He probably followed his elder brother into this interest as well, though both Solandt brothers would suffer from sea sickness throughout their career, sticking to work on land or in the air whenever possible.[57]

Conclusion

Solandt's early years shaped his future outlook on science and problem solving and provided a nurturing environment for his intellectual and personal development. His father's absence was as instructive as his mother's support. Good men like Reverend Solandt provided for their family by hard work and service and solved problems on their own. Omond had great respect for his father and his abilities, and their professional lives had many parallels. His mother's encouragement and constant eye on his development made for a childhood rich in challenges and achievement. Solandt would thrive on both. But it was Don who led him into his interests with science, tinkering, and nature. When the family moved to Toronto, he still followed Don's lead, but soon developed his own personality, interests, and track record for excellence.

Solandt entered university as a superb student whose interests were as varied as his mind was formidable. He exhibited a quiet but aggressive enthusiasm for learning, but did not let his abilities ruin his social life through conceit. Nor did he suffer foolish situations that could be changed by his own efforts. At the University of Toronto, Solandt honed these skills and attitudes into a distinguished academic career, one that laid the groundwork for his future success in government science.

3 Protégé:
Omond Solandt and Charles Best in Toronto and Leningrad, 1927–1935

Omond Solandt's momentum for academic achievement continued at university. For nearly a decade, he thrived in the intellectual environment of the University of Toronto (U of T). He made money for school using his technical skills, first as a radio operator for the Department of Forestry, and later as a flying observer for the Ontario Air Service. It was at U of T that Solandt became the chief protégé of one of Canada's most famous scientists, Doctor Charles Best, co-discover of insulin. Best replaced Don Solandt as the chief influence on Omond.

Solandt inherited Best's research methodology and made it his own, as demonstrated through his academic articles of the early 1930s. Solandt was also well served by Best's numerous contacts in the medical and scientific elite in Britain and Canada. Despite a nearly fatal attack of bulbar polio, Solandt finished his master's thesis under Best's tutelage in 1932 before preparing for further training in Britain. Best thought high enough of the young Solandt to take him to Leningrad and Moscow in 1935 for the Fifteenth Annual Physiological Conference. Here, Solandt would experience a nation that made state science a pillar of its society and would witness the great distance between reality and illusion in Soviet society.

The University of Toronto

After considering geology, Solandt followed his brother's path and began his bachelor of arts in the biological and medical science at U of T in the fall of 1927, reuniting with Chute. Both took an advanced medical science program that gave them solid grounding in physics, chemistry, measurement, and other subjects.[1] Solandt made the most out of his undergraduate years. His interests outside of studies remained legion: writing and editing for the student *Medical Journal*, membership in the small but active Science Society, and a full social life with university dances, since there were more girls in this honours science program than in others. But he could not do everything he wanted. "The P and M course [Physiology and Medicine] was so arduous that none of [us] took a very active part in the committees," Solandt recalled, "but we did participate in many of the activities."[2]

Swimming and squash were added to his love of football, which he continued to play. Unlike that of his father, the star university athlete, Solandt's gridiron prowess started to fade as more students began to match his size and strength. He was demoted from senior intercollegiate to the second team in his final year, nicknamed the "orphans" (they played in the Ontario Rugby Football Union).[3] Despite this setback, he enjoyed playing the game and meeting the different characters on the team. Here, Solandt was coached by his future civil service colleague, Lester "Mike" Pearson, whom he had already met through Pearson's cousin. "I saw a fair amount of Mike Pearson during the war and was fairly closely associated with him and his various jobs in Ottawa during the 10 [*sic*] years that I was Chairman of the Defence Research Board."[4] Solandt, who would spend the Second World War working for the British Army, would not be a complete unknown to the personal networks of power built by the Canadian civil service mandarins during the Second World War.[5]

Radio Forestry and the Ontario Provincial Air Service

As mentioned, Solandt worked as a radio operator for the Ontario Forestry Branch, 1928–9. In 1930, he became an observer for the Ontario Provincial Air Service.[6] Flying fascinated him and remained a lifelong interest both personally and professionally. He had enjoyed watching airplane flights at Leaside Aerodrome in Toronto in 1921 and took his first flight in a "1023" at the Potomac River at 14th Street Bridge, Washington, DC. He began flying training at de Lesseps Field, Toronto, in late 1927, the year of Charles Lindbergh's epic cross-Atlantic journey in the *Spirit of St Louis* on 21 May. Cultural historian Jonathan Vance argues that Lindbergh's achievement shifted Canadian perceptions of pilots as reckless, Great War daredevils pursuing hedonistic thrills. "No longer was flying a pointless pursuit, merely cheap entertainment.

Now it took its place with the great epics of the age, its practical value demonstrated."[7] One can imagine this achievement of wonder and practicality appealing to the inquisitive and pragmatic Solandt.

In 1930, Solandt became an observer for the Ontario Provincial Air Service (OPAS) at Caribou Lake near Armstrong, Ontario. He flew with OPAS pilot Harry McCoy in a Cirrus Moth, an HS2L Curtis Flying Boat (former submarine-hunting aircrafts in the U.S. Navy),[8] a Fokker Universal, a Super Fokker, and others.[9] Observers flew with the pilots, their main job to patrol for and sketch fires (there was no general air photography yet). With the exception of sketching, these jobs were easily done by the Moth pilots. Solandt took every possibility to fly, but the lulls in action, passive nature of the work, and lack of independence ruined his interest.[10] He never flew solo. The lessons cost thirty dollars an hour, and Solandt, no matter how hard he worked during those difficult years after the Depression, could never afford the final fee to achieve his pilot's licence.[11] Fiscal limitation on his possible achievements was a theme that would stay with Solandt his entire life.

Solandt and Charles Best

But the focus of his efforts was towards his degree, for which he received top honours. Some teachers were more interested in their research than working with students, but there were also legends "able to convey to the students the spirit of inquiry that motivated them."[12] Professor Lash Miller taught a year's worth of chemistry using copper sulphate (blue stone) as a model, to good effect.[13] British-born physicist John Satterley's "Liquid Air" lectures drew big crowds, as he flash-froze goldfish and shattered them with a hammer. His very brief course on the theory of measurement and his introduction to calculus were among the best math courses Solandt took and served him well in his operational research days ahead.[14] Professor Lawrence Irving, a young American general physiologist whose main interest was diving mammals, was also a favourite. In addition to Best's influence, Solandt attributed his future skills in operational research as deriving from Irving's training. Don, Omond, and Chute would later all become involved with Irving's research at Woods Hole and at the Marine Biological Laboratory at St Andrew's, New Brunswick. Solandt also fondly recalled an eccentric biology professor, name unknown, who lived a simple life in the country, studied wildlife preservation, and had an uncanny knowledge of birds and the animals of southern Ontario.[15]

U of T's Physiology Department still basked in the glow of the discovery of insulin by Frederick Banting and Charles Best in 1922. Most of Solandt's course work in physiology was taught by Best's colleague Norman Taylor. Taylor was thorough, as he and Best were in the process of finishing their famous textbook *The Human Body and Its Functions*.[16] Solandt and his class would be the guinea pigs for its application.[17] Occasionally, the famous Banting attended meetings of the Physiological Journal Club, where it seemed he and Best were making an effort to get along. They never appeared to be great friends. Best's pride was always wounded for not receiving the Nobel Prize with Banting. Banting, a complicated and troubled man, maintained an often dark temperament.[18] Solandt never regarded Banting as a good scientist, in part because the quality of the students he trained was sub-par (see chapter 4). During the war, when Banting briefly stayed with Solandt during a trip to Britain, Solandt's opinion of the man improved.[19]

Solandt graduated with a B.A. in biological and medical science from Victoria College with Gold Medal standing in June 1931. Long before, he had decided to follow his brother into research in physiology in September 1931. "My intention was of course to stay out only a year and to go in medicine in 1932. Laurie Chute had made the same decision so we both started work in the Physiological lab in the summer of 1931." The work was done under the supervision of Charles Best.[20] Best had begun mentoring Don Solandt and now began with Chute and Solandt in their final year. For Solandt, Best would be a supreme mentor. Best's career was skyrocketing when the Solandt brothers arrived at U of T. In 1922, when Banting and Best's discovery of insulin was international news, Best became friends with eminent British physiologists Professors H.H. Dale and A.V. Hill. They convinced him to complete his MA at U of T, then come to England for final graduate work with Dale.[21] While in England, Hill and Dale introduced Best to Britain's medical and scientific elite. Among others, he met Sir Edward Mellanby, Nobel laureate for the discovery of the vitamin D deficiency that caused rickets and future wartime secretary of the Medical Research Council (MRC), and Cambridge physiologist Sir Alan Drury, future founder and director of the Lister Institute. Drury, like Dale and Hill, would become a lifelong friend, and both he and Mellanby would become critical to Solandt's career and wartime research work.[22] In 1929, Best returned to U of T to be appointed professor and chairman of the Department of Physiology, largely on Dale's recommendation. He was only thirty years old.[23]

After mentoring Don, Best took an active interest in the younger Solandt. During their final undergraduate year, both Omond and Chute participated in Best's special Friday classes on physiology, which only the best students attended. Best excelled as a teacher when the class was small and the learning was hands-on. The class would work in the lab instead of the classroom, and conduct experiments on each other as well as Best himself. This novel approach was common in experimental physiology at the time. Solandt would develop a similar personal approach to operational research.[24]

After a summer of research with Best in 1931, Solandt and Chute visited with the Hellyer family on their farm. The Hellyers were friends of the Chutes, and Laurie and Omond had enjoyed previous visits and horseplay with young Paul Hellyer, future minister of national defence, and his sister.[25] The fall 1931 trip, however, was no joke. Both boys got sick and Solandt developed a deadly case of bulbar polio.[26]

In Toronto, Doctor Armour at Toronto General Hospital would not commit to the polio diagnosis that Solandt had given himself.[27] Polio was considered a children's disease, and Omond was in his early twenties. Nonetheless, he suffered paralysis in his hands and, most dangerously, in his chest. In the years before the Salk vaccine, bulbar polio was usually fatal.[28] For three months, Edith Solandt cared for her son, day and night, massaging his shoulders, chest, and back to ease pain and tension on his lungs. Omond always attributed his survival to her tireless efforts.[29] By Christmas, he had sufficiently recovered to return to his studies. But damage had been done. His pectoral muscles were wasted away, ending his sports activities, and he suffered permanent loss of strength in his hands. Chute remembered Omond fighting to relearn how to write with a massive pen that required his whole hand to use. These hardships never stopped Solandt from returning, as soon as he was able, to his aggressively active work schedule at U of T, conducting research for Best that would form the basis of his master's thesis.[30]

Solandt attributed much of his future success to the teachings and support of Charlie Best. Best believed in the patronage of key students and worked hard to provide them direction, assistance, and opportunity. He would often spend an entire day going over a student's paper to help him improve.[31] When Solandt finished his MA, Best took special care in planning his postgraduate career, telling him to go to Cambridge for senior graduate work in clinical cardiology under Best's friend, Sir Alan Drury.[32]

Best's professional ethos on research in experimental physiology was also critical to Solandt's skills in medical and operational research work, and became a foundation for his approach to managing scientific enterprises in war and peace.[33] Researchers had to know of the key arguments and refutations in the scientific and medical literature to see how their own work fit within a greater context. Genius in science lay in the proper questions that opened up knowledge. Scientists had to be sceptical and wary of their own intuition and perceptions of answers before experiments were conducted. Meticulous measurement was also demanded so researchers could have total confidence in experimental data within the limits of technique – a conclusion Best no doubt based on his experience with Banting during the discovery of insulin.[34] Interpretations might change, but solidly measured data formed the bedrock of analysis. Best defined a hypothesis as "the story" of the research, a story that tied all that had been known on the subject towards the goal of the experiment, whose answers could not be predicted and required experiments. As knowledge increased, the story would change, so researchers were required to be both sceptical and flexible.[35]

Research and Graduate Work

Best's research ethos was a perfect fit for Solandt's meticulous intellect, and soon the student made the approach his own. Future Cambridge colleague Doctor William Feldberg was always confident that Solandt's research was 100 per cent reliable and could be published without fear of error. "You could always trust his work." This was no small praise, given the occasional eclectic nature of some scientists.[36] According to H.A. Sargeaunt, who worked with Solandt in operational research during the war, failure to reach Solandt's high standards for accurate data usually meant getting a lecture on why the offender "should have studied with Charley Best."[37]

Solandt's 1932 research became the basis for his MA thesis, "The Duration of the Recovery Period, and the Respiratory Quotient of the Excess Metabolism, Following Strenuous Muscular Exercise."[38] The thesis attempted to repeat and replicate some experiments conducted by Best and A.V. Hill in Britain. Solandt conducted the work with Jesse Rideout, an MA graduate in physiology who had worked as Best's research assistant for years. Rideout was supposed to lead the research, but Solandt's self-acknowledged "aggressive enthusiasm" kept him in charge of the project. Rideout, whom Solandt respected but felt lacked confidence, allowed the unspoken transfer of leadership with no ill will. They stayed friends for years afterwards.[39]

Solandt conducted experiments with human subjects to find and possibly quantify the "fuel of life": the "fuel oxidised to supply the energy for muscular contraction. It was hoped that the nature of this fuel might be indicated by the respiratory quotient [RQ] of gaseous exchange observed during exercise."[40] Using a Douglas Bag, a cheap but difficult measurement device that required painfully careful application, Solandt investigated the level of RQs in three subjects under various degrees of stressful exercise. These technological limitations were lamented but necessary, given the funds available.[41]

Throughout preparation of the thesis, Solandt kept a close eye on not only the measured values witnessed in his subjects' respiratory gases, but also on their observable state during the experiments. In the appendix, he discussed an exercise experiment done with subject "T.B." on 1 December 1931. The subject's performance was weaker than previous experiments, and his face twitched as if it were in pain. After the experiment, he complained of a toothache. "The results of this experiment were not included with those from other exercise experiment. They are presented here in order to show how atypical the results of a metabolism experiment may be when the subject is disturbed by pain."[42] Solandt's work in physiology, the systematic study of the function of living beings,[43] had to take into account situational as well as biological factors to understand how his subject operated. Such training would serve him well throughout his varied career.[44]

In the end, Solandt's MA experiments found no definite conclusion on the value of RQs in determining the nature of the "fuel of life."[45] He never considered the work a fascinating point of entry into research, but working under Best was a boon. With his MA completed in 1932, Solandt prepared for his medical education at Cambridge with Sir Alan Drury by working with Best. He did this for the remainder of the year at the Department of Physiological Hygiene in the School of Hygiene building which also housed the Connaught Labs where Don would later do much of his critical wartime research. "I worked both on the estimation of Chlorine and tissues," Solandt recalled, "and during the summer I continued some of Dr Best's own work on continuous injections of Heparin into dogs which gave great promise of helping in the treatment of coronary diseases in humans."[46] Indeed, after insulin, heparin would be Best's next and last great research project. Solandt earned his master of science in medicine in 1933, worked with Best until 1936, and soon departed U of T for Cambridge.[47]

Solandt's Intellectual Development

Solandt's professional interests during this period would serve him well in his war work, but he sustained interest in respiratory quotients.[48] In fact, Solandt's first published article for the university's undergraduate *Medical Journal* challenged the modern and orthodox view of RQs as a valuable way of measuring metabolism. He first evaluated all the literature concerning RQs, compared it to the results of recent RQ experiments (though not his own), and concluded that questioning their value ten years ago may have been "rank heresy," but recent experiments negated such dogmatic thinking. This was a fairly bold statement for a student working on RQ theories based on his mentor's work, but Solandt rarely lacked confidence.[49] Next, Solandt and Rideout published an article, based on their attempts to recreate Hill and Best's work, for the prestigious *Proceedings of the Royal Society*, where Hill was secretary. Hill received and approved the article, which was Solandt's first major publication and the last he would write specifically about RQs.[50]

Solandt maintained an interest in the respiratory system under duress and exertion and the nature of blood throughout his academic career. He wrote and co-authored articles on the relationship of strenuous exercise and sugar content in blood,[51] before taking a keen interest in the effects of and treatments for cyanide poisoning, a growing problem as Canada became more industrialized.[52] Of particular interest was the relation between cyanide and carbon monoxide poisoning, since both produce asphyxia. Research was being conducted on the possibility that methylene blue, used to treat cyanide poisoning, might have applications for carbon monoxide poisoning as well.[53] Such knowledge was invaluable to his future work on carbon monoxide poisoning in tanks.

Never satisfied with book learning alone, Solandt, along with his brother and Chute, worked in experimental physiology with Lawrence Irving at Woods Hole and at the Marine Biological Laboratory at St Andrew's.[54] This research involved the nature of nerve impulses in dogfish, more in line with Don's interests,[55] and the unique nature of blood and the respiratory system in seals.[56] This last article investigated the possible role blood played in a seal's ability to dive for considerable stretches of time. The writing of the final paragraphs, in which this theory is discounted, again reveals the systematic way in which physiologists viewed the body. The blood could not provide any "considerable extension of the time of diving. The capacity of blood for oxygen is so small compared with metabolic needs that the blood is not a significant reservoir of oxygen. It is only the system of respiratory transport for the tissues. But in its function as a transportation system the properties of blood can determine how rapidly the oxygen debt which is incurred during diving can be discharged and how soon another dive can be made."[57]

This professional development was intended to increase Solandt's skills to be a good doctor of medicine as he continued his education at Cambridge. It was fortuitous that they would also provide him the tools to investigate the physiological dimensions of tank-operator problems during the Second World War.

The Fifteenth International Physiological Conference in Leningrad and Moscow, 8–18 August 1935

In 1935, Best offered Solandt the chance to join him and his wife on a trip to the Fifteenth Annual Physiological Conference in Leningrad and Moscow. They would travel from London with Solandt's brother Don and Don's future doctoral supervisor, Nobel laureate A.V. Hill. Hill was secretary of the congress and Don was presenting a paper.[58] Omond agreed, relishing the chance to travel to the exotic and mysterious Soviet Union.

The congresses began in Basel, 1889, and had grown in size and stature in the following years. Canadian representation began at the fourth conference in 1898. International scientists and their achievements were showcased. H.H. Dale, Banting, as well as Best, who attended in 1926 with Dale, had all presented papers. In 1929 the congress became truly international with Soviet participation at the thirteenth congress in Boston, where I.V. Pavlov was the star of the meeting. The congress visited Solandt's old haunts of Woods Hole and Banting and Best's lab.[59] In 1932, the Soviet Union was chosen for the next congress. The Organizing Committee was Russian, but Hill took on the responsibilities of secretary.[60] The congress would he held primarily in Leningrad, with a sojourn to Moscow, from 8 to 18 August 1935.

Solandt and company left London on a small Russian passenger ship to Leningrad. Every day, the passengers were summoned on deck for the captain's daily lecture on the virtues of communism, delivered in good English. The British expected the USSR to be unhygienic and warned Don and Omond not to touch anything. The Solandt brothers expected it could be no worse than Britain, where they had seen the only cases of typhoid in their young careers. The forward-looking Soviet Union might even be cleaner.[61]

The opening session began on 9 August at the Uritsky Palace, formerly the meeting place of the Duma and the Provisional Government of 1917 and now the Leningrad Soviet. "Members passed through a succession of four doors with armed guards into the great Soviet auditorium, which was decorated for the occasion with palms and flowers. Each seat in the auditorium was equipped with a head-phone, and by placing the plug in the appropriate socket one could listen to the proceedings in either Russian, or English, or French, or German."[62] Attendance was estimated at 1,200 to 1,500. While 485 papers were presented, the congress was also "a demonstration to the foreign members of a large number of facets of life in modern Russia."[63]

Solandt recollected primarily the spectacle and scale of Soviet architecture and hospitality. The Leningrad Council Hall was a scale model of the Parthenon, lined with tables holding mounds of salmon and caviar, as well as copious supplies of alcohol. The two minister's sons drank cherry brandy cautiously and enjoyed the late and elegant dinners on the weekends. Pavlov, nearing the end of his life, was the star attraction. Solandt thought he was praised by Soviet doctors in particular because he did not buckle under to the Bolsheviks and was worshipped as a national hero. The Solandt brothers wore badges that said they were "Friends of Pavlov," which got them access to anywhere they wanted to go. There were also major dinners in Moscow at the Kremlin, with concerts at both places, where some members met with foreign minister V. Molotov. Best introduced the brothers to all the important members of the physiological world. The papers were good, but the highlight was meeting Pavlov and seeing his famous lab.[64]

Solandt found the USSR a strange society with an oppressive atmosphere. Everywhere there was a feeling of being watched. Despite the spectacle of the conference, most Soviet technology looked ragged, old, and dangerous. Peasant women in "rags without shoes" worked along railroad track from Leningrad to Moscow. After the congress Don and Omond travelled to Finland by train. When they crossed the border, a cheer came from the train, most of the passengers glad to be out of Soviet Russia. After travelling to Scandinavia, they returned to London, and soon Omond was home and preparing for Cambridge.[65]

Solandt always appreciated Best's offer. "It was an outstanding example of Charley's generosity and thoughtfulness," Solandt later remarked. "He talked Don and myself into going and facilitated our registration, but the main thing was everywhere we went, he made sure we met all the big shots. We went with him twice to [Doctor Ivan] Pavlov's lab."[66] Later in life, Solandt well recalled the value of working with "big shots" in any field of inquiry, especially in government science. To have any push or pull against the inertia of a bureaucracy required the influence of the most powerful players of the game. The Leningrad journey also showed that Solandt was no shrinking violet. At the age of twenty-six, he had mixed with Nobel laureates and medical legends with confidence, thanks in part to the support of Charles Best. This confidence remained when he entered the world of military affairs and dealt with generals, admirals, and air marshals.

Conclusion

For almost a decade, Solandt made a name for himself at U of T. He had flourished academically and become the protégé of one of the school's most famous scientists. Using the research methodology and contacts of his mentor Charles Best, Solandt became a star pupil on the cusp of a promising medical career. His publication record showed a fierce mind for research, coupled with a wide interest base. Under Best's direction, Solandt decided he would finish his studies at Cambridge under Sir Alan Drury.

It is easy to see in hindsight that Solandt had all the makings of a solid researcher. His technical skills maintained their proficiency with summer work for the Ontario Forestry Branch and the Ontario Air Service. His meticulous research and a fierce yet practical intellect would pay great dividends during the Second World War. But military affairs were the furthest thing from Solandt's mind. His professional destination was clinical medicine.

At the Leningrad Congress, Solandt heard the opening speech of Pavlov. At one point, the elder statesmen of experimental physiology directed his speech specifically to "the importance of such assemblies for the young generation, for the young scientists." Pavlov had found such assemblies of great value as a young man, working with naturalists and physicians. "Our government spends great means on scientific work and draws large numbers of the young generation into scientific research. To see the world's scientific work personified should be highly stimulating to young people."[67] He called for continued international approaches, the comradeship of working towards "the rational and final unity of humanity."

War, however, would change this. Scientists would then become enemies, and appreciations of each other's work would be effected. "I can well understand the greatness of a war of liberation," Pavlov said, "yet, at the same time, it cannot be denied that war is essentially a bestial method of settling difficulties, a method unworthy of the human mind with its unlimited resources."[68]

These feelings echoed Solandt's approach to science and warfare during both the Second World War and the Cold War. Solandt did not relish war. Its destruction and chaos were an affront to a doctor who had experienced the good that could be accomplished through reason, cooperation, leadership, and knowledge. But in facing the aggressive and duplicitous nature of Nazi Germany and, later, the Soviet Union itself, Solandt saw the duty of scientists to provide their national government with critical aid against deadly foes.[69] It is one of the great ironies of Solandt's impressive life that his first experience of state science, and its relation to warfare, was in the heart of his nation's adversary during his days at the DRB. He would be the only member of the post-war Canadian chiefs of staff to have personally witnessed the grim environment and technical and scientific prowess emerging from the shadowy world of Stalin's Soviet Union.

4 Commencement: Solandt, Cambridge, and The Ontario Polio Epidemic, 1936–1939

The last phase of Solandt's education was filled with the same enthusiasm and vigour of his Toronto years. He began his Cambridge studies in clinical research in cardiology and pathology in September 1936 on the Ellen Mickle Scholarship. He was made a member of the Medical Council of Canada on 2 July 1936.[1] Don had finished his graduate work with A.V. Hill, where he met and then married one of Hill's other students, Barbara Mary, and returned to work with Best. For the last time Omond would be following his brother's path. Soon, he would outrace him.

Solandt continued to build his reputation as a fiercely meticulous researcher. He succeeded on his first attempt to pass the exams for the Royal College of Physicians, but faced his most difficult professional work during his internship at Toronto General Hospital in 1937 during the worst polio outbreak in Canadian history. His performance during the crisis demonstrated many markings of Solandt's greatest abilities as a manager of a complex workplace under dire circumstances. Within the world of clinical research, his star was on the rise. The Second World War, however, would redirect Solandt's career and life, ending forever his original goals of a life at Cambridge.

These were hard years. On 6 August 1936, Reverend Donald Solandt died after a short illness at the age of sixty-five.[2] Never an emotional man, Solandt likely found it best to honour the memory of his tireless father by continuing the family tradition of hard work, service, and achievement. He would also, unintentionally, follow his father's lodestar of leaving his original profession to apply his multitude of skills towards war work and state service. It would also be during his Cambridge days that he would meet and soon marry his first wife, Elizabeth Maud McPhedran.[3]

Cambridge Days

For his first year at Cambridge Solandt stayed at London House, a residence for graduate students from the Commonwealth that Don Solandt had used while working with Hill. Solandt enjoyed meeting the many South Africans and Australians who used London House, and would stay there during his second stay from January 1938 until September 1939.[4] Solandt only had one social rule during that first year: he would not join the Canada Club or spend his time cloistered solely with other Canadians. These expatriates complained about missing home instead of embracing Cambridge life. Solandt would do the opposite, with no regrets.[5]

His initial work began at the London Hospital.[6] The rest of the year was spent conducting physiological research at Cambridge with Drury, who was the last of Solandt's great mentors.[7] The son of a successful railroad man, Sir Alan Drury was a respected physiologist whose personal friends included Best, Dale, Hill, and E.D. Adrian, future senior professor of physiology at Cambridge. He had served in the Royal Army Medical Corps in India and Mesopotamia during the First World War before continuing his education at Cambridge, and, during the 1920s, worked for five years with Thomas Lewis, secretary of the Medical Research Council, using the Einthoven string galvanometer in clinical and experimental investigations of auricular fibrillation and flutter. He also studied with Professor Einthoven in Holland before returning to England, where he survived a dangerous bout of tuberculosis. From 1927 until the outbreak of the Second World War, he worked at Cambridge on a host of collaborative research efforts with British and international scholars, Solandt among them.[8]

Solandt enjoyed his time under Drury, who had a solid reputation as a medical scientist. Drury treated him as a member of the family during the first homesick years in Cambridge, and Solandt would often vacation with him and his family on their barge, the *Dutchman*, along with fellow Cambridge student Tenny Spooner, who would also become a wartime colleague. As part of his degree requirement, Solandt needed to serve a hospital internship. After the initial lonely year at Cambridge, he chose to go back home, and from 1 July 1937 to 30 June 1938 he was a member of the Toronto General Hospital Staff, Junior Rotation Service.[9]

The internship was initially disappointing. Solandt, who was a surgical intern, received no surgical time. Instead, he was tasked to administer anaesthetic and conduct blood transfusions. According to Solandt, the woman who had been responsible for surgical anaesthetics had killed a few people with the procedure, and the doctors no longer trusted her.[10] Blood transfusions were still in their infancy as well, and many doctors were happy to let Solandt administer the procedure. He soon had a reputation as a minor specialist on transfusions that kept him in demand and out of surgical training. The training was fortuitous for a future director of a wartime blood bank. During his tenure at the Southwest London Blood Depot, Solandt would be shocked that no one outside of the teaching hospitals had been trained for or had done transfusion.[11] He made the best of his internship, but the lack of surgical training frustrated him. This feeling was soon quashed as Solandt confronted an old medical adversary sweeping through the province with fatal speed.

The Ontario Polio Epidemic of 1937

During the summer of 1937, Ontario suffered its worst outbreak of poliomyelitis, the second of four polio epidemics in the nation's history. It would be eighteen years before the Salk vaccine was introduced. Until then, polio remained one of the most feared diseases in North America, attacking primarily the young almost annually.[12] The Canadian west suffered an epidemic in 1927–8, with 182 cases in British Columbia, 313 in Alberta, and 434 in Manitoba. Over the next two years, Ontario and Quebec would suffer over 1,000 cases each before dropping to half that number in 1936.[13] But in the summer of 1937, polio returned in strength to affect 2,546 people in Ontario alone. This accounted for 64 per cent of the nation's polio cases that year.[14] The mortality rate was 4.3 per cent and paralysis rate 56 per cent.[15]

At the time, Solandt's recovery from polio was thought to make him immune, and for this reason he was given full responsibility for the polio ward at Toronto General.[16] Solandt recalled that general fear of the disease also whittled away the number of nurses and physicians who would risk working at the hospital.[17] Solandt and fellow interns worked twenty-four-hour shifts during the epidemic, two hours on then two hours off. Solandt received a pithy honorarium of one hundred dollars.[18]

Death from polio generally resulted from "bulbar" cases, paralysis of the throat or lungs, which Solandt had managed to survive. When prophylactic sprays developed in the United States failed to stem the tide of the disease in the early 1930s, the bulbar cases became a top priority, and for them, respiratory technology was critical. Philip Dinker had invented the "iron lung," an electric tank respirator, in 1928 at Harvard University. "It was essentially a metal tank into which all but the head of an individual was sealed. A motor, or hand crank, operated a set of bellows, and since the head remained outside the lung, the negative pressure inside acted like a human diaphragm and forced the lung to expand and contract like regular breathing."[19] Toronto's Hospital for Sick Children (HSC) bought the nation's only iron lung in 1930. During the 1937 epidemic, more machines were bought, and the staff at HSC also constructed their own respirators for other bulbar cases during the crisis. The majority of these respirators were used at HSC and Toronto General.[20]

Desperation led to use of risky technology. Solandt, as head of the polio ward, was involved in the experimental use of a corset-styled respirator created by Ed Hall. Hall and William "Bill" Franks got their mentor Frederick Banting to petition Toronto General's professor of medicine Duncan Graham to try the respirator on bulbar patients. Graham had clashed with Banting during the early days of insulin and had doubts about the man and his students. He finally agreed to the trials but told Banting that Solandt was responsible for the patient's welfare. If, at any time, the patient was threatened or it appeared the respirator was not working, the patient would be put on the old respirator, no questions asked. Hall and Banting agreed.

Solandt was informed of these machinations after the fact, souring his already low opinion of Banting. Still, he worked with Hall and Franks on the tests. A paralysed patient who had been in good spirits agreed to the procedure, so long as they kept an eye on his vitals. But once he was in the contraption, his face turned blue. Solandt immediately ordered him back in the mechanical "dinker" and said the experiment was over. Hall was furious, claiming it had worked, and demanded more trials. But Solandt would not budge, and later figured it was Banting's clout and not the scientific merit of his students that allowed the experiment to be tried. His opinion of Banting as a scientist sank even further. Hall's conduct confirmed for Solandt that Banting's students never measured up to those of Best. The experience also demonstrated the danger of patronage instead of scientific merit being the deciding factor in medical policy. This lesson would stay with him throughout his career in government science.[21]

A year later, Solandt and some senior colleagues wrote a paper on their experience and efforts during the epidemic. It explored how they handled the influx of sixty-six cases of acute polio, during which time the hospital was quarantined for three weeks.[22] The paper showcased and analysed the methods they used to organize the patients, their treatment, and post-hospital recovery. They also revealed aspects of Solandt's management style.

Patients were classified by the nature and location of their paralysis. Treatments were designed to meet the specific kinds of paralysis, ranging from blood serum injections to slings and cradles for affected body parts. Convalescent blood serum was also used in pre-paralytic cases, but when compared to others in similar condition to whom no serum was given, the percentage of those who became paralytic was almost identical (65 to 69 per cent).[23]

Nine deaths occurred. "One case was a pure bulbar, seven were bulbospinal, and one was a spinal type." All died within two weeks of the onset of the illness, and six cases died within three days of admission.[24] Seven of these died during respirator treatment, but not from lack of effort.[25] The Toronto General team organized a strict system to provide for the respirator cases. Their care and treatment, in and out of the respirator, was measured and recorded "not only to ensure adequate attention to the individual but in order that the entire group might receive attention."[26] This systematic method of care to maximize the efficiency of the respiratory schedule suggests the application of scientific management, the time-motion methodology developed by American efficiency expert Frederick Winslow Taylor.[27] Solandt was familiar with time-and-motion studies, and there is little doubt that the efficiency of scientific management would have appealed to him. This may have been Solandt's first exposure to "Taylorism" in an operational setting, but it would not be the last, as time-and-motion studies would be integral to his work in operational research (see chapter 7).

For Solandt and his teammates, therapy in this climate required thorough analysis of not only physiology but also of the patient's mental health. They devised activities geared to patients' interest to keep them mentally stimulated and alert: book-rests were used on respirators, radios were brought in, and mirrors were employed so that patients could see the whole ward and not feel claustrophobic. Teachers even braved the disease to instruct the young. According to the team, emotional well-being was good and constant.[28]

It would later be a pillar of Solandt's management style that any organization had to be run efficiently so that research could be conducted without harming the primary objective. Clearly, this approach was first learned during the polio crisis of 1937. As soon as the ward was run effectively, Solandt and his team conducted research. Their tests dismissed both the notion that the sense of smell was associated with the disease, and that serum treatment had a preventive effect on paralysis. The relative merits of different kinds of physiotherapy were recorded and analysed. They conducted six autopsies of the fatal cases, recording that emphysema had not occurred. "Of the paralyzed patients who survived 40 per cent have completely recovered or have obtained good functional use of their limbs. In 26 per cent a good functional result is anticipated within the next six months. In only 34 per cent of the cases is the ultimate result in doubt."[29]

While the epidemic was tragic, the work of Solandt and his team was both hopeful and forward-looking. They believed that the knowledge gained, spread, and built upon this tragic episode could provide solutions for the future. Solandt's constructive attitude of support for research during a crisis would continue throughout all his work, from blood-substitute research at the Blood Depot, to his examination of casualties in the wake of the atomic bomb. Knowledge gained today was the key to tomorrow's solutions, even under duress.

By the time Solandt returned to Cambridge in 1938, his experience in medicine was not purely academic or experimental. It was also pragmatic, tactile, and tinged with mortal consequences for failure. His meticulous mind applied efficiency and rigour to management as well as research, and with positive results. Two years before the Second World War changed his career trajectory towards state science, Solandt had witnessed the value of his systematic examination of human concerns in a dangerous environment, where the cost of failure was in lives risked or lost. The lessons learned from his experiences and success during the polio epidemic formed a pillar of his science management style.

Solandt before the War, 1938–1939

After the polio epidemic, Solandt returned to Cambridge, resumed his work with Drury, and continued to publish articles.[30] He wrote on the defining factors involved in lung collapse, thoroughly familiarizing himself with the current literature on the subject. Indeed, Solandt himself often emphasized the need to include all relevant data, and noted that new data had to be accounted for if current theories and practices were to remain valid, or, if necessary, change. Solandt would make similar arguments on the successful application of new solutions based on operational research (OR) for the Army. Without knowing why current methods were used, the innovator could not effectively provide an alternative to the organization.[31]

Solandt's formal education ended and his young career began in medical research in 1938. In the fall, Solandt was employed at London Hospital to finish his studies before taking the dreaded exams for membership in the Royal College of Physicians. He succeeded on the first try, an uncommon feat. On 28 May 1938 he was made a member of the Ontario College of Physicians and Surgeons.[32] On 12 May 1939 he received his MA in pathology from Cambridge.[33]

At Cambridge, Solandt continued his research on blood pathology with Drury,[34] as well as work on mammalian physiology to do with blood sugar,[35] and learned a few lessons from outside his studies that stayed with him well into his career in defence matters. First, the University of Toronto's standard of education was of the same calibre as that of Cambridge, though Cambridge provided better support for their students. Second, Cambridge students had been crammed with knowledge at an early age from a variety of fields, "from multiplication tables to Greek," all keys to advanced study. Third, Cambridge also tolerated and accepted the inclusion of "peculiar people" whose eccentricities often indicated future brilliance. This suited Solandt just fine. His general interest in people of all walks of life allowed him to keep a keen eye on their relative merits and detriments. One of his greater strengths would be his ability to find the right people for the right job, regardless of the traditional chains of command or operational environments or if they were strange, "eccentric" people.[36]

Unlike his lonely first year at Cambridge, Solandt's return in 1938 was marked by a burgeoning social life. Chute was teaching, and Solandt added new cronies to his cadre. Cambridge chemist Owen Wansbrough-Jones was a good friend and future colleague in defence research. Lawyer T. Thomas Ellis "Tel" Lewis was a dear friend and arranged for Solandt to be accepted as a member of Trinity Hall, a right usually reserved for law students. Solandt also made friends with German expatriate W.S. Feldberg, and the two collaborated on research projects to good effect. Through Drury, he was reintroduced to senior medical science elites like Mellanby, and met his future wartime boss at the Army Operational Research Group, physicist John Cockcroft.[37] Solandt was a great admirer of Cockcroft, a gifted scientist and manager of science who could manoeuvre through the corridors of power in Britain as easily as a university laboratory.[38]

With his education complete, Solandt was hired by Drury to work at the Physiological Department's laboratory. Here, against protocol, he struck up a friendship with Adrian by initiating conversation with the "senior" physiologist. They soon became friends and, not long after, Solandt became a lecturer. He did not want the job, but had no choice of turning it down, as it was a rare honour for such a young doctor. Solandt always felt his time at Cambridge exposed him to the best medical professionals in the world, particularly the "invisible college" of electrophysiology that included British and American physiologist such as Hill, Dale, and Adrian, as well as American Ted Bronk, whom Don had worked with after his MA and who later, as scientific advisor to the U.S. Air Force in Washington, worked with Omond during his DRB days. Solandt included Don among these professionals, one who would have surpassed them all if not for his ill health.[39]

Conclusion

The summer of 1939 was spent working and vacationing with Drury, who took his son, Solandt, and Spooner on a trip through the Dutch canal system on his barge. Solandt and Graham McCullough, a senior lecturer in pathology, also went to Belfast in August before having to return to Cambridge. When Solandt returned in early September to prepare for the autumn term, the German Army had invaded Poland. Days later, Britain was at war.[40]

By the fall of 1939, Solandt was professor of mammalian physiology and about to begin a career in clinical research. His work at Toronto General had provided him the opportunity to demonstrate his natural skills in managing and research during a crisis. The pay was rotten, but the experience invaluable. Solandt now knew that if he had to, he could manage patients, staff, and research under duress, and that he had done it efficiently and well. He established order in chaos to serve the needs of others and get the job done. This attribute was a hallmark of his legacy in defence research.

It is difficult to assess how Solandt's medical research career might have gone if not for the war. Feldberg postulated that there was a fifty-fifty chance that he would have become a great research scientist, but since he left the field so early it was impossible to tell. As previously noted, Feldberg felt Solandt's work was always 100 per cent reliable, there was never a fear of error in his work. Many scientists did not share this methical work ethic.[41] Indeed, his intellect was extraordinarily clear and rational rather than creative. While the lack of creative insight might have crippled his achievements in medical research, his ordered mind would be invaluable to government science projects, where there was a natural tension between the exploratory, experimental, and trial-and-error method of science, and the budgeted, goal-oriented needs of the state. But 1939 marked the end of Solandt's strictly medical career forever. Solandt's qualifications, skills, energies, and confidence were soon in high demand. Over the course of six years of war, Solandt's successes took him from running a blood depot in London, to the atomic battleground of Hiroshima and Nagasaki. All of these future successes, however, rested on the foundation of professional skill Solandt forged through his university education and experiences during the Ontario polio epidemic of 1937.

5 First Taste of War:
Solandt and the South West London
Blood Depot, 1940

The German invasion of Poland on 1 September 1939 found Solandt preparing for another year of research and teaching of mammalian physiology at Cambridge. With war declared, he attempted to enlist in the Royal Canadian Army Medical Corps, but was told he would have to return to Canada to join. Solandt decided to wait for the Army to arrive in London instead, and with this decision changed the course of his career and life forever.

After an erroneous letter was sent out recalling medical students back to early class in September, Solandt instructed civilians on physiology through first aid. Soon after, the British Medical Research Council (MRC) selected him to direct the South West London Blood Depot. The British government had started preparations for collecting and maintaining a supply of blood in case of war since the Munich Crisis of 1938. The MRC was responsible for creating, staffing, and maintaining the organization that would become the four London Blood Supply Depots. MRC membership also included Mellanby (the secretary during the war), Drury, and Hill, and they selected Solandt to replace Doctor J.O. Oliver as director of the South West London Blood Depot.[1] Drury was likely Solandt's strongest backer.[2]

For the next year, Solandt would organize the depot's materials and staff to service the blood needs of hospitals in his sector, and stamped his own personality onto the depot's organization. Efficiency and excellence were his watchwords as he prepared and then led the depot against the looming threat of the Luftwaffe during the Battle of Britain. When time could be spared, research was promoted, including Solandt's own contribution to the emerging problem of "crush injuries" resulting from German bombings. Clearly, he built upon his ideas and experiences of management first demonstrated during the polio crisis. The reward this time would be greater. By the end of 1940, the MRC would select Solandt to establish his own laboratory and charge him with the responsibility of solving tank problems for the Armoured Corps. As such, his directorship of the Sutton Blood Depot was his first major success in government research and science management.

The Depots

The depots served civilian blood needs primarily through hospitals, though the Royal Air Force and Royal Navy, each without a sufficient independent blood service, were also recipients of depot blood. The depots also cooperated and coordinated with emergency organizations to cover the growing needs of the whole country. While maintaining the blood needs of the nation was of prime importance, research was also conducted on drying blood serum, improving whole blood transfusions, and more.[3] Each depot collected, stored, and supplied the blood needs of London hospitals. By July 1939 the embryonic shape of the depots was in place. The bleeding program commenced 1 September 1939. The depots were staffed with a mix of paid employees and volunteers who were trained as nurses and drivers. Depot directors had four or five other medical officers who bled donors, gave transfusions on request, and engaged in research. Research assistants with science degrees were often attached to depots for research work. Each had ten to twelve full-time nurses, doing a variety of the tasks, including bleeding, transfusions, and research.[4]

Depot policy was initially centralized under Professor Topley and then Drury. But with the unique demands of each depot arising during the Battle of Britain, Drury allowed directors the flexibility to make their own policy, and so they developed in different ways. This made administration difficult, but the practical and academic benefits grew. Successful methods of organization were communicated from depot to depot through the MRC's new Blood Transfusion Research Committee, set up in 1940 under Drury, to facilitate efficient coordination. Solandt, as director of Sutton, was a member. It was his first introduction to the British committee system.[5] While running the depot, Solandt witnessed Drury's management system. His ability to provide the depots with some guidance but also the freedom to conduct their operations as best they saw fit would be mirrored in Solandt's work at AORG and DRB.

The New Director

Oliver had assembled an effective establishment and prepared the depot when Solandt arrived, but left before the excitement of the summer. With the collapse of French resistance in June 1940, Germany could focus its power on Britain, now fighting with only the Commonwealth at its side. During this desperate year, Solandt would make his name directing the depot from 1 May to the end of December 1940, servicing the injured civilians during the entire Battle of Britain (July to October 1940), including the worst of the Blitz.[6]

His arrival also marked a shift in professional conduct. Doctors Patrick and Maggie Mollison had started work at the depot under Oliver. They noted that as soon as Solandt arrived, everything began "tightening up." Solandt wanted details on every aspect of the operation: donors, bleeding techniques, research, transport, bomb shelter preparations, everything. His first major policy shift was reorganizing the transport system because he felt it was inefficient. During the Blitz when blackouts were normal and teams of drivers had to make it to and from hospitals in the dark with no street signs to guide them, inefficiency was dangerous. Solandt also ended the misuse of depot transports for personal use. Patrick Mollison noted Solandt was not mean in this regard. He had a knack for making the perpetrators feel ashamed for their misconduct, given the serious business at hand. Both Mollisons recalled he was well liked by the staff and, though quiet, took great pride in their efforts and company. He was loudest when introducing important visitors to his team, always proud to celebrate those who had done good work, making the team feel their contributions were valuable and noticed.[7]

"He worked obscenely hard," Patrick Mollison remembered, and set high standards.[8] He had no tolerance for people not doing their best, and Maggie Mollison felt many wanted his praise to avoid his disapproval. Solandt's leadership was both quiet and forceful. Maggie Mollison once demurred from approaching members of the Rotary Society for bleeding, and Solandt, quiet but firm, made it clear that it was an important job, and failure was not an option. She appreciated the push. He might call his team out for doing a poor job, but he was also loyal to them.[9]

Clearly, Solandt continued the approach he had learned from Best and employed at Toronto General. He read as much as possible on blood transfusion and resuscitation from papers from the Royal Society, whose office he frequented, and during this reading turned Patrick Mollison on to RH immunization issues that Mollison would later focus on. Solandt did not participate in this research, but he kept on top of Mollison's efforts.[10] But his greatest gift was efficient management of the operational elements of the depot. Patrick Mollison was particularly impressed by how Solandt's meticulous mind could get to the bottom of every operational concern. Maggie Mollison was similarly impressed with his ability to size up just about any problem and work towards effective solutions. This ranged from "diagnosing" plumbing problems himself to avoid paying for a plumber (as he had diagnosed radio problems in the Forestry branch as a younger man), to reorganizing the team's fleet of vans for optimal efficiency. Both approaches would be hallmarks of his OR career. Patrick Mollison recalled Solandt's shock that the team did not know how to fix their own cars or change tires, which he had been doing since his teenage days. Whatever the problem, Solandt attacked it with the same gusto that had driven him to the top of his classes in Canada and led him to Cambridge.[11]

Solandt encouraged and took an active interest in the perfectionist scientists of his laboratory. Instead of involving himself directly, he provided guidance and support and was always interested in their work. He often referred to his role as that of a "maestro" with an excellent orchestra that needed the best instruments. Sterilization was critical, and Solandt and the lab chief both struggled to find the best ovens for the procedure, finally working with a consumer baking company to create them. Solandt recalled the ridicule he received for not using conventional lab ovens, but had the last laugh when their blood had the lowest infection rate of any other depots. "If the unconventional worked," he argued, "you used it."[12] Mollison was also impressed that Solandt seemed to know all the key people in physiology, both the established and the up-and-comers, which allowed him to make contact with the right people. Doctor R. Mayon-White was closest to Solandt at the depot, no doubt because he was among the sharpest members of the team. He was the only depot member Solandt would bring with him to Lulworth in 1941.[13]

A Depot at War

The Battle of Britain found the Sutton Depot in good shape, but their initial conduct under fire was less than stellar. Solandt wrote his mother in August 1940, when the big news was the air raids. He had seen part of the Croydon raid while with Doctor Todd, a colleague on staff at the local LCC lab in Sutton, overlooking the local aerodrome.

> We watched the planes approaching but timidly retreated to the air raid shelter and heard the row from there. It was all over in about three minutes and we came out to watch the fires. The siren sounded about ten minutes later and we again had dinner there. They have a very spacious shelter opening from inside the house. It seemed silly to miss the fun at the time but we have since heard of people in the neigbourhood who were hit by machine gun bullets even though we were two miles from the battle. The Spitfires were right on the raiders tails and the air was probably full of bullets. As you probably heard, none of the raiders got back.[14]

Another raid the following Friday did not leave Solandt in good spirits. He was at the MRC, away from the depot, when bombers dropped their ordnance north of the city, the nearest bombs two miles away. Then dive-bombers appeared, machine-gunning the streets. "It was our first big test and we did not come off very well." He had told his secretary about his schedule, with the order that if there was trouble to call him immediately, but she left without telling anyone where he was. "We did not have much blood on hand but had extra plasma but in the excitement no one thought of it." The only one who could have thought of it was stuck in an air raid shelter and the warden would not release him. Obstacles blocked the depot's trucks everywhere. It took two hours to get their blood to one hospital. "Most of the hospitals ordered twice the blood they needed. We sent 150 bottles of blood and 50 of plasma and probably not more than 100 were used. We have 200 or more fresh bottles on hand now so there will be no more trouble this week end."[15] He assured his mother that Sutton had no real strategic value to the bombers, "and it is not even sufficiently closely packed to making bombing civilians profitable." He signed off saying that they would likely be attacked but hoped it could be avoided.[16] Even in trying to soothe his mother's concern for his safety, his pragmatism would not lie still.

Solandt's View of the Depot

Just as he had during the polio epidemic, Solandt wrote about his experience in an article for both British and Canadian medical journals. He emphasized the role of the South West London Blood Depot in an effort to raise awareness of the work and how it was done.[17] The article reveals not only his opinion of the import and function of his own depot, but also of civil defence. Solandt would never forget what it meant to try to conduct research in what Anti-Aircraft Commander General Frederick Pile referred to as Britain's only "public theatre" of operations.[18]

Solandt argued that the MRC's pre-emptive measures before the war were now justified by the depot's successful track record.[19] Solandt's depot was originally to supply hospitals in Sectors 8 and 9 (South West London and all of Surrey), which contained more than seventy-five hospitals between them. The depot had also taken over the supply needs of Sussex, which had thirty-four hospitals. By 1941, 50,000 donors were located in the outer suburbs of London and towns and villages outside the London area, forty miles from the city, as part of the dispersion strategy against the bombings. Enrolment of donors was continuous, but the donors were to be scattered. Solandt argued this scatter strategy was useful. It helped "to have a widely scattered donor panel in case war conditions should make bleeding impossible in the area immediately around London."[20] Six to seven hundred donors were bled each week after the bombing started: 40 per cent at the depot, the remainder at various bleeding centres. Donors were called by postcard, and emergency measures for quick contact were also created. Blood reserve stocks remained adequate at the time of writing, and the depot staff was run so efficiently that it never needed to bleed at night and rarely on Sunday. "This is one great advantage of the Depot system, particularly when night transport for donors is made very difficult and even dangerous by blackout and barrage."[21]

Solandt also described the technologies and methods of bleeding, and the effect on the donors.[22] At the end of bleeding, a sample of blood was put in a test tube and stored on the side of the pint bottle. This was to be used for the Kahn test and for regrouping by blood type. "No blood is released for distribution until the results of these tests have been reported."[23] Almost all blood sent out was "Universal Donor" (group O) to minimize chances of haemorrhaging that occurs when different blood types are transfused. No error in grouping had been detected or even suspected.[24] Blood was stored in large fridges at 4–6 degrees C. With addition of glucose, the blood could be stored safely for three weeks, but the older blood was to be used only in emergency cases. Experiments had shown that stored blood was of almost identical quality to that of fresh blood for up to two weeks. "When the demand for blood is fairly large it is possible to keep a stock of 150 to

200 bottles of group O blood on hand all the time and yet have none of it over a week old."[25]

As noted, Solandt changed the delivery procedure during his tenure to increase efficiency and effectiveness. Originally, hospitals had a stock of blood that was replenished, but this led to great waste and the dangerous use of old blood often stored in less than ideal conditions. Hospitals were then encouraged to have stocks of serum or plasma for emergency use only and to call for blood as soon as casualties arrived. This greatly reduced wastage. A night and day delivery system was maintained with "a fleet of light vans," and service was rendered usually within an hour. A sub-depot system existed with an alternative communication system in case phones were unavailable.[26] Solandt's efforts to increase efficiency paralleled the methodology of organization he had applied during the polio outbreak, proving itself once again.

As the bombing increased during the Blitz of September 1940, so did depot responsibilities. Solandt described the many other jobs of the depot, such as the maintenance of transfusion kits and preparing plasma from whole blood for treatment.[27] This latter point was critical to the war effort. During the 1930s, many physicians had found that plasma and blood serum could be effectively used to treat patients suffering from traumatic shock just as well as whole blood. Unlike different blood types, plasma and serum could be used with any patient for this effect, and could also be stored longer than whole blood.[28] Solandt had been introduced to these efforts while working with Best and his brother at U of T. Indeed, Best and Don Solandt would spend much of the 1930s and the war years working together on blood concerns, including the formation of Canada's own and first nationwide blood donor service and the storing of dried plasma.[29]

Solandt's depot had also created a Mobile Resuscitation Unit to work in the field. The depots originally had no mandate to administer blood or in the assistance of wounded.

However, during the clinical trial of fresh and stored blood (*Lancet*) the Depot staffs began to give transfusions, and this side of the work has been considerably extended. Experiences in air raids soon showed that many of the smaller hospitals were in great need of assistance in resuscitation of badly shocked casualties. The Depot now maintains a day and night transfusion service, with never less than one doctor ready to go out and give transfusions and to supervise the resuscitation of cases of shock. In addition, many transfusions are given to ordinary civilian patients.

The Depots are gradually building up a reputation for careful transfusion technique and are frequently called upon to transfuse specially difficult cases.[30]

Research and "Crush Injuries"

Every task took second place to serving the blood needs of hospitals, but any lull was taken up with research. As a member of Drury's Blood Transfusion Research Committee[31] of the MRC, Solandt tabled a report outlining the research of his depot. Most research investigated shock among air-raid casualties, though many members of the staff were engaged in other work. Patrick Mollison worked on the survival of transfused erythrocytes, which were preserved in various solutions. Doctor E.F. Auebert and his team were finishing work on suppression of agglutinins on pooling, serum, and plasma. Research at the depot was under Solandt's direction. Doctors van der Watt and Pierce had made measurements of venous pressure in blood donors before, during, and after bleeding. Doctor R. Mayon-White was engaged in controlled experiments on donors to see whether iron therapy is necessary to accelerate the restoration of the haemoglobin after bleeding and, if iron was required, to determine the effective dosage. Miss I.M. Young assisted Doctor Mollison and conducted her own experiments on the relationship between mechanical and saline fragility of stored blood.[32]

Incredibly, despite his responsibilities of running the depot, Solandt produced two articles of note based on his experience during the Battle of Britain. The first warned Britain's medical community about the toxic effects of sodium alginate, a substance many thought might serve as a substitute for plasma or serum in treating shock.[33] The second dealt with the appearance and rise of a shock condition previously unrecognized and named "crush injuries." They occurred in persons who had been buried for several hours beneath masonry or heavy debris, usually with a limb pinned for more than a few hours under the debris. The patient would show "initially some redness and blistering of the compressed limb, which later swelled. As fluid accumulated in the limb, haemoconcentration occurred and the blood pressure fell. The blood pressure level could usually be restored to normal by plasma infusions, but soon signs of damage to the kidneys appeared and about a week after the injury the patient might die in uraemia, with hypertension and with an increased content of potassium in the blood."[34]

The concept of crush injury came from observations made by Doctor E.G.L. Bywaters and his colleagues at the British Postgraduate Medical School. In March 1941, data on some sixteen examples were published.[35] Solandt and Mayon-White contributed one article in an MRC pamphlet on crush injuries published in 1941.[36] Their article provided a case study concerning a girl pinned under debris in October 1940. The horror of the scene is dealt with in objective and clinical detail. "A girl aged 11 was pinned under bomb debris for about three hours on a night of October. Members of the rescue party stated that she was lying on her right side, body tilted head downward at an angle of 45 degrees, the legs acutely flexed at the hips and knees. In this position she was compressed, though not actually crushed, by the dead bodies of her parents; a heavy beam lay upon her left thigh."[37] Solandt administered aid the night after she arrived in hospital. She had smoky red urine, positive for blood.[38] She appeared to have uraemia, the accumulation in the blood of nitrogen-bearing waste products (urea) that are usually excreted in urine. Her condition deteriorated quickly. She was awake and irritable until an hour before her death at 4:30 p.m., a week after her admission. Attempts were made to change her blood, including the injection of plasma. Unfortunately, nothing relieved her, but the post-mortem examination revealed the kidneys to be large and pale. Under a microscope, they showed granular debris within the tubules, indicating the relationship of kidneys to "crush injuries."[39] Treatment was still experimental when Solandt wrote his article, focusing on correcting urinary output by means of heat to the loins, saline dilution of plasma protein, increasing blood volume and pressure through the use of serum, and the use of diuretics such as caffeine. Solandt left this field of inquiry before a treatment could be found.[40]

Conclusion

The uncertainty and constant peril of the war environment created new demands on Solandt, the likes of which had only been hinted at during the polio epidemic. Where sixty-six polio cases had arrived over a few months in Toronto, more than fifty casualties could arrive at hospitals in a day, demanding blood that the hospitals did not store. For Solandt, the depot served a vital function in the medical arm of the war when the casualties became that high. "Even the most adequately staffed hospital cannot cope with such a rush and at the same time call up and bleed a large number of donors. It is in such emergencies that the Depots have proved indispensable. In several cases 30 to 50 bottles of blood and similar amounts of plasma have been delivered to one hospital within a few hours. Calls for 20 bottles of blood at one time are not at all unusual, and have always been filled promptly and without difficulty."[41]

Until his work for the London depot, Solandt had not considered the nature of modern warfare. Indeed, he carried a general disdain of it and the soldiers who toiled in it. Such a view shifted when war was not theory but an environment in which he would have no choice but to work in. And the reality was telling. In 1940, he witnessed the fury of modern war as it was waged on civilians, including the young girl who died of crush injuries despite his best efforts to save her. Such events touched him in deep ways, not just regarding war's destructive nature, but also the courage of those who confronted it. To the end of his days, Solandt remembered the selflessness of the British during the air raids. "Anyone who has seen people pinned under the wreckage of bombed buildings or who has worked in a hospital ward full of those who have been mutilated by bombs will know what I mean. Almost without exception the injured asks only for the bare necessities and keep their pain and their troubles to themselves[,] feeling that there must be others who need help more than they do. This rugged disregard of personal problems is by no means confined to those who are actually injured."[42]

From the Battle of Britain onward, Solandt remained impressed by the courage of the British civilians he encountered while working at the depot. "The London donors have shown a magnificent disregard of personal safety in coming to be bled in spite of air raids. Even in the heavily bombed districts there has been no decline in donor response. Many donors have come to be bled within a few hours after the destruction of their home by a bomb."[43] Solandt also loved to share the anecdote of picking up an elderly woman at 4 a,m. She was her way to her son's house, because hers had just been bombed. Solandt expressed regret, but the woman didn't feel that way. "She said that the bomb had been aimed at the Hawker factory and it was much better that it was merely her home that had been destroyed."[44] Such pragmatic selflessness no doubt appealed to him.

Working with civilians in the battleground of London moulded Solandt's future opinion of civil defence and of modern warfare. As chairman of the DRB, where he was directly involved in managing Canada's scientific effort on atomic, chemical, and biological warfare, he would often recall his depot experience to make his opinion on conventional and unconventional ordnance perfectly clear. For Solandt, there was no good or better or morally superior way to harm human beings in warfare. They were all horrible. "Many propagandists would have us believe that there is something particularly vicious and inhuman about the anti-personnel effects of atomic, biological and chemical weapons. They suggest by inference that old-fashioned bombs and bullets produce results which are by comparison humane and not too unpleasant. I am sure anyone who has seen wards full of people maimed and mutilated by high explosives would firmly reject this idea. I can imagine no injury more horrible than some that I saw during my year with the Blood Transfusion Service."[45]

By the end of the Blitz, Solandt's depot had performed well. The style of management he had employed during the polio epidemic had proved just as effective at the depot. By maintaining high standards, fighting inefficiency, and personally supporting quality work, he had managed the depot through the dangerous year of 1940. His personal leadership kept all the variables of the depot in line. Solandt did not suffer fools gladly, nor did he tolerate graft or anyone not working to the best of his ability. His intellect and efficiency were occasionally intimidating for the staff, especially upon his arrival, but soon the team found in him a leader who worked harder than anyone to make the depot effective and supported their research and tasks with praise, if they were willing to meet his standards.

His success was soon noted. In late 1940, MRC Secretary Mellanby would select him to create and lead the Physiological Laboratory at the Armoured Fighting Vehicles School in Lulworth, Dorset. The British Army had witnessed problems of tank crews passing out during firing exercises and asked for scientific and medical aid in determining its cause. Solandt was replaced at the depot by Doctor J.F. Loutit.[46] At Lulworth, Solandt would make his name not only in managing a laboratory, but as a pioneer in the emerging applied science known as operational research. Both experiences built upon his emerging capacity to conduct and manage science at the behest of the government, first tested at the Sutton depot.

6 Tank Doctor:
Omond Solandt, Director of the Physiological Research Laboratory, Lulworth, 1940–1942

From late 1940 until the end of 1942, Solandt concerned himself with the dynamic relationship between man and machine in armoured fighting vehicles (AFVs) as the first director of the Physiological Research Laboratory at the AFV Gunnery School, Lulworth, Dorset. It was quite a change and challenge. Solandt had no knowledge or experience of tanks, and only a general pre-war impression of soldiers that was not flattering. What he did have was his medical background, his life-long interest and training in mechanical arts, and a thirst for solving complex problems. Within a year, Solandt established himself as an expert on tank design and operation. His laboratory was the premier research centre for solving problems with tank design in the country. His efforts at Lulworth would earn him a U.S. Medal of Freedom with Bronze Palm, and contributed to his receiving the Order of the British Empire (OBE) in 1946.[1]

At Lulworth, Solandt had two main objectives. First was conducting and directing research on physiological concerns regarding AFVs. Second was delivering these reports to the MRC's Military Personnel Research Committee's (MPRC) Sub Committee on Armoured Fighting Vehicles. This body consisted of members of the MRC, the War Office, the Royal Army Medical Corps (RAMC) and, most importantly, the Ministry of Supply. Tank design was their responsibility, facilitated by their Directorate of Tank Design (DTD). It was up to Solandt and other members of the subcommittee to persuade DTD of the validity of their work at Lulworth. The official history of British tank development at the Ministry of Supply includes no mention of Solandt or the MRC's AFV laboratory, its reports, or subcommittees.[2] While the MRC official history includes a general report of Solandt's work, this chapter will add substance to previous sketches and discuss the friction Solandt encountered between scientists, government, and industry. At Lulworth, Solandt learned the difficulties of changing government policy based on scientific research. While it was initially frustrating, he soon became adept at it.

At Lulworth, Solandt also became an early supporter and, indeed, pioneer in the use of operational data from the battlefield for wartime scientific research. He championed the MRC's decision to send field teams into battle areas, and selected his friend Laurie Chute, a captain in the Royal Canadian Army Medical Corps, to conduct research on AFVs in the battlefields of North Africa in 1942. In turn, Chute's in-theatre data were used to make the Lulworth teams' work more realistic and harder for the military to dismiss as mere theory. By championing the use of battlefield data, Solandt earned his early reputation as pioneer in the wartime science known as operational research (OR).

The Origins of the Lulworth Laboratory

At a general meeting of the MRC on 22 November 1940, the War Office expressed interest in physiological investigations of tanks and their crews. The MRC agreed to appoint Solandt to do the research on the council's behalf at his present salary of $950 per annum,[3] and secured agreement for his appointment with the War Office on 4 December 1940. After he visited the requisite sites of his lab and met with authorities, Solandt's initial mandate was to study tank problems under "practical conditions" at the Gunnery School at Lulworth and with a field unit at Luton to study the layout for the A.22 tank, or Infantry Tank Mark IV (the "Churchill").[4] By 9 December, with unanimous approval, Solandt was secured as a fulltime research physiologist for armoured vehicles.[5]

The Physiological Research Laboratory was constructed at the Gunnery Wing of the Armoured Fighting Vehicles Training School at Lulworth in late winter of 1940 and early 1941. Solandt's team consisted of Professor I de Burgh Daly, Doctor E.E. Pochin, G.L. Brown, and Doctors J.A.B. Gray, J.A.V. Bates, and P. Hughes Jones. Solandt took Doctor Mayon-White from Sutton and secured old friend Captain Laurie Chute to join the team.[6]

The MRC's research-management approach of the lab consisted of providing Solandt with cheques to use at a local bank to secure necessary materials and housing. If Solandt could not satisfy his needs, he referred to Doctor Landsborough Thomson, the MRC's chief manager, who had an uncanny knack for finding equipment fast. This management system, like Drury's approach to the depots, was predicated on two principles to be effective: that people were both sensible and honest. It worked remarkably well and Solandt would employ it later at AORG and DRB.[7]

Carbon Monoxide and Ventilation

The military's main concern was tank crews losing consciousness during gun trials. After disabusing them of the notion that it might be a neurological problem, Solandt focused his efforts on measuring the effect of carbon monoxide from gun fumes, since he believed it was the most likely culprit. This subject dominated his time at Lulworth.[8] As Solandt wrote at the time, carbon monoxide (CO) is a colourless and odourless gas that is lighter than air. It is formed whenever carbon is burned with insufficient oxygen. It can affect living tissue by combining with iron contained in the respiratory enzymes within the cells, or combining with the blood's haemoglobin. Normally, oxygen forms a compound with haemoglobin in the blood and then is carried from the lungs to other tissue. CO combines with haemoglobin in a similar way, and haemoglobin has a greater affinity for CO than oxygen. It is this combination of CO and haemoglobin that poisons the body's ability to transport oxygen, so the person suffering from CO poisoning is suffering as if experiencing severe anaemia. The effect and rate of the poisoning is determined by the concentration and duration of exposure.[9]

Solandt maintained the research methodology that he had mastered under Best. He had produced articles on the danger of CO in Canadian industry, but he still thoroughly reviewed British and American medical literature on CO poisoning in the 1930s,[10] then got up to speed on CO concerns in the military. Explosives produced CO,[11] and exposure occurred for both sender and receiver. Gunners, especially of automatic weapons, were at small risk from the CO produced by their cartridges if the weapon was fired in the open. The risk increased when the weapon was fired from within an enclosed space, such as a pillbox or AFV.[12]

CO was only one part of the complex system of concerns for AFVs. Solandt's March progress report to Colonel Bond, head of the AFV Directorate at the War Office, on the Mark IV Cruiser used a "diagnostic" or systematic approach to understanding the operation of the tank and its crew as part of the same system. The following points were ranked according to importance, the first four being most critical.

1. Ventilation in tanks
2. Inter-communication in tanks
3. Visibility from tanks
4. Interior lighting from tanks
5. Entrance and exit from tanks with special reference to the rescue of wounded crew
6. Seating and driving positions in tanks
7. Noise and vibration for tank crews.[13]

Ventilation of the auxiliary turret of the Mark IV Cruiser was the greatest concern. Without field tests yet done, Solandt offered some hypothetical directions for change. A positive pressure system in which the indrawn air could be passed through a gas filter offered the crew the greatest possibility of achieving "maximum protection" and was found effective in trials against poison gas. No trials had yet been done regarding gun fumes and heat, which Lulworth would rectify. If this system was not practicable, localized ventilations around guns might be a substitute that would allow for maximum protection, combined with a large and controllable draught from the engine fan through the fighting chamber. When gas was encountered, the cooling ventilation could be sealed. Gun ventilation could still render weapons fireable, but the penetration of poison gas into the fighting chamber would be delayed because of the slow rate of air change. This plan depended upon the invention of some suitable method for gun ventilation.[14]

Before he had even heard the term *operational research*, Solandt wanted field tests of these points to be as realistic as possible. "Any method suggested should be tested under service conditions to determine the concentration of fumes in the fighting chamber and the effects of such concentration on the crew."[15] The report also discussed work needed on communications, visibility and lighting, and evacuation techniques, all done by specialists and all with the goal of helping the designers know what was required for a tank to be physiologically and systematically efficient.[16]

Early Friction

Solandt made a bad impression during his early days at Lulworth. He noted to his immediate superior, E.A. Carmichael, secretary of the Body

Protection Committee, that if current tank trials on the Mark IV at Farnborough proved it was not yet safe to use, breathing masks with tubes drawing air from outside might be a short-term solution, if modified to be used with the Rawdon-Smith intercommunications system. Solandt also requested closer liaison with the Ministry of Supply's DTD if any of Lulworth's research was to be turned into effective tank design.[17]

Carmichael had no concerns about Solandt's report, but many about his attitude. He wrote Mellanby in early April that Solandt had not yet produced worthwhile results because he was "dashing about the country," discussing his research with tank people. A probationary period to get Solandt grounded was discussed. Carmichael also agreed with Doctor Brown that Solandt did not have much "push" in the political arena of policymaking,[18] likely because, instead of sticking to the lab, Solandt was "interviewing people connected with tanks" and expressing "opinions without experimental facts to back them up."[19] Glowing reports from Libya were arriving with no concerns about gas fumes attached. Carmichael feared that the lab's work would fall on deaf ears if he did not have better reports, especially about turret research. "I told him that he must have experimental facts to back up his opinion that the turret is too small for efficient work and have further suggested to him that he carry out experiments which might indicate the minimum dimensions for an efficient working turret." Carmichael believed Solandt was capable, but "his constant cry to be on committees, etc., disappoints me. I have done my best with the tank people for them to have him on the relevant committees, but until he has shown his value it is no easy matter to press for co-operation on the like of Tank Users Committee. Were he to deliver some goods first, then this difficulty would melt away."[20] Carmichael feared "it may be necessary to look for someone else" if they did not persuade him to focus on research.[21] He failed to mention, however, that the suggestions he had made to Mellanby on the future of research at Lulworth had largely originated from the report and letter Solandt provided him on 5 March 1941.

After talking with Mellanby, Solandt agreed to settle down. Still, he was adamant that unless there was a current presence, meaning himself, on these committees, the results would not make much difference. The experience in Libya had put the "tails of the tank people up very high," he told Mellanby, "and it was a difficult matter to persuade them that further investigations might reveal means of great improvement." Even after being told to calm down, his confidence in his argument never wavered.[22]

Carmichael's criticisms were not without merit. Solandt was aggressive and confident, in contrast with his relative inexperience in British military and government circles, leading to doubts on the threat of CO poisoning. But Solandt's approach, like everything else in his professional life, had a fierce logic. He studied as much of the relevant literature as possible, then moved on to practice. Since tanks were not readily available for research, he talked to everyone involved in tanks and their design. Solandt feared talking to no one, especially engineers, given his technical training. His ignorance of tanks could best be corrected by talking with tank designers. Learning directly from them how they did their job, before he made any suggestions on how to change their work, was a keystone to his operational research approach.[23] He also clearly understood the need for networks of influence if his team's work was to translate into practical value, especially before the Sub Committee on Armoured Fighting Vehicles existed. Solandt knew his influence was limited but paid close attention to where it could be effective. A year of successful and important research and results stymied any further criticism of his approach by Carmichael. Soon, Solandt's requests to have closer liaison directly with DTD were acknowledged and supported by Mellanby.[24]

Early Work of the Phsyiological Laboratory

The Lulworth team's earliest reports in May 1941 were made in conjunction with Doctor C. Lovett-Evans from the Chemical Warfare Experimental Station, Porton, who had also been working on chemical warfare gassing trials for AFVs.[25] The joint gassing trials indicated that the greatest producer of CO in the Infantry Tank Mark III was the Besa 7.92 mm medium machine gun. The trial was done to assess risk from CO poisoning when the Besa gun was fired in an Infantry Tank Mark III with the tank stationary and the engine off. Nine belts were fired (2,025 rounds) in thirty minutes with mild stoppages upsetting the firing rate. At the twelve-minute mark, four belts had been fired, and the trial was paused when the CO level became dangerously high. A further four belts were fired before the end of the trial. See table 6.1 for results.[26]

They concluded that the Infantry Tank Mark III had unusually large ventilating apertures in the main turret and it might be possible to fire the Besa gun with the engine off without actually asphyxiating the crew. Still, the CO concentration during the firing of the eight belts (1,800 rounds) in thirty minutes would render the crew unfit for action if the

Table 6.1. Early Tests on CO Concentration at Lulworth

Crew	CO concentration in continuous sample 0–35 minutes	% saturation of blood with CO after firing
Loader	.20	35
Gunner	.09	22
Driver	.06*	16

*Sample taken from 15–35 minutes only

firing were prolonged or repeated in less than three to four hours. "The results of this trial emphasize the conclusion that the Besa gun should never be fired in closed space without adequate mechanical ventilation." A series of Porton reports was used to bolster the point.[27]

Solandt's first major report on CO poisoning was released on 11 June 1941. It filled the gap of the previous trials, measuring how breaks from CO saturation would affect the crew. Unable to obtain their own tanks for trials, Solandt and Mayon-White followed Best's method: they used themselves as test subjects. They exposed themselves to concentrations of CO for three periods of a half-hour, followed by breathing pure air for three and one-half hours, to stimulate the effects on tank crews should they be engaged in three short, sharp engagements during a day of fighting. The concentration was based on those to be used during gassing trials. They lay down for a half hour before each exposure to CO. Following the rest, they breathed a CO-air mixture made up in a 1,000 litre Douglas bag. The same mixture was used for all three exposures. The subjects breathed the mixture through a small oro-nasal mask fitted with valves. Blood samples were taken immediately before and immediately after the gassing. Between exposures, Solandt and Mayon-White carried on with regular lab work and ate a meal as normal. No muscular effort of any severity was induced. Solandt breathed in a heavier concentration than Mayon-White (.29 per cent to .26 per cent) to start off with and had a much harder time in doing simple tasks, suffering from vertigo and nausea as well as fever throughout the experiment.[28]

The results suggested that a half-hour exposure to .015 per cent CO could be repeated every four hours without producing serious symptoms. However, "it must be emphasized that the concentration of the CO would produce a very serious result if the three ½ hour exposures were separated by a shorter interval or were continuous. In experiment

II [the one that featured Solandt] the limit of useful fighting efficiency was reached after the second exposure. Had the three exposures in this experiment been continuous the subject almost certainly would have died." The effects of CO would also likely increase if the subjects had been physically active.[29] The report concluded,

1. A person is not seriously incapacitated for light work by carbon monoxide poisoning unless the blood saturation exceeds 25 per cent.
2. An exposure to CO that produces 15 per cent blood saturation in thirty minutes can be repeated every four hours without producing serious symptoms.
3. An exposure of CO that leads to a blood saturation of 25 per cent in thirty minutes will lead to definite symptoms of poisoning if repeated after four hours.
4. These conclusions are valid only when the subject is at rest. The blood saturations that can be tolerated during exercise is known to be considerably less.
5. From a limited experience of gassing trials in tanks, it would appear that a concentration of less than 0.10 per cent CO in air produces blood saturation of 15 per cent in thirty minutes.
6. It must be emphasized that the concentration of CO used in these experiments can tolerated only for very short periods. Even 0.10 CO may prove fatal if breathed for several hours.[30]

In June, Solandt collated the laboratory's efforts in a host of areas into "Some Notes on Armoured Fighting Vehicles" for Carmichael. The chief concerns were almost identical to those of his earlier report, a credit to his initial hypotheses.[31] Ventilation remained the most critical, involving complex considerations of fighting in war-gas environments, gun fumes, and use of engines. Research on the mode and entry of CO in different tanks was next, followed by comparing different designs for those with less protected fighting compartments against engine exhaust. Future experiments would seek to determine the presence and concentration of CO in AFVs of all types and under different circumstances, determine the source of CO, devise methods to overcome these dangers, and determine the effect on the efficiency of the human after frequent exposure over a short period of heavy concentration.[32] Solandt also laid out the research agenda for the rest of the lab.[33]

The Military Personnel Research Committee's Sub Committee on Armoured Fighting Vehicles

These were the primary concerns of the Solandt's lab when the MPRC formed its Sub Committee on Armoured Fighting Vehicles in July 1941. Solandt had worked on committees since Sutton, and before July 1941 had also dealt personally with DTD members Misters A.A.M. Durrant, A.T. Sweeney, and W.J. Mesher, as well as Carmichael on night-fighting issues, a subject Solandt had written reports on.[34] Carmichael made no ill remarks about Solandt's research approach, though he maintained Solandt should get away from the politics involved in dealing directly with the Ministry of Supply. Instead, Solandt used the subcommittee as his arena to change tank design.[35]

The first meeting of the MPRC Sub Committee on Armoured Fighting Vehicles occurred at the London School of Hygiene on 7 July 1941. Mellanby chaired the committee and Carmichael was secretary, with representatives from Lulworth, DTD, Porton, the Royal Army Medical Corps, and interested parties from the Commonwealth nations and the United States.[36] Solandt led the discussion on tank ventilation and CO poisoning. He discussed different pressure systems, the effect of CO poisoning on psycho-motor mechanisms including judgment and other psychological effects, and the need for work on CO fumes in armoured cars, HQ vans, and other enclosed vehicles. Solandt and Doctor Craik also discussed vision and auditory problems associated with operating tanks. The committee concurred with the research, and the reports were to be sent to the requisite authorities at the Ministry of Supply.[37]

While Solandt waited for his research to turn into policy, he led and managed his team while keeping abreast of similar work done throughout the services and civilian industry. He made contact within the Admiralty, the RAF establishment at Farnborough, the U.S. Armoured Fighting School at Fort Knox, the telephone service committee on communications, and more. He usually led discussions of his team's efforts at the subcommittee, primarily on ventilation, but also allowed key staff members such as the vision specialist Poachin to make their case in their field of expertise. He also kept a close eye on DTD, such as notifying Carmichael that a Colonel Ross of DTD should attend the next meeting of the Sub Committee since he was "sympathetic" to the work at Lulworth.[38] However, Solandt's influence seemed limited. When Solandt realized that his team's research was not being fully considered by DTD, he made his case known.

The Greatest Nemesis: The Besa Machine Gun

Solandt's greatest frustration was his research regarding CO production from the Besa machine gun. By August 1941, when Solandt submitted his first report, it was clear that this weapon was the greatest CO problem on AFVs.[39] That fall, deflector bags were used to reduce CO when the machine gun was at the maximum rate of fire.[40] By December, comparative field trials had shown the Browning machine gun generated much lower concentrations of CO than the Besa.[41] By February 1942, discussions of gas-proofing tanks led Solandt to again warn the subcommittee about CO poisoning in tanks, especially those sealed to fight in chemical warfare. Lovett-Evans feared for the British forces in the North African theatre, since the desert was a prime environment for CW. The subcommittee formed an informal investigative team to research these problems, including Solandt, Lovett-Evans, MacLeod, Ross, and Carmichael. As the result of concerns about toxicity of fumes from gun trials at Larkhill, Solandt led a discussion as a consultant on design changes. If his advice was taken early enough in the design phase, he hoped, positive change could occur.[42]

By spring 1942, there were some positive results. An effective working relationship between Lulworth, Porton and DTD on gassing concerns had emerged, especially on gas-proof clothing. But Solandt pressed on about CO contamination in turrets. Colonels Timmis and Franklin from DTD finally agreed that they would inform the authorities of the need of a General Electric Company (GEC) fan, and the committee backed the report and made sure the ministry received it. Lulworth's report on CO poisoning in the Crusader Tank stated that the Besa gun could not be fired at operational rates without putting the crew at risk without adequate ventilation. The GEC fan was also needed in the Churchill III tank. Timmis and Franklin were to make this clear to the ministry. For the next year, Solandt argued that the Besa and six-pounder Quick-Fire (QF) gun were among the chief CO hazards in the Churchill tank if ventilation or weapons choice did not change, but his position met resistance. Franklin argued against the Browning because of difficulties in procuring them and their ammunition. Timmis argued for the operational merits of the Besa. Still the MRC forwarded the report to the Army Council in the hopes of getting a decision.[43]

By May 1942 Mellanby wanted a more forceful "push" regarding Lulworth's reports and the committee's recommendations to change tank design. Solandt condensed his team's reports into a formal memorandum that outlined sources of fumes and their character, discussed safety standards and its relation to the General Staff's requirements of rates of fire, and reviewed weapons options. The key noxious gas concerns were with the Besa, six-pounder, three-inch twenty cwt, and twenty-five- pounder field guns. If they were to be used, effective ventilation was critical, and specifications for ventilation required were soon to be generated. Mellanby agreed with Solandt. While DTD should be responsible for seeing that manufacturers met the vent requirements and undertook the acceptance trials, Solandt, while not directly responsible for acceptance trials, was to be in close touch with them. The committee agreed that Ross, Timmis, Lovatt-Evans, Bedford, and Solandt would report on specifications for ventilations of tank design to remove noxious fumes.[44]

But as the lab continued to report its research over the next year, and the committee continued to affirm their work, there appeared precious little interest from the Ministry of Supply. During the 6 July 1942 meeting, Lord Falmouth, an accomplished engineer and committee member for the Ministry of Home Security, wanted to know if any action had been taken after all Lulworth's work. Anxious that the authorities were now alive to the dangers, Solandt said changes to design were at hand, such as the addition of appropriate doors in the bulkhead, alteration of exhaust fan outlets, and use of improved fans. Design of future models had been studied and advice given and accepted for improved and, he hoped, adequate ventilation. But opposition from DTD remained.[45]

One reason for the Ministry of Supply's general resistance was their overriding concern for tank production to the exclusion of all other factors. From 1939 until 1942, quantity trumped quality, as tanks were desperately needed after the tragedy of Dunkirk and before the start of desert warfare in North Africa. Institutional inertia and the general distrust of outsiders giving advice to private industry contracted by the ministry for tank design and production would compound distrust of the tank doctors at Lulworth.

Solandt and the Need for Operational Data

By late 1942, however, quality, not just quantity, was now on the ministry's agenda. Support for change grew. One component of this shift in perception was the integration of battlefield experience into Lulworth's field trials and experiments. Solandt was an early and strong proponent of applying operational data to government research. It allowed for more accurate and relevant experiments and trials, and this improved accuracy gave Solandt's arguments greater gravitas on the subcommittee. It was easy for the Ministry of Supply to dismiss or ignore the Lulworth results when Solandt was the subject of the experiments. But Solandt and his laboratory were harder to dismiss when their arguments were based on the operational experience of the British Army in North Africa.

The need for operational data was championed at MRC. Mellanby had witnessed the need for such information during the First World War, where the gathering and then implementation of timely casualty data was critical to saving lives.[46] By 1941 Mellanby and Carmichael prepared to send a Colonel Marriott, RAMC, to Egypt to gather data relevant to MRC research with War Office approval. Solandt supported the decision and consulted Marriott about his specific tank concerns.[47] From 16 March to 18 June 1941, Marriott travelled over 2,500 miles across the Western Desert, Canal Zone, and Palestine conducting preliminary research and analysis of subjects including the "Physical Efficiency of Tanks" for Solandt.[48] The data centred on the Seventh Armoured Division, and Marriott separately interrogated seven combatant officers and three medical officers from it.[49]

Tanks, he reported, were closed for only about a half hour at a time because of the heat and a commander's need for full range of vision because vision equipment was inadequate. If more than two belts of Besa were fired at one time, stuffy and malodorous conditions followed, and there were cases of dizziness and one of lachrymation, but no reported cases of passing out, perhaps because the turret was open.[50] The greatest complaints were with night vision when tanks were on the move, poor field of vision due to bad telescopes and periscope, the vulnerability of the tank floor, and the difficulty of removing wounded. Communication comfort and needs meant that little clothing was worn and helmets were never used. This was a preliminary view of how tanks operated in reality, not theory.[51]

The subcommittee praised the report on 18 August 1941, and Mellanby worked with Carmichael and Solandt to prepare a request for more observer missions. In a draft entitled "The Need for Scientific Observers in Battle Areas" the report argued that scientists at the front could obtain data that had been absent from the official records. "Statistics about injuries to despatch riders, to specialized troops such as personnel of armoured fighting vehicles ... have not been obtainable from the official archives." All field research needed to be sent back fast and be put into action if it was to be of any real value.[52]

Solandt believed the value of scientific observers would come from sending out "two or three junior observers under the direction of a more senior person. The junior observers who would be attached to armoured regiments, must obviously have some scientific interest or training." However, a typical regimental medical officer would not be sufficiently interested in the work to do the job properly. Junior officers could be sent to Lulworth to observe the laboratory's objectives and operation so they could provide Solandt with data relevant to their needs. In the field, the observer would need months for observation, research, and analysis, but the "results would be worth the effort."[53]

By 29 September 1941, the War Office supported the subcommittee's desire for more information from officers with experience in actual fighting conditions and arranged for Solandt to interview officers returned from the Middle East,[54] but this was not good enough for Solandt. He wrote Mellanby emphasizing the need for field research. "If there is a possibility of sending someone to the Middle East, I think that it would be a good plan to send some Army man here for training with a view to going to the East. Captain Chute would be very willing to go but I do not see how we can spare him."[55]

Mellanby agreed to take Chute for an overseas fact-finding mission and increase Solandt's staff. He sent his memo on scientific observers to the War Office in December 1941. When reports came back from North Africa on the need for research on burn casualties in January 1942,[56] the MRC was prepared to send a team of scientific observers to collect data that would lead their research at Lulworth and elsewhere. With their agreement, the next MRC team to collect battle data from North Africa was formed. Chute, on Solandt's recommendation, would go for the Lulworth team. [57]

The mission to North Africa was designated No. 1 Medical Research Section (1 MRS), commanded by Lieutenant-Colonel W.C. Wilson, a professor at Aberdeen. Wilson would work alongside Solandt's old Cambridge friend, Major E.T.C. Spooner, a bacteriologist.[58] Wilson and Spooner studied disease, burns, and psychological and physiological effects of weapons, including wounds-shock and body armour. Chute, now promoted to major, worked exclusively on the experience of armoured vehicles and their crews, producing eleven reports on the subject. As Ronnie Shepard has shown in his analysis of Solandt's contribution to operational research, Chute's reports became the bedrock for Solandt's research agenda at Lulworth.[59]

Chute arrived in North Africa in April 1942 and submitted reports throughout the year until returning home after being injured in the winter of 1943. Unlike Marriott, Chute collected data from the front lines as well as the rear areas of battle, and encountered general resistance for his efforts from the General Staff and the Medical Corps in North Africa.[60] But the mission was a success from the MRC and Solandt's point of view. Chute's data provided the evidence Solandt needed to challenge assumptions about how tanks and their crews actually operated in battle, giving his trials the weight of operational experiences.

Solandt's own perspective of tanks as a man-and-machine system was validated by these reports. Poor telescopes and periscopes meant the turrets were often left open so that commanders could have a full range of vision. This exposure led to a higher number of commander casualties proportional to the rest of the crew. Open turrets likely mitigated the ventilation concerns, with a high price. Marriott's preliminary data on leg burns were contested by reports that most burn injuries in tanks occurred on exposed areas because crews generally wore as little as possible there.[61]

Solandt argued throughout the mission in Libya for the value of Chute's work and the need for more data on the actual conditions of men in tanks during operations. It was difficult to discuss requirements without knowing the operational reality of men in tank battles. General Richardson of the War Office said Chute's information would be returned as fast as possible. Timmis argued that there had been no complaints of crews about heat in tanks or of heat exhaustion, but Chute's reports made it clear this was not a validation of the status quo, but likely a result of the open hatch, the absence of full uniforms, and ventilation provided by the tanks while moving. Trials done by Chute in Libya with full gas-protection equipment in a sealed tank made it clear that dehydration would likely cripple most tank crews in a gas environment if they wore the current gas-proof gear.[62]

Solandt used Chute's report on casualties for his own analysis, tying them all to the importance of improved tank design, especially in regards to casualties. There was sufficient evidence to deduce that many tank commanders were the sole casualties because officers kept their head out of the tanks. This exposure was caused by the need for a maximum field of vision, which was denied by current periscopes and telescopes.[63] Crews also preferred to have the hatch open to allow speed of escape in case of fire. Solandt pressed for more analysis on damage to tanks, casualties among the crew, and the effectiveness of armour-piercing missiles. He felt the answers would aid determination of the vulnerability of certain tank sections.[64]

The evidence on commander casualties convinced War Office representative Major Hartford of the value of Chute's research. The vision and cupola work of Doctors Poachin and Craik was now critical, and there was general agreement that the more operational data Solandt could get his hands on, the better.[65] The Sub Committee held a special meeting regarding Chute's findings on 1 October 1942, in which everyone praised his work. Slowly, the committee members from the Ministry of Supply succumbed to the persistent pressure of the Chute reports, now that lives were clearly at stake if turrets, periscopes, and ventilation concerns were not redressed.

Still, the ministry's view of Lulworth was coloured by systemic lack of interest or knowledge of their work. In August 1942, Doctor G.P. Crowden at the London School of Hygiene was contacted by the assistant director of tank design (turrets and ammunition). Crowden had worked with them before the war on turrets and heat exhaustion. Reports coming back from North Africa had made these issues pressing again, and "we are anxious to avail ourselves of the best possible advice in this connection." Crowden informed Mellanby, "It is amazing to think that the writer of the letter did not know of our work at Lulworth on tank ventilation and associated problems."[66] Mellanby agreed and wrote the assistant director: "Your Department of Tank Design has already got a laboratory, dealing with questions of ventilation in tanks, in being at Lulworth. It is the Physiological Research Laboratory of the Medical Research Council and is under the charge of Dr Solandt."[67] In September 1942, Mellanby pushed the Ministry of Supply from the top, contacting Oliver Lucas, controller-general of research and development, who diplomatically forced the issue about Solandt, his lab, and their reports. Despite the presence of DTD representatives at committee meetings, "there is a feeling that much of its work has not been fruitful, either because of lack of interest or lack of drive of those responsible." Important work from Lulworth had been done on CO, the Besa machine gun, and proper ventilation to protect fumes, as well as night vision and optics for gunnery. This work had been accepted only in a piecemeal fashion. "I am sure, however, that if you would kindly look into this matter, you would find that a great deal more could be done to improve tank efficiency from this particular angle."[68] Mellanby emphasized Solandt's leadership as critical to the lab's success. "Dr Solandt has been in charge of this work and he has apparently been regarded as so successful by the War Office that they have asked him to become Director of Gunnery Research at Lulworth."[69] By November, the Ministry of Supply created their first ventilation panel, which was concerned with gun fumes and chemical warfare, its members selected from DTD, Porton, and the MRC. Before he left working for the MRC, Solandt's long-time goal of close cooperation between scientists and government agencies on tank problems had been established.[70]

Conclusion

Solandt was succeeded at Lulworth by Doctor Brown, who inherited an innovative and valuable research arm for Britain's war effort. Sadly, Brown failed to push Lulworth as hard as his predecessor, but Solandt had no time to worry about the failings of his successor, for new challenges lay ahead. By late 1942, Solandt's contributions at Lulworth, his innovation, research, and leadership, had been noticed by the War Office, who appointed him to work for them in what would become the British Army Operational Research Group (AORG). Solandt would maintain close contact with the Lulworth staff and his membership to the AFV subcommittee, but his career soon took him beyond the concerns of AFVs alone. The ventilation difficulties were soon ended with proper ventilation in British AFVs and the use of the less CO-heavy U.S. Sherman tank.

Solandt's two years at Lulworth were marked by struggle and success in translating effective research into changes in government policy on tank design. He left just as the MRC began to force the issue with the Ministry of Supply, though British desire for the U.S. Sherman tank curtailed some of their enthusiasm for changing their own tanks. Solandt left Lulworth a very different manager of science than he had been at Sutton. The skills he had honed under Best were still employed, but he had made them his own in two ways. First, he applied them to solve complex military problems across a spectrum of different fields (casualties, battle-damage analysis, vision and mobility, etc.). Second, he applied them to the management of these problems across a spectrum of professions and disciplines, which included maintaining close contact not only with the relevant literature, but also the main practitioners. This meant developing professional relationships and contacts with scientists, engineers, government officials and the military. As his career flourished with the opportunities generated by the war, Solandt's own research began to take second place to management and leadership of scientific establishments. His ordered and efficient intellect found fertile ground in working as the translator of government need into scientific research, and pushing that research to change government policy. This facility would become another pillar of his success as a state scientist, witnessed as he rose to become superintendent of AORG.

7 Managing Science: Omond Solandt and the British Army Operational Research Group, 1943–1945

For Solandt, the final war years were filled with promotion and frustrations. In 1943, he began conducting research on tank problems for the British Army Operational Research Group (AORG). Within a year, his capability to manage diverse people and science organizations led to his selection as deputy superintendent and later full superintendent of AORG at the rank of full colonel in the Canadian Army. Solandt had tried to join the Royal Canadian Army Medical Corps while in England in 1939, but was told he would have to come back to Canada. As the Canadian Army was on its way to England, he thought he would stay put and wait until they arrived. From 1940 on, however, the Canadian authorities made it difficult for him to join. Finally, Solandt spoke with General A.G.L. "Andy" McNaughton about this problem. McNaughton was a supporter of Solandt's tank research and informed Solandt that joining the Canadian Army would mean he would likely be removed from his important position at Lulworth. By 1943, Solandt convinced McNaughton that his work at AORG would be greatly facilitated if he was in uniform and would also reduce his payable tax and thus increase his salary. McNaughton agreed and in 1943, Solandt became captain. Within days he was made a lieutenant-colonel and later full colonel.[1]

Solandt's diverse intellect and interest in science were well matched during AORG's period of diversification. In May 1945, Solandt led a fact-finding mission on the use of military science in India and Burma. During that trip, he was selected to be scientific advisor (SA) to Lord Louis Mountbatten's South East Asia Command (SEAC) with the rank of brigadier. The dropping of the atomic bombs at Hiroshima and Nagasaki ended the war in the East and denied Solandt this appointment but produced a rare opportunity to study the casualties of the atomic bomb in Japan for the War Office (see chapter 8).[2]

These accomplishments came with both friction and failures. Some at AORG disliked his leadership style and preferred his predecessor, South African physicist brigadier Basil Schonland. Solandt also came into conflict with Schonland over the use of OR sections in Normandy when the South African was SA to 21st Army Group. Indeed, Solandt failed to replace Schonland at 21st Army Group and was denied the position of deputy scientific advisor (DSA) of the War Office when no suitable replacement could be found to lead AORG.

These were tough blows for the young and ambitious Solandt, but his confidence never weakened. AORG was an even greater test than Lulworth on how to run a unique, varied, and dispersed government science operation. He accounted for himself well, learning by doing and building on his past successes. His years at AORG also revealed the organization's inefficiencies and the limits of Solandt's influence within the British government. All of these lessons were parlayed into the creation of the Defence Research Board.

Origins of Operational Research (OR)

OR was a form of applied science that became a critical and innovative component of Britain's war effort. A scientific offspring of radar research from the mid-1930s, OR began as scientific analysis of strictly technical issues. This approach expanded to include review of military operations in action and providing suggestions for improvement. By 1941, Professor Patrick Blackett, the noted physicist then in charge of Coastal Command's OR Section (ORS), wrote the first useful treatise on OR's purpose.

> The object of having scientists in close touch with operations is to enable operational staffs to obtain scientific advice on those matters which are not handled by the service technical establishments. Operational staffs provide the scientists with the operational outlook and data. The scientists apply scientific methods of analysis to these data, and are thus able to give useful advice. The main field of their activity is clearly the analysis of actual operations, using as data the material to be found in an operations room, e.g. all signals, track, charts, combat reports, meteorological information, etc. It will be noted that these data are not, and on secrecy grounds cannot, be made available to the technical establishments. Thus such scientific

analysis, if done at all, must be done in or near operations rooms. The work of an Operational Research Section should be carried out at Command, Group, Station or Squadron as circumstances dictate.[3]

Solandt's work at Lulworth was not OR as Blackett defined it. Solandt's own definition of OR was the application of scientific principles to solving complex problems of men and machines in operation and in field trials. Most work at Lulworth was of the latter variety, though the Marriott and Chute missions had convinced Solandt of the primary value of operational data to conducting effective OR. His success in this field at Lulworth led him to do OR work specifically for the Army, beginning in January 1943.[4]

The Army Operational Research Group

OR in the Army began with Blackett, who was made scientific advisor to General Frederick Pile, commander of Anti-Aircraft Command, in August 1940. Here he formed an ORS named Anti-Aircraft Command Research Group (AACRG), but known as "Blackett's Circus." By studying AA teams in action and deducing how to efficiently change their training and operation, they improved anti-aircraft fire with gun-laying radar, predictors, and searchlights even after Blackett left "Ack Ack" Command in March 1941. Its success led to OR sections being stationed at most command HQs in Britain.[5] The "Circus" also worked with the radar school formed in Petersham on the edge of Richmond Park under the directorship of radar pioneer Doctor J.A. Ratcliffe.

Sir John Cockcroft, superintendent of Air Defence Research and Development (ADRDE) in the Ministry of Supply, had selected Ratcliffe for the radar school. The "Circus" and the radar school were amalgamated into Cockcroft's organization but under Ratcliffe's direction. The group was formally recognized as the Operational Research Group of the Air Defence Research and Development Establishment: ADRDE (ORG) in June 1940.[6]

In July 1941, Cockroft selected his friend Lieutenant-Colonel and later Brigadier Basil Schonland, an eminent South African physicist and lightening expert, to lead ADRDE (ORG) when Ratcliffe joined the Telecommunications Research Establishment (TRE) in Malvern. While Cockroft would be chief superintendent of ADRDE, Schonland would have direct access to key commanders such as Pile. Schonland's original mandate focused on radar and anti-aircraft needs of the Army, including training and equipment, as well as general applications of science towards Army needs. He inherited Ratcliffe's two OR sections, one dedicated to radar under Doctor Maurice Wilkes, and the second on gunnery fire-control problems under Doctor L.E. Bayliss. Major Patrick Johnson, a physics professor, friend of Schonland's, and future rival of Solandt's, led the radar school.[7]

By late 1942, ADRDE (ORG) had sections for a wide range of concerns beyond air defence. On 16 January 1943, Schonland and his staff were separated from ADRDE and on 1 February were named the Army Operational Research Group (AORG), maintaining their equipment and facilities at Richmond Park, but now headquartered at Ibstock Place, Roehampton, and had OR sections spread across England. AORG was now responsible to both the War Office and the Ministry of Supply. Schonland was responsible to the scientific advisor to the Army Council (SA/AC) in the War Office. Sir Charles Darwin was briefly SA/AC, but Sir Charles Ellis, a friend of Cockcroft and Schonland, held the post for the remainder of the war. Administration and technical efficiency rested with the Ministry of Supply and its controller of physical research and scientific development (CPRSD), Doctor Paris. Solandt's previous difficulties with the Ministry of Supply meant he likely had no illusions of the difficulties he would face when he succeeded Schonland.[8] AORG's duties when Solandt joined were

1. to investigate the performance of selected types of service equipment under conditions obtained in field operations;
2. to collaborate with design establishments in studying the performance and use of early models of new equipment;
3. to investigate methods of using selected equipment;
4. to analyse statistically the results of selected tactical methods, whether they involve the use of technical equipment or not;
5. to advise the War Office and Commands upon the experimental planning of troop trials of equipment or not;
6. to be represented by observers at troop trials; and
7. to carry any scientific investigations that may be approved by SA/AC or the CPRSD.[9]

Operational Research Section 4

Ellis had visited Solandt at Lulworth in December 1942 and offered him the chance to start a similar OR section within the ADRDE

(ORG).[10] Solandt agreed and briefly worked for both organizations before Doctor Brown took over at Lulworth.[11] Solandt's new section was designated Army Operational Research Section Four (AORS 4) "Armoured Fighting Vehicles, Field and Anti-Tank Artillery." Its own lab was also at Lulworth. Solandt was also tasked to create an OR section at Larkhill concerned with anti-tank weapons. As such, AORS4 was split between AORS4 (a) "Tank Gunnery and Armour and Mobility of Tanks" at Lulworth and AORS (b) "Anti-Tank and Field Gunnery" at Larkhill.[12]

Solandt never forgot the Ministry of Supply's resistance to his research. Now he fought the same fight with the Armoured Corps. In early 1943, Solandt prepared a lecture for the Armoured Corps specifically on tank operational research. Tank OR focused on operational use of tanks with a view to possible improvement in three areas:

1. the tactical handling of tanks,
2. the design of tanks – both in minor details of design and general specifications, and
3. the training of tank crews.[13]

Critical information was still lacking on a host of questions regarding tank gunnery (What are the main targets? What limits accuracy? What limits range? Are there discrepancies between success in trials and battle?), armour (What weapons are responsible for most of our tank casualties? What is the best balance between armour and weapons on the tank? How vulnerable is the track?), and reliability (What are the most common defects in each type of tank? What percentage of tanks can be kept in the front line? Where are the tanks when not in the front? How can the return to battle be accelerated?).[14]

In the lecture, Solandt demonstrated the value of operational data versus intuition with the case for flame-proof clothing for the armoured units in North Africa. Reports from Libya in 1940 claimed that tanks were prone to fires when damaged, and these fires resulted in many casualties. The government translated this to mean most casualties were from burns, so there was "an orgy of designing clothing that would protect crews against flaming petrol even though none of the reports had specifically mentioned petrol burns and the design of modern tanks is such that the crew are unlikely to get splashed with burning petrol." Solandt and his lab invested time and money on clothing designs. But Major Chute's reports from the battlefield proved that few casualties came from severe burns. Minor burns were due to the lack of clothing worn on exposed areas and could be reduced if light clothing was designed. The wasted time and energy on the wrong solution could have been avoided if more research had been done on the initial problem. In short, the best questions were not asked at the start.[15]

Even while at ORS 4, Solandt's work on gun fumes was not taken seriously by the Armoured Corps. Solandt warned them of the danger, but Colonel O'Roarke, head of gunnery, volunteered to be a guinea pig to prove CO poisoning was not an operational threat. Instead of continuing the argument, Solandt agreed to the test, taking an oxygen mask and canister to the trial. O'Roarke fired continues rounds of the Besa while inside a Churchill tank. Eventually, he stopped. Solandt ran to the tank, canister in hand, and opened the hatch. O'Roarke was unconscious. Solandt administered oxygen to O'Roarke while he was still in the tank, knowing full well it was impossible to pull him out while immobilized. The risk of CO, and the difficulty of removing wounded from tanks, was made clear. When O'Roarke awoke, he backed Solandt's research.[16] When AORG was formed in February 1943, Solandt had built a reputation at Lulworth, Larkhill, and the War Office that if he said something, it was right.[17]

Solandt maintained a steady hand on research, contributing seven reports while directing ORS4.[18] Physicist H.A. "Tony" Sargeaunt, who would succeed Solandt as superintendent of AORG, worked with him at ORS4.[19] Sargeaunt doubted the tank section did much good changing minds on tank design, but he remained impressed with Solandt's leadership. Solandt walked a fine line between being a scientist and an administrator. He got on well with an Army that was sceptical of scientists telling them how best to do their business, and in this regard, Solandt did very well. Solandt always liked meeting new and different people. His quiet personality was not hard to bear, but his vast intellect and interests allowed him to talk to most professionals about their work in some detail. This earned their trust,[20] and in doing so Solandt built bridges between scientist and soldier.[21] His OR section shared a "coffee house" with the MRC scientists at Lulworth, where ideas were exchanged and problems and solutions discussed. They also shared a mess with the officers so that they could socialize and become accustomed to each other. Solandt encouraged this multidisciplined approach, a hallmark of other wartime science efforts such as the "Sunday Soviet," a gaggle of scientists who informally discussed war research and helped solve each other's problems through personal contacts. Solandt enjoyed keeping his section organized and respected. His meticulous intellect and acute pragmatism provided him a compass for seeing flaws in research proposals, steering the team away from dead ends before they started. His command of factual data was vast, so he was hard to fool. This point enforced the value of his judgment to the team. What humility Solandt had was reserved for his brother Don. According to Sargeaunt, Omond spoke of Don as if he were a Sherlock Holmes of science.[22]

Solandt maintained his aggressive presence on committees, especially those regarding tank design, though he again met opposition. The factor in tank construction that decided every other, Sargeaunt argued, was the size of the gun mounted on the tank, its recoil, and ultimately the size of tread required. Design should have started with treads and moved up. Instead, they started with suspension and thought about the weapon last. Solandt, he believed, was the first to spot this problem. He wrote a very forceful memorandum arguing that successful tanks were always mounted with big tread tires like the German Panther and Tiger, but the Churchill had terribly small treads and this was part of its defect. Solandt brought this to a committee meeting with the Ministry of Supply under the chairmanship of Duncan Sandys, who was prime minister Winston Churchill's son-in-law. As the Churchill tank was just starting its commission, Sandys ordered all copies of Solandt's critical memo burned. Another lesson learned on the risks of patronage and politics in science. Most of the reports were destroyed, though many survived because those who read it knew Solandt was right.[23] It would not be the last time Solandt was forced to destroy good research to serve political ends, and it no doubt displeased him.[24]

Solandt had his detractors among scientists, too. Physicist and mathematician Neville Mott, a future Nobel laureate, had been scientific advisor to Pile at AA Command when Cockcroft and Ellis asked him to work on tank issues. Mott found the work useless without operational data. His hatred of the work was "not helped by my dislike of the man in charge and probably his for me."[25] The feeling was mutual. Until the end of the war, Solandt employed Mott as an advisor and "human calculator"[26] to conduct complex calculations on the lethality of weapons.[27] But Mott's theoretical genius, Solandt felt, was mired by lack of pragmatism, so crucial to government science. Mott was a typical, wandering-intellect boffin.[28]

After the war, Solandt highlighted several projects of AORS4 as worthy of note:

Tank Gunnery
1. The value of zeroing, and devising, zeroing drills when shooting is not possible, which led to a study of the variation in jump of tank guns
2. An investigation of the errors of visual estimation of opening range – leading to work on the value of range-finders in tank gunnery
3. The accuracy of central laying with different types of aiming at moving targets
4. Methods of correcting fire – in A.P. [armour piercing] Shooting the "lines of site" method of correction and direct correction. In H.E. [high explosives] Shooting, the general theory of bracketing has been developed, tested, and applied to the production of bracketing drills
5. The experimental work on drills to be employed with ricochet airburst fire[29]

Tank Armour
1. Theoretical studies were made of the best armour distribution on tank hulls and turrets, together with work on vulnerability of tanks, both allied and enemy, to A.P. fire.[30]

Tank Mobility
1. The production of Tank Going Maps in conjuncture with the Met. Office, reliable data on the performance of tanks under given conditions have been obtained in various trials. Instruments and methods of measuring the condition existing in tank mobility trials have been developed.
2. Methods of reconnoitring ground to estimate the going conditions of tanks.
3. The general theory of the relations between the factors which influence the going, including the design of suspension and tracks, is being developed.[31]

It is unclear how well these lectures were received, but what is clear is that Solandt continued to employ his diagnostic and systemic approach to dismantling complex problems with AFVs and their human occupants, Soon he would leave behind this research for the new challenge of greater leadership within the organization.

Schonland and Solandt

By the spring of 1943, Schonland was suitably impressed with Solandt's efforts to promote him to deputy superintendent of AORG. They were an odd duo, the South African physicist and Canadian physiologist. Schonland, at forty-seven, was the senior statesman of science with a proud military record, respected international reputation, and patrician demeanour. Solandt, thirteen years younger, was an aggressive former wunderkind with no military experience, whose biggest scientific achievements had begun to burgeon only during the war. And here they were, two "colonials" running a new, unique, and complicated British wartime science establishment without a map. Sargeaunt detected a mild tension between them because of Schonland's military bearing and Solandt's forceful attitude. This personality clash led to only one major confrontation, after Schonland left and Solandt became superintendent (see below). Otherwise it never interfered with their work.

Solandt would work with Schonland to establish ORS Sections 5, 6, and 7, but the remaining two sections were formed under Solandt (see table 7.1). Schonland was a good boss and Solandt respected his skills and approach to leadership, which in many ways mirrored Solandt's own abilities. Schonland would discuss concerns, offer critical suggestions, but leave the implementation of solutions to the section chiefs.[32] Schonland provided the strategic vision for the organization, and Solandt was responsible for getting the section heads to keep in line. Being AORG's taskmaster likely made Solandt less popular, as well.[33]

Table 7.1. Operational Research Sections of the British Army Operational Research Group

AORS1: Anti-Aircraft Defence, Radar, and Searchlights
AORS2: Coastal Radar and Gunnery
AORS3: Signals in the Field
AORS4: Armoured Fighting Vehicles, Artillery and Tank Gunnery
AORS5: Airborne Forces
AORS6: Infantry Weapons and Tactics
AORS7: Lethality of Weapons
AORS8: Mine Warfare, Assault Equipment, Tactics and Flame-Throwers
AORS9: Time and Motion Studies
AORS10: Battle Analysis

Source: Austin, *Schonland*, 251

As deputy, Solandt participated in the SA/AC "control meetings," where Ellis provided general direction for the research at AORG.[34] For Solandt, Ellis was a good man and fine administrator but too easily dominated by others. Solandt attended meetings on diverse topics from artillery doctrine to plans for sending OR teams into the field.[35] Solandt's only criticism of Schonland was the South African's utter lack of interest in administration. Solandt was horrified when he discovered the best scientists at AORG were making peacetime university wages far below that of many officers. Solandt always made a point of increasing material support of his best teammates. So Solandt visited physicist C.P. Snow, in charge of scientific staff at the Ministry of Labour, but Snow had no knowledge of the salary situation. Solandt and Snow worked to improve wages, though Snow resisted Solandt's pressure to double anyone's salary within a year.[36] Maintaining control of staff salary and hiring policy would be one of his legacies at the DRB, largely rooted in this AORG experience.[37]

Deputy Superintendent Solandt Abroad

While deputy, Solandt also acted as an international liaison for AORG. He was invited to the United States by the Committee on Medical Research in Washington, on a trip that lasted from 12 November 1943 to 4 January 1944 and focused on AFV research done at the Armoured Medical Research Laboratory, Armoured Board and Armoured School, Fort Knox. Solandt also briefly visited prominent war science leaders and organizations in the United States and Canada, including Vannevar Bush, chairman of the National Defence Research Committee (NDRC), Doctor Fry at Bell Telephone Labs, and Doctor Gregg at the Rockefeller Foundation. Solandt also visited the Ballistics Research Laboratory, Aberdeen Proving Grounds, and the Chemical Warfare Development Lab, MIT. But these were not just observer missions, for Solandt also lectured the combined Army, Navy, and Air Force group and had meetings with Doctor C.J. "Jack" Mackenzie, head of the NRC, Brigadier Meakins, Royal Canadian Army Medical Corps (RCAMC), and Doctor Frank Rose, scientific advisor to the chief of the General Staff (CGS).[38] Interestingly, Solandt's visit corresponded with AORG's top-secret radar trial "Exercise Feeler," proposed by Schonland as part of preparatory work for Operation Overlord. Schonland had been "bigoted" (selected to contribute) to the plans of D-Day under a committee of scientists approved by General Eisenhower to investigate the equipment needs of the invasion, chaired by Watson-Watt and including Cockcroft, but Solandt was not a member of this inner sanctum.[39]

Becoming Superintendent of AORG

In February 1944, Ellis asked Schonland to join 21st Army Group as SA to Field Marshal B.L. Montgomery for the Normandy invasion. Schonland agreed, and Solandt, only thirty-five, succeeded him as superintendent of AORG in March. L.E. Bayliss and Humby acted as his deputies, though it is unclear if theirs was an official position.[40]

Solandt and Schonland were very different leaders. Everyone respected Brigadier Schonland for his service, reputation, and well-established scientific acumen. Mott respected Schonland as a contemporary physicist. Lieutenant Michael Swann, later Lord Swann and head of the BBC, who worked at AORG and later joined an OR section in Normandy, found Brigadier Schonland a bit frightening for a mere lieutenant, but very approachable, tough yet fair, always engaging and with charm. For both Mott and Swann, Solandt never matched up. Swan recalled the young Solandt as less patient and more critical, even if he was usually right. Many believed Solandt never had Schonland's "old firm hand" in the War Office over Ellis. Some believed he was easily coerced by the likes of Ministry of Supply mandarins such as Lieutenant-Colonel Nigel Balchin and others.[41]

Schonland's boots were hard to fill and Solandt had to fight for respect since his youth and reputation could not compare with that of his predecessor. But his efforts and engaged personality earned him many supporters. E.A. Treadwell, who had worked with Don Solandt in Hill's lab as a teenager before the war, worked for both Schonland and Solandt at AORG's soil analysis lab. If Schonland was to visit, Treadwell recalled, they were given advance warning and cleaned things up so that it was fit for a brigadier's inspection. But they were often unsure of what exactly he wanted from them, leaving the lab with little direction.

Solandt, on the other hand, would arrive unannounced to see the lab in action. After the shock of this approach wore off, the staff found Solandt a keen supporter of their work, so long as the research was in step with AORG's strategic outlook. Such an approach meant Treadwell and company always had to be on their toes and working at their best, never knowing when the "boss" might drop in. Treadwell, like Sargeaunt, felt Solandt's broad knowledge and appreciation of science and technology made him one of the best superintendents AORG ever had.[42] Sargeaunt argued Solandt's great work as deputy had been in administering and managing the growing sections of AORG. Indeed, he was better suited to this than Schonland, whose keenest interest was in radar and AA work of AORS1 and 2. Solandt's wider interests and sense of efficient organizational management "tightened" AORG's scattered sections and provided him an insight into their research that was, as Tredwell described, personal and enthusiastic.[43] Solandt also had no illusions about the abilities or machinations of Balchin or other War Office mandarins. Perhaps Swann mistook his quiet Canadian demeanour for naivety. Omond Solandt was many things, but he was rarely naive.

Solandt and Managing AORG

Solandt maintained Schonland's ethos that AORG's priorities were determined by Army needs, and operational needs always came first. If problems arose in the deserts of North Africa, or with AA coastal gunnery against the enemy, they became a priority.[44]

Almost all of AORG's research during the Solandt era followed the same general procedures. Observing came first. "Never uncritically accept the users' diagnosis of the source of his problems" since he is often wrong. This was a double-edged sword. The military wanted the best possible equipment and procedures for its use, but they resisted anyone outside their profession telling them how to do their business. As such, Solandt believed one had to know the original reasons how and why the Army used their equipment. Otherwise they would resist any attempt at change.[45] Next was collecting quantitative data, which often meant measurements of things that had not yet been measured, and rough quantification was better than no numbers at all. The next stage was experimentation to confirm or dismiss a hypothesis or to test a remedy. Solandt maintained his role as maestro with his section heads, providing guidance and support, and leaving them to decide the best way to play.

The Value of Eccentrics

Solandt's team was a diverse lot, in training and background. Most of the early OR pioneers were physicists like Cockcroft, Blackett, and Watson-Watt. Demand soon dried up the supply of the country's physicists, so more esoteric and diverse scientists were employed in OR work. Solandt was among a small cache of OR pioneers from the medical sciences, like zoologist Solly Zuckerman, who became famous for his bombing surveys, or biologist Cecil Gordon, noted for his work on optimizing the RAF's air escort maintenance and deployment.[46]

While not lacking physicists, AORG held an eclectic range of scientific and technical personnel. G.D. Kaye was an actuary who served as a living computational device for Mott and Frank Nabarro's lethality of weapons studies.[47] Theo Fabergé, the grandson of the famous Russian jeweller Carl Fabergé, was one of Solandt's best machinists, despite a violent temper that raged in Treadwell's lab.[48] Of all his scientists, Solandt held Frank Nabarro in the highest regard. The work of Nabarro, one of Mott's best students, on the lethality of weapons was among the best research AORG produced during the war. Solandt also had a soft spot for Nabarro's outsider status in England as a South African and a Jew.[49] Indeed, Solandt championed outsiders and genuine eccentrics whose disposition avoided conformity. As he said in 1954, eccentrics were often the sources of innovation. "One of the very real dangers to our North American civilization is our worship of conformity. In almost every walk of life the person who conforms most pliably to the accepted standards of dress and behaviour is most likely to succeed. We must recognize that this enforcement of conformity will finally result in universal mediocrity. New ideas, especially in human relations and often even in science, come from those who refuse to conform."[50]

This was an enlightened and useful position for such a ruggedly non-eccentric man. Others might have found working with such an eclectic mix of strange and interesting people akin to herding cats. Indeed, managing various scientists and assorted tinkerers required firm guidance and a flexible mind. Solandt provided both, even for those who didn't care for him. Treadwell felt this was Solandt's greatest gift. Few doubted the direction Solandt was taking them in.[51] Solandt also worked hard to get promotions and increased pay for those he felt were doing excellent work, often having to raise an uproar in the War Office to get his way.[52] Simply put, Solandt backed hard work, even if it meant confrontation with his superiors.

For his subordinates, however, Solandt had one terrific weakness compared to Schonland: he lacked access to Prime Minister Churchill and his Cabinet.[53] Schonland was a friend of former South African prime minister, General Jan Smuts, and Smuts was a good friend of Churchill. Many felt this political patronage helped gain support for

AORG.[54] Solandt knew from working with Best the value of working with the policy "big shots" or "operators," so Solandt cultivated powerful friendships and connections as best he could. Cockcroft, his old friend from Trinity Hall, was one ally. Solandt's friendship with Watson-Watt, scientific advisor to the Air Ministry, allowed him to bypass bureaucratic red tape. Sargeaunt recalled Solandt's strong relationship with Colonel Bond and the Directorate of Armoured Fighting Vehicles, themselves powerful members of the War Office.[55] Solandt secured support from Brigadier Robertson, deputy director of research (DDR), who was interested in Nabarro's work.[56] Still, he was outside the inner ring of power in Britain.

Work at AORG

It is clear from his correspondence with Schonland that Solandt took an active interest in every section, visiting the dispersed teams when he could, keeping abreast of their developments and needs. His wide interests and strong intellect were engaged across a spectrum of military concerns, from infantry battle analysis to continental air defence against modern rockets.

AA Command had done the best with Army OR. Later in life, Solandt stated that AORS 1 and 2 had advantages the other sections did not: they had got in at the beginning; modern electronics had revolutionized every aspect of AA gunnery that the need for scientists was clearly seen and validated; and in General Pile they had a powerful supporter and utter convert to the value of scientists in war. "There is nothing more sobering for an advisor than to know that his advice is quite likely to be accepted and acted upon." Pile's support for science, more than any other reason, convinced Solandt of the need for direct access to the minister of national defence.[57]

The lion's share of AORG's focus was in preparation for or support of the Normandy invasion, including bombing analysis and weapons studies as well as improving training exercises and many other topics.[58] From September 1944 on. AORG also became involved in continental defence against German V1 and V2 rockets. Before the V2 sites were overrun, Solandt recalled, "AORG had tested a system that anti-aircraft command was prepared to try that gave the same calculated probability of a kill against the V2 as the best that the anti-aircraft could do with a slow flying bomber at 10,000 feet at the beginning of the war. Unfortunately permission to fire the guns against the V2s was never obtained."[59] Also by September 1944, Nabarro's team had done excellent battle analysis of minor battles in Italy. The War Office wanted the same for Overlord.[60] Solandt also had high regard for Bayliss's AA work and that of his section, which was filled with scientists, civilians, men and women, and often a complete Army survey regiment with recording vans under his command.[61]

AORG and the OR Sections Abroad

Before Schonland left for 21st Army Group, he initiated OR sections that would be used in the upcoming battles in France. AORG provided most of the scientists. The exception was physics professor Major Patrick Johnson, who had been working at Petersham. Johnson, a friend of Schonland's, would lead No. 2 ORS. The teams were picked by Schonland, who consulted with Johnson and Solandt.[62]

The relations between AORG and ORS after D-Day produced the only major controversy during Solandt's tenure as superintendent.[63] Schonland had promised Solandt that when the time was right, AORG members would be allowed to visit and conduct research in the field. Schonland would have to request the chiefs of staff to invite the AORG members when the time was right. Solandt maintained close contact with Schonland throughout the Normandy campaign, and he was happy to send any staff Schonland might require, knowing full well he might lose some of his best scientists.[64]

Months passed. No request for AORG scientists appeared and no ORS reports were being sent to Ibstock Place. By August, Schonland argued he did not have the authority to send the reports and that Ellis or a senior rank should be contacted to initiate any AORG staff heading to 21st AG.[65] By September Nabarro and others were rankling Solandt and Schonland about getting data from the field. What limited information Solandt got from the battlefield came from personal correspondences with Sargeaunt, who was in No. 2 ORS. Still, the War Office pressed Solandt to use more operational data at AORG.[66]

With pressure coming from above and below, Solandt pushed Schonland to request AORG staff to visit France to study German long-range guns and fire control.[67] In a rare flare of aggravation, he also sent an undiplomatically phrased letter in early October 1944. No real OR had yet been done in 21st Army Group, Solandt claimed. It was a poor choice of words: Solandt's desire for operational data, the bedrock of OR, the kind that generals and politicians could not dismiss, got the better of his tact.[68]

Schonland, fighting illness, was responsible for both OR sections. He responded with ire: "What the hell is biting you?" He then read Solandt the riot act about the positive work done by each OR section, the difficulties that they and he had faced over the past two months in an Army that did not want them, and stated that it was possible only now to get an analytic team from AORG with him.[69]

Brian Austin's biography of Schonland does not follow the dialogue beyond this point, and leaves an unbalanced impression of Solandt's conduct.[70] But Solandt continued the conversation, his own anger and frustration evident, though kept in check. Schonland, Solandt reiterated, had told him AORG would be "the direct servant" of 21st Army Group" through him. Solandt was to visit every month, and most of the senior AORG people would be invited over for various jobs. "You personally told most of them this. I did not believe this ideal could be achieved for some time but some of the others did." As such, Solandt's own patience was tried by scientists expecting him to fulfil Schonland's promises.[71]

What little information AORG had received had been about Johnson's poor conduct and interference with Schonland's attempts to support the ORS teams in the field. "Since then there has been no news from and no reply to an intervening letter asking if you would object to us trying to get Varley over to look at German coast guns," Solandt said. "Can you blame us for being a bit fed up?"[72] Solandt had managed to stay in the loop with Sargeaunt, though Swann had never returned his letters.[73] Then Solandt got to the heart of the matter:

> Briefly, it is just what always bit you so hard when you were here. The overseas people expect us to supply our best people on no notice at all for jobs that we know nothing about and we then hear nothing at all about anything that goes on. You know that if you were in my place you would be madder than I about the inefficiency and muddling of Sc.1., which blocks what little contact we might have with you e.g. Chapman says that you have written him telling about Sargeaunt's work and what he needs in the way of assistance. You did not mention this in your letters to me although you presumably want a person from AORG and, in fact, under the present scheme the man will remain on our staff on loan to you. You know from bitter experience that things do not pass through the WO with any speed or certainty.[74]

Despite this confrontation, peace was restored. Solandt, once he finally received a large batch of ORS reports, was very impressed.[75] He revealed that his initial negative impressions of the teams were based on Schonland's own letters before any reports were available. "Maybe you did not realize how depressing your first letters were."[76] By March 1945, Solandt had established effective liaison with the ORS teams, but little else. For this he blamed Johnson, who replaced Schonland as SA at 21st Army Group. Johnson had really let AORG down, never forming a strong working relationship between the ORS and AORG.[77] Solandt hoped the increase in AORG staff overseas might force Johnson's hand, though his opinion of Johnson never changed. For Solandt, Johnson was a charming man, and a fine teacher, but a poor leader and worse evaluator of quality research, and could be absolutely "pig headed and reactionary."[78] The feeling was mutual. Whenever Solandt visited Johnson, he "seemed more of a bore than ever."[79] Schonland became the repository for their mutual distaste, getting an earful from both men throughout his last days with the British Army in the field. Despite the cross words, Solandt and Schonland maintained a positive and collegial relationship long after the war.[80]

Failed Appointments

Writing to Schonland in August, Solandt had revealed that Chapman, Ellis, and Kennedy at the War Office had "elected" Solandt to succeed Schonland as SA to 21st Army Group when Schonland left. Balchin was selected to be the next superintendent of AORG.[81] Schonland, who left 21st Army Group in late October 1944,[82] doubted this would happen, and he was right.[83]

According to Solandt, when he visited Schonland and Colonel Francis W. Guingand, the chief of staff for 21st Army Group in France in 1944, both wanted him to be the next SA.[84] Schonland biographer Brian Austin disagrees. Schonland argued after the war that as SA had to be both soldier and scientist to be effective in an operational environment. Austin then extrapolated that Solandt, while contentious and sincere, was both pedantic and verbose, and senior military officials would soon lose patience with him.[85]

Like Carmichael's criticism years before, this view has merit but is easily contested. Solandt was aggressive and did not suffer fools gladly, but he also knew how power operated. His lack of actual soldiering experience would no doubt have made him even more of an outsider within the staff system, but there is no hard evidence to suggest that he would have floundered at 21st Army Group. Chute's own success in

North Africa indicated one did not have to be career military to succeed at OR in an active theatre of war. Solandt got jobs done, no matter how different the task from his actual training, and was a sage and practical problem-solver. These are desired commodities in an army headquarters as much as laboratories. He likely would have done a better job than Johnson, who attained the post and did little with it, according to Austin and Solandt.[86]

With this post denied him, Solandt's ambition turned to the War Office, where his skills were well known. By late January 1945, the deputy director of science (DDSc) wrote a scathing report on the malaise infecting AORG's research and Ellis's War Office staff, himself included, and suggested as part of cleaning up the mess that Solandt be made deputy scientific advisor (DSA). If Solandt was not available, "you must get someone equally alive, enthusiastic and competent."[87] Doctor Chapman then resigned as deputy scientific advisor to the Army Council, and Ellis asked Solandt to fill the position. Solandt agreed and maintained his duties as superintendent of AORG while Gough at the Ministry of Supply attempted to find a replacement.[88] In a strange parallel with his last days at the Defence Research Board, Gough argued that no one could really replace Solandt and refused to consider Bayliss or Humby, the only two serious candidates.[89] A severe case of flu cramped Solandt's own efforts to secure the position. By February 1945, Balchin secured the post.[90] Solandt respected Balchin but knew that Ellis was "putty in his hands."[91]

India and Burma

His ambitions twice checked, Solandt had little time to be frustrated. In early 1945, convinced that the war in Europe would soon be over, Ellis and Paris decided to send joint representatives to India and Burma to work with South East Asia Command (SEAC), and Solandt was the man for the job.[92]

He left on 1 May with Major Macklen of Ellis's staff and spent a month travelling through Egypt and India. Their primary task was to survey the OR sections in Asia and provide advice on expanding and improving them. Solandt also attended a conference held by Brigadier Welch on future work of ORS and user trial establishments (UTE) as well as post-war plans (20 May). He witnessed trials by UTE on their method of work and experiment, then visited the Ordnance Laboratories at Awnpore (21–2 May), and returned to various sections of GHQ in New Delhi before heading home on 29 May.[93] Here, Solandt gathered data on how OR and the UTEs operated in an environment so different from that of Europe.[94]

Two reports were produced. Solandt authored "Science in the Army in India," while both he and Macklen contributed to the formal report "Army Operational Research in South East Asia." While generally positive about OR in India, Solandt's paper critiqued AORG's structure, a result of its ad hoc, wartime creation, and provided suggestions on future OR and other science organizations within the military. Clearly, Solandt had well-considered ideas about war and post-war research organizations.[95]

According to the formal report by Solandt and Macklen, OR in SEAC should also be organized to avoid the many organizational pitfalls in AORG. India should have an SA directly under the commander-in-chief and have access to all branches of the Army, and not focus his efforts on research. His main duties would be close contact with scientific progress in universities, industry, and the service, and offer scientific advice to the CinC. He would have a staff that included "forward thinkers" to assess feasibility of certain new scientific ideas, and maintain a staff of specialists within key scientific disciplines. He should also have an Army OR group as a self-contained organization under SA direction, similar in structure to AORG, which would include field sections. Also required were sections on recruitment of personnel and structure specific to the Indian environment.[96]

The report emphasized a future organization's need for independence and freedom of thought. These subjects appear in Solandt's other post-war science writings, so such themes within the report likely came from him. If independence was not achieved, their research would suffer from the same political ruin that Solandt had experienced on MRC committees and at the Ministry of Supply. So the SA's organization should be responsible to an independent body such as the Privy Council. While these suggestions regarded the Army, "many of the problems affecting the future of the Army cannot be separated from those of other services. Close collaboration between the scientists in the three services should be championed. An inter-service research team to study problems of mutual interest should be formed. Personal contact between scientists in different services should be encouraged."[97]

The formal report outlined the history and structure of OR sections within the Allied Land Force South East Asia (ALFSEA). It praised the OR units, whose problems were particular to their theatre. Still, they suffered difficulties similar to those of other OR units in the field. For instance, Brigadier Welch had never met the Army commander. There was little inter-service work and too rigid control of the scientists. There was also an OR Division at HQ of Supreme Allied Commander South East Asia Command (SACSEA) under Mister T.W.J. Taylor that coordinated OR and scientific advice on planning problems. Direct contact with senior commanders was crucial to the success of OR.[98] "The division consists of a head with a small personal staff and attached representatives from the operational research organizations of the three services. The Division coordinates Operational Research in the three services and also acts as a means whereby the results of Operational Research are brought to bear on future planning. Through this division operational research has a direct channel of approach to the Supreme Commander which does not exist in any other theatre. This facilitates the effective use of the results of operational research."[99]

Though he did not know it at the time, Solandt's conclusions on the future use of science by the military paralleled the emerging vision in Canada for post-war defence research. When Solandt returned to Canada in 1945, he brought with him these conclusions, where they would become the strategic and structural framework for the Defence Research Board.

From AORG to SEAC

While in India, Solandt met Lord Louis Mountbatten at SEAC HQ Kandy[100] and his scientific advisor Doctor Taylor. Mountbatten "asked for me to go out to replace Taylor. I have agreed to go and hope to go out there about Sept. 1." The War Office agreed and Solandt was to be made brigadier.[101] Before leaving, Solandt also contributed to the discussions of the future of AORG, arguing against the machinations of Gough and Ellis.[102] He "violently" opposed the splitting up of separate sections between the War Office and Ministry of Supply, but to no avail. For this, he blamed Balchin, whom he accused of building an empire in government without any direction to shape it.[103] Solandt also contributed reports to the post-war use of OR in the Directorate of Scientific and Industrial Research,[104] and in creating an industrial physiology research group for the MRC. Both emphasized the need for OR-styled teams in-being to quickly conduct research.[105]

During the final AORG "Conversazions" (*sic*)[106] Solandt led two tours of the establishments for "the common herd" as well as the elite. During the former, he made a "very unpopular speech emphasizing the need for independence in operational research," and its arguments echoed those of his India report. The future organization needed the backing of top people, like General Pile. It required young scientists full of enthusiasm "prepared to attempt the impossible and free from pre-conceived ideas and from the dogma of the army." All branches of science should be represented and encouraged, and they should foster an independent and impartial attitude. AORG had often acted as arbiter between weapons designer and user and had never been accused of having its own agenda. "Independence is absolutely essential to the maintenance of impartiality" because the Ministry of Supply "allow[ed] us freedom even to the point of criticizing other departments of the ministry and partly to vigorous ways in which Brigadier Schonland[107] repelled any attempts to encroach on our freedom."[108] As Solandt summed up his thoughts of future Army science management, "I think that these three essentials must be maintained in any post-war organization. You can supply the first essential – support in high places. The second can be maintained by making arrangements for a steady flow of young scientists from the government research establishments, the universities, and from industry. The third can probably best be maintained by having the work of scientists guided by a committee appointed by some independent scientific body such as the Royal Society.[109] As he knew it would, Solandt's speech upset both Ellis and Gough so much they made "mildly rude" comments about it, but he was not concerned. After all, he thought he was right.[110]

Conclusion

For Solandt, the war ended with disappointment and opportunity. SEAC cabled him after VE Day to depart for the Far East no later than 1 September, but on 31 August this order was cancelled. Atomic bombs had been dropped on Hiroshima (6 August) and Nagasaki (9 August) and by 15 August the formal war with Japan was over. Solandt had few regrets in his life, but failing to serve as SA to Mountbatten, a theatre commander who championed science, was one of them.[111]

This disappointment was fleeting. Ellis informed him that the War Office was sending a mission to Japan to investigate the effects of the atomic bomb, and Solandt's experience and medical training made him an ideal candidate for the team. Would he like to join? He had also been contacted by the Canadian government regarding the possible opportunity of becoming scientific advisor to the new chief of the General Staff in Ottawa, Lieutenant-General Charles Foulkes. Would he accept the job?

There was no question about Japan. It was a rare and important opportunity to witness the first atomic battlefield in history. He was less certain about Canada. He wrote Schonland, "If I go back to Canada I will be sorely tempted to give up Physiology and stay on the military or administrative side of science – what do you think of it? I have grave misgivings but feel that I will be better qualified for administration than for research after another year of it. I would very much value your advice on what sort of job you think that I should look for."[112] Schonland's response is not known, but Solandt did choose Canada as the starting ground for his post-war career.

For six years, he had applied his mind and energies to the challenge of directing science for war. His reward was a rise in prominence within British scientific, military, and government circles, the likes of which he had not attained in his chosen profession of physiology. While his pioneering work in OR was heralded, his years at AORG solidified his reputation as a highly effective, if aggressive, manager of government science. It was in this role, rather than research, that his career would reach its apex. His experience in Japan, one that allowed him to become among a handful of people to examine the actual effect of atomic weaponry, shaped the trajectory of that apex. When the mission was over, he would spend a decade not only managing science, but preparing Canada for the realties of nuclear war.

8 Atomic Battlefield: Solandt and the British Mission to Japan, November 1945

From the summer of 1945 onward, Solandt's work in government service would involve atomic affairs. As a member of the British Mission to Japan (BMJ), Solandt entered an elite group of experts who had studied the actual effects of the atomic bomb. Solandt's primary goal was to study the human and structural damage of this new, deadly, and mysterious weapon. His secondary goal, much less dramatic, was to obtain information on Japanese technology for the Canadian government. It was after the mission to Hiroshima and Nagasaki that Solandt finally agreed to the Canadian government's offer to be the "scientific advisor" to the chief of staff. Thus Canada, at the birth of the atomic age, obtained a rare pillar for its post-war defence posture: an expert on the reality of atomic warfare.

Solandt's experience in Japan informed his judgment on all subsequent atomic matters, and his meticulous intellect kept it safe from hyperbolic assertions. Japan's lack of any serious civil defence structures and the quantitative and not qualitative reality of the atomic bombs was emphasized in his contribution to the BMJ's final report. Such rationale came from a man whose intellect was at the forefront of his emotions. For his entire life, Solandt found fulfilment in work, challenge, and achievement in professions of service rather than idle recreation. But by the time he had walked the ashes of Hiroshima and Nagasaki, even he was emotionally numb.[1] While he always maintained a congenial, if occasionally aggressive, demeanour, Solandt's feelings had retreated further behind his intellect before arriving in Japan, for the war had made such numbness necessary. Solandt, dour to start with, became even less emotional as he witnessed the carnage of modern war. He had pulled bodies out of the wreckage made by German bombs. He had watch some die in the streets and in hospitals. He toured the battle-fields of Europe to see the Old World ravaged by modern arms. It was a sight he never savoured. But to have worn his emotions on his sleeve would have curtailed his greatest asset, his precise and rational intel-lect. Still, Solandt was not devoid of awe at the catastrophic power of the atomic bomb. He just had no flair for describing it. Instead, Solandt told the public and friends that John Hersey's brutally frank account of the devastation at Hiroshima summed up best what he experienced as a member of the British Mission to Japan. In short, the world had wit-nessed the dangerous and terrible birthing pains of the atomic age, and there was no way to turn away from its awful realities.[2]

Solandt and the Manhattan Project

Later in life, Solandt claimed that he had caught wind of the Manhattan Project while serving in England, primarily through Sir John Cockcroft, Solandt's friend from Cambridge and former boss at AORG. In the later stage of the war, Cockcroft became the chief scientist of atomic research in Canada at the Montreal Laboratory and the reactor program at Chalk River.[3]

> We became close friends and met regularly up to the time of Sir John's death in [1967]. Sir John was one of the pioneers in Nuclear Physics that lead [sic] to the bomb. [Peter] Kapitza who later led the Russian work on the bomb had worked with Cockcroft in Cambridge well before the war. With this background it [is not surprising] that I heard enough discussion of a possibility of the explosive release of energy from the nucleus and also vague references to a curious industrial enterprise [called] Tube Alloys to make me realize that work on an atomic bomb was going on.[4]

Brian Austin disputes this point. There is no direct evidence that Schonland himself knew about the atomic bomb until Hiroshima and Nagasaki. The chances that Solandt would know this tightly kept secret before Schonland is unlikely, especially since Schonland and Cockcroft were better friends.[5] It is reasonable to assume that of the two, Schonland would have had his confidence more than Solandt.

However, according to Guy Hartcup and T.E. Allbione, Cockcroft's biographers, Solandt's early liaison mission to the United States and Canada in November 1943 also included participation with Cockcroft and Watson-Watt's mission to the United States concerning radar development in November 1943. Before returning to England, this mission, Solandt included, visited Montreal, where the Canadian, British, and U.S. governments had set up a laboratory for plutonium production for the atomic bomb. Indeed, it was here that Doctor C.J. Mackenzie, acting president of Canada's National Research Council, offered Cockcroft the position of director of the Montreal Laboratory, which he subsequently accepted.[6] This mission was not included with Solandt's AORG report of his trip and is the earliest indirect evidence that suggests Solandt, like a detective hunting for clues, may have realized an atomic weapon was being developed. What he could not have ascertained was the impact these new weapons were to have on his own career and his vision of their role in war and peace.

The United States Strategic Bombing Survey

The United States Strategic Bombing Survey (USSBS) originated on 3 November 1944. President Franklin D. Roosevelt authorized the request of secretary of war Henry Stimson for the creation of a survey mission to conduct a "scientific investigation of all the evidence" of strategic bombing in the European Theatre that would provide the data needed for assessing post-war strategic planning. USSBS came under leadership of Colonel Franklin D'Olier, president of Prudential Life Insurance Company. His chief aides included future American security-policy gurus George Ball and Paul Nitze, and Canadian-born Harvard economist John Kenneth Galbraith. The survey was completed in September 1945.[7]

After the atomic bombings of Hiroshima and Nagasaki, president Harry S. Truman authorized a similar mission for the Pacific. In a 15 August 1945 letter to D'Olier, Truman suggested to the chairman of USSBS that "It would be similarly valuable for our post-war planning and future policies to have the same kind of impartial and expert study of the effects of our aerial efforts in the war against Japan. This study would include the effects of *all types of air attack.*"[8]

This would include the use of the atomic bombs. D'Olier accepted and with Nitze as his vice-chair began to reassemble a team for Japan. In the fall of 1945, the U.S. chiefs of staff invited their British counterparts to send their own survey mission to Japan with a focus on the effects of the atomic bomb. The British had participated in the European survey[9] and gladly accepted the offer. The British had contributed significantly to the U.S.-led effort to create the atomic bomb and were especially keen to assess the scope of atomic damage to an island nation for their own post-war strategic planning.[10]

The British chiefs of staff quickly assembled the two RAF field teams that would serve in the Pacific, collectively known as the British Mission to Japan (BMJ). Despite the military designation, the members were largely from the Ministry of Home Security, all experts in assessing bomb damage. BMJ would be led by Professor W.N. Thomas, a civil engineer from Cardiff University, Wales. Thomas was a leading authority in structural damage caused by air attack and had led many bomb-damage assessment teams for the British Ministry of Home Security during the Blitz.[11]

Solandt represented the War Office in the mission and led RAF Field Team No. 1, Group 2 (see table 8.1) He was also the only member of the team with medical background, which he found odd.[12] While instructed to investigate structural damage to military institutions, Solandt recalled "there were no military installations in either Hiroshima or Nagasaki so I spent most of my time studying casualties from a medical point of view."[13] Solandt had not conducted his own research on casualties since the "crush injuries" studies at Sutton, relying instead on the deployed research teams from the MRC or AORG. In Japan, he would finally have the chance to lead a research team in the field.

The British mission was under the auspices of the USSBS. During the mission, Solandt received a directive from the U.S. chiefs of staff on the nature of relations between the two bodies. It emphasized the collaborative nature of the enterprise and included the following points:

> 3. The activities of the British party will, therefore, be governed by the terms of reference of the United States Strategic Bombing Survey. You should consult with Professor Thomas, head of the Civil Defence party, arrange with M. Franklin D'Olier, the Chairman of the U.S.S.B.S., how make the best use of you particular qualifications, bearing in mind ... the special importance of studying the results of the two atom bomb incidents at Hiroshima and Nagasaki.
>
> 4. You should not make any inquiries into research, constructional or operational aspects of atomic bombing that are outside the scope of the U.S.S.B.S terms of reference.[14]

Clearly, the United States was concerned, even among the allies that had helped build the atomic bomb, to keep its secrets locked away for the time being.

Table 8.1. Task Organization for RAF Field Teams Attached to the United States Strategic Bombing Survey (Pacific)

RAF Field Team No. 1, Group 1		
Officer in Command: G/Capt. William N. Thomas		
Name	Rank	Service
Bronowski, Jacob	W/Comdr.	RAF
Burn, Douglas	W/Comdr.	RAF
Hawker, James B.	S/Leader	RAF
Elder, Henry	F/Lieut.	RAF
Badland, Percy A.	SDO	RAF
Bevan, Ronald	SDO	RAF
Pavry, Francis H.	SDO	RAF
Young, Oliver C.	SDO	RAF
Walley, Frances	SDO	RAF

RAF Field Team No. 1, Group 2		
Officer in Command: Col. O.M. Solandt		
Name	Rank	Service
Dark, Albert E.	G/Capt.	RAF
Mitchell, Fredrick G.S.	G/Capt.	RAF
Evans, David	Lt. Comdr.	RN
Whitehead, Robert G.	S/Leader	RAF

Source: UT OMS, B1991-0015/011, file 6, "US Strategic Bombing Survey (Pacific) Operations Order No. 34," Tokyo (29 October 1945), 1, Annex A

En Route

The mission left Britain on 11 October and arrived in Washington two days later. Here they were to get a week's full briefing on what to expect in Japan from the U.S. Army Corps of Engineers, the organization that had provided the backbone to the Manhattan Project/District that created the atomic bomb. In 1975, Solandt recalled that during that week in Washington they had in fact been "briefed by experts in the Manhattan project from General Nichols[15] on down."[16] But this statement is contested. In 1980, Solandt remembered that the meetings with the engineers never materialized and "we left Washington with little or no information on the atomic bomb."[17] It is unclear just how much verifiable evidence the United States provided the mission before it arrived in Tokyo on 27 October.[18]

The mission was briefed on the effects of the atomic bomb from Japanese physicists and doctors who had been working with the American survey teams, though what was said is not clear.[19] The mission's next activity was an overview of the bomb damage of Japan. The mission spent an entire day flying in a DC-3, examining other bomb sites in Japan to assess the differences between them and the devastation of Hiroshima and Nagasaki, but from the air only an expert could tell the difference between conventional and atomic bomb damage. The real work would have to be from the ground.[20] Still, what they did see was "almost overwhelming. Each one of these cities had been extensively devastated."[21]

On 31 October, the mission, now situated at the British Consulate, began its work in Nagasaki. The USSBS provided lodgings and jeep transportation but left the British mission largely to their own devices. As mentioned, there were no military institutions of significance in Nagasaki, so Solandt turned the team's attention to casualty analysis.[22] He was joined in this effort by Doctor Jacob Bronowski, a renowned statistician adept at the mathematics of casualty surveys who had worked for the Ministry of Home Security. Bronowski led RAF Field Team 1, Group 1. Together, they would investigate the number and nature of casualties from the atomic bomb in Nagasaki (2–13 November) then Hiroshima (14–24 November), Bronowski concentrating on the former and Solandt on the latter (see table 8.1).[23]

First Impressions

Walking through the world's first atomic battlefield was a unique and life-changing experience for both men. Bronowski, who was also a literary scholar, recorded his impression.

On a fine November day in 1945, late in the afternoon, I was landed on an airstrip in Southern Japan. From there a jeep was to take me over the mountains to join a ship which lay in Nagasaki Harbour ... suddenly I was aware that we were already at the centre of damage in Nagasaki. The shadows behind me were the skeletons of the Mitsubishi factory buildings, pushed backwards and sideways as if by a giant hand. What I had thought to be broken rocks was a concrete power house with its roof punched in. I could now make out the outline of two crumpled gasometers; there was a cold furnace festooned with service pipes; otherwise nothing but cockeyed telegraph poles and loops of wire in a bare waste of ashes. I had blundered into this desolate landscape as instantly as one might wake among the craters of the moon.[24]

While Bronowski waxed poetic, Solandt had the tricky job of driving the jeep through these "craters of the moon." "I realize that I was probably the more practical of the two, since when we were out together I drove while he planned. Incidentally, driving a large ¾ ton Dodge 4 x 4 in a ruined Japanese city requires as much concentration as writing poetry."[25]

Solandt's first impressions were much less poetic, even dour. But the awe of Nagasaki was not lost on him. He wrote to his family,

> The damage is quite up to advance notices. The bomb lit in the northern part of the town where most of the factories are and completely devastated an area about one mile [sic] and 2 ½ miles long ... I've been spending the last few days looking at the casualties that remain. They are a sorry looking lot since most of them have had practically no treatment. Fortunately most of the bad cases are dead and the others are getting better ... The attitude of the Japs is quite amazing. They treat the Americans as welcome guests not as conquerors. The Americans in turn are very friendly especially to the kids. The kids are very attractive and spend most of their time talking to the troops. Kids about Sigrid's [Solandt's eldest daughter] age are to be seen carrying small brother or sister about on their backs all day – no wonder they grow up small and tough ... I spent many several [sic] hours [in the previous days] looking at damage in the residential quarters – I have never seen such filth in all my life. It is a mystery how anyone survives. Every place is just full of dirty little kids with running noses and impetigo and probably fleas. I had a bath immediately upon returning.[26]

Only Solandt could write this account of witnessing the birth of a new era of science and warfare. This was the precise mind of both a doctor and scientist, who found solace that most of the bad cases were "already dead," amazed at both the toughness and filth of the Japanese people in the wake of the atomic bomb.

Information Collection

The goal of the mission was to find out the number and nature of Japanese deaths in relation to the distance and effect of the atomic bomb and what structures were in the path of the explosion and its radiation.[27]

Japanese hospital records on casualties were scattered throughout the region. Solandt worked with the U.S./Japanese Joint Commission for the Investigation of the Effects on the Atomic Bomb, under the direct authority of General Douglas McArthur, supreme Allied commander in the Pacific.[28] Using this limited information source, as well as information garnered from his interpreter, Solandt gained a framework for the atomic bomb's effects. Casualties had been produced in four general categories:

1. Ordinary thermal burns due to direct radiation from the heated air around the explosives;
2. By irradiation with gamma rays and other short-wave radiations;
3. By blast; and
4. Secondary injuries due to being hurled against buildings or hit by buildings. Burns due to fires should be included in this group.[29]

Solandt learned from the American/Japanese team that out of a population of 250,000 there were 21,000 killed, 1,000 missing, and 40,000 injured. All those out in the open and within 1,000 metres of the centre died, though the cause of death was difficult to determine.[30] Those 1,000–2,000 received heavy doses of radiant heat and gamma rays. Thick clothing could screen against the former, but the latter could penetrate up to 15 inches of concrete. Blast damage had created minimal casualties at this range (10 per cent) and screening determined the nature of the casualties. The lightly clothed suffered severe flash burns and died before the radiation could kill them. Those 2,000 metres away or less clothed generally survived the flash-burns but showed symptoms of irradiation, including epilation (hair removed from the roots). Solandt likely noted the similarity with burns of tank crews and the value of even light clothing in avoiding serious injury.[31]

The research was fascinating. While investigating some cases of structural damage, Solandt came upon two Franciscans from Quebec, Father Prudent Moffette and Brother Moreau. They had been in Japan for seventeen years and were interned when the war broke out. Nagasaki, they claimed, had over 16,000 Roman Catholics and 8–10,000 were killed.[32] Solandt also interviewed an eyewitness of the blast from the *Tsuruoka Maru*, a 10,000-ton vessel that was 4,000 yards away.

Three of them were on the bridge at the time – watched the parachute come down and one went to get glasses to observe them more clearly.

All were looking toward [the] bomb when it went off. They saw a bright white light – heard one loud bang followed by four smaller ones which they think were not echoes. The ship rocked violently but was not in any danger of capsizing. They did not emphasize any feeling of wind. Several others on the ship had mild flash burns on exposed skin. A few had blisters on their forearms. Next the whole area was covered with a white smoke which rose forming a bright fiery column which ended in a mushroom shaped head of white smoke. The bright white smoke rose and was followed by bluish smoke and then flames. The flames appeared within 1–2 minutes of the first flash. One man on the ship had his hat charred by the flash – they disagreed about flame on the hat. Trees on the east of the harbour were seen to burst into flames within a minute or two. They also [saw] several small vessels between them and the bomb burning.[33]

All of Solandt's writings at Hiroshima and Nagasaki were written with a cool, pragmatic, and clinical eye for detail and research. Still, his letters home and remarks on the casualties were not without notes of wonder, both for the weapons and the Japanese who had suffered under them.

Creating the Report

The entire mission was an admittedly rushed affair. The British were eager to get as much first-hand data as possible to make their own independent analysis on the nature of atomic weapons so that they would not have to rely strictly on U.S. data. In preparing his reports, Solandt repeatedly warned the British chiefs of staff that this was an initial report, that the assertions were preliminary at best and would be superseded by the final report.[34] Solandt and Bronowski created at least three drafts before completing "Casualties Due to the Atomic Bomb at Hiroshima and Nagasaki" in late November. This paper would be included in the final report of the BMJ, *The Effects of the Atomic Bombs at Hiroshima and Nagasaki*, produced in England in 1946.[35]

The most important and repeated point made by Bronowski and Solandt was that the extensive damage done to both Hiroshima and Nagasaki by the atomic bomb had to be contextually understood if its effects were to be rationally considered. They were both adamant that everyone reading their report be reminded that Japan had constructed almost no structural civil defence measures in either city. If they had, the devastation, while great, would not have been as complete. "The rescue services were so slow in coming into action after the bombing and were so inadequate that no records were kept of the dead, injured, missing and evacuated people. The hospitals were so swamped with work that their records are of little value. The Japanese did make scattered attempts to keep records beginning a few days after the bombing but these were very incomplete."[36] Such views are not surprising for men who worked so extensively in British civil defence, where the value of preparation and safety measures helped curb the effects of German bombing and air raids.

The mission ended in December of 1945, and Bronowski wrote the final report of their efforts as Solandt made his way back to Canada. The report itself contended, much like the final USBSS reports, that the revolutionary characteristic of the atomic bomb was quantitative: it had performed the tasks of many tons of bombs with significant efficiency, but it should not be considered a war-winning weapon. Proper civil defence measures would mitigate some of the worst damage experience at both Hiroshima and Nagasaki, and both cities could have contended better against the weapon with proper defence materials in key areas.[37]

Solandt maintained this appraisal and the importance of civil defence when he assumed his position as director-general of the Defence Research Board later that year. While this view changed in emphasis as the nuclear weapons advanced into the 1950s, this core conclusion from his month in Hiroshima and Nagasaki remained (see chapter 14).

Solandt's Canadian Mission

While Solandt's work with the mission was as part of the British post-war effort, he was still in Canadian uniform and also operated as an agent of Canadian national interest. Before leaving for Japan, he was instructed by the Canadian Ministry of Munitions and Supply to "collect opinions concerning the advisability of sending Canadian technical and scientific representatives to Japan to study recent developments there."[38]

His six-page report was done in two days, but provided an overview of Japanese strengths and weaknesses in agriculture, industry, medicine, and maritime and military science and technology. In the process, Solandt secured permission from U.S. Colonel W.S. Wood of the Military Intelligence Section, supreme commander of the Allied Powers in the Pacific, to allow for a Canadian mission to be sent for a more thorough investigation. According to J. Hancock, Canada had a long interest in Japanese agriculture and maritime technologies. At the end of the war, Ottawa's participation in the Pacific war allowed it to gain access to these technologies, and to punish Japanese war criminals as part of the International Military Tribunal for the Far East. Solandt helped secure the former before leaving Japan.[39]

Conclusion

Six years of war had a profound impact on Solandt's life and career, changing it in ways he had never considered and providing opportunities he had never dreamed. He began the war a mammalian physiologist, set on a life of medical research. He ended the war a colonel in the Canadian Army, experienced in the management of military research, and an expert on the impact of atomic weapons. At war's end, Canada selected him to be the first director-general of a peacetime research establishment, and they got a rare and powerful asset to their post-war goals.

Here was a man who had pioneered a new field of applied science and had either researched or managed military research concerning tanks and chemical warfare, continental defence against modern rockets, and the effects of atomic weapons on both structures and people. Solandt ended the war a seasoned and insightful state scientist. He had definite opinions on how science should be managed, promoted, and led within government to serve military needs. From Sutton to Ibstock Place, he honed his skills of managing, leading, and inspiring teams of scientists, doctors, engineers, and assorted eccentrics towards positive results for a military client. He had witnessed the importance of connections with industry that his team at Lulworth lacked. At AORG, he had seen the essential value of support from senior military and political officials in translating good research into practical results and cutting through red tape. He had also seen the limit of his ascension in the ranks of "operators" within the British political and scientific world. Now, after walking the battleground of Hiroshima and Nagasaki, he had realized that government-led scientific effort had taken on a terrible urgency and import. For the next ten years, he would apply all he had learned in war to mould a new defence research organization in his home country of Canada. After six years in Britain, Solandt arrived in Ottawa a stranger. By 1956, he would leave government service as a "wise old man" of science and warfare at the ripe old age of forty-seven.

Donald McKillop Solandt, ca 1905. Photograph by Charles H. Boyes, Kingston, ON. B1994-0020/001P(03).

Edith Young Solandt. B1994-0020/001P(04).

Donald Young Solandt.
Photograph by Milne, Toronto.
B1994-0020/001P(22).

Charles Best, 1938.
Photograph by Ashley &
Crippen, Toronto. A1978-
0041/002(26)01.

Omond Solandt, with and
without hat, London, 1944.
B2012-0014.

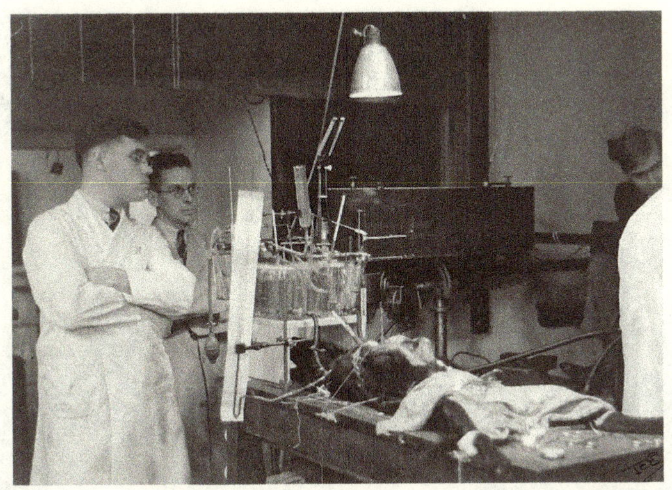

Solandt attending a dissection at Cambridge University, ca 1939. Photograph by J.O. Edgcumbe. B1994-0020/001P(07)01.

Omond Solandt shaking hands with Brooke Claxton: "To Dr O.M. Solandt, O.B.E., with warmest appreciation of his service to Canada in war and peace. Ottawa, December 1, 1948 [Signed] Brooke Claxton, Minister of National Defence." B2012-0014.

"NATO unclassified. 3 Jan [19]56: Dr O. Solandt, chairman of Canadian Defence Research Board, accompanied by Mr A. Hartley Zimmerman, chairman designate of the Board, today (5 Jan 56) visited Supreme Headquarters Allied Powers Europe and were received by Field Marshall Lord Montgomery, Deputy Supreme Allied Commander Europe and by other senior officers." *Left to right*: Dr Solandt; Lt Gen C.V.R. Schuyler (SHAPE chief of staff); FM Lord Montgomery; Mr Algrain (assistant scientific adviser to SACEUR); Mr Zimmerman. Official SHAPE Photograph. B2012-0014.

Meeting of the Defence Research Board at Suffield, Alberta, June 1951B2012-0014.

Solandt, cover of *Saturday Night*, 15 August 1950. Photograph: Canadian Army. B2012-0014.

Solandt and friends. *Left to right*: (W/O D.C. Brown, Solandt, G/Capt Dark, [Helen Hamilton?], S/Ldr. Whitehead; attached to the card "The Troika," personal direction Helen Hamilton, Washington, DC, No. 537. B2012-0014.

Medical Registration Certificate, 14 June 1940. B2012-0014.

Map: "Hiroshima, Hiroshima Prefecture, Honshu, Japan." Emergency provisional edition, August 1945, with concentric circles radiating from Ground Level Zero superimposed, and annotated by OMS with red pencil lines and numbers [for photographs?] documenting the extent of damage inflicted. B1991-0015/024(64).

Solandt posing for a practice with the Senior Intercollegiate Rugby Team in Varsity Stadium, 1930. His rugby years were 1928–30 and 1931 (Juniors, I; Senior Intercollegiate, II; and Senior Intercollegiate and Orphans, IV). B2012-0014.

Victoria College graduation, 1931. Photograph by George Freeland, Toronto. B2012-0014.

Alpha Omega Alpha Honorary Medical Fraternity, 1935–6. Photograph by George Freeland. B2012-0014.

Medical Society Executive, University of Toronto, 1935–6. Photograph by George Freeland. B2012-0014.

Solandt, graduation in Medicine (MD, 1936). Photograph by George Freeland. B2012-0014.

Solandt with fellow students (doctor unidentified); Solandt looks quite young, so it may have been taken while he was doing his Bachelor of Science (Meds) in 1932–3. The alternative would be while getting his MD, 1933–6. Photograph by Park Bros, Toronto. B2012-0014.

Toronto General Hospital house staff, 1937–8 (while Solandt was doing his residency). Photograph by Freeland. B2012-0014.

T-holders of Medicine, 1931. Solandt is at the centre of the back row. This was his first *T*; he received his second in 1936. B2012-0014.

Solandt, undated. B2012-0014.

9 The Only Man for the Job: Omond Solandt and the Origins of the Defence Research Board, 1943–1946

On 28 December 1946, Solandt accepted the offer to become the first director-general of defence research for the Defence Research Board (DRB). For ten years, Solandt served as the senior military science tsar in Canada, applying all he had learned during the war about the good, bad, and inefficient nature of government science in Canada's emerging Cold War strategy.

However, the DRB did not originate with Omond Solandt, but with senior Canadian government science and military officials concerned with maintaining and reorganizing the defence research assets Canada had generated during the war. C.J. Mackenzie, president of the National Research Council (NRC), C.D. Howe, minister of munitions and supply, and Colonel Wally Goforth and his team debated and then laid the foundation for post-war defence research. Incredibly, the majority of Solandt's conceptions for a viable peacetime defence research organization either mirrored or supported the needs and vision of his Canadian colleagues.

For Solandt, those early years were ones of unrelenting pressure and monumental work in a country he had barely seen in a decade. Solandt returned to Canada as an outsider to the Liberal Party's power elite. Still, his demeanour and intellect moved easily alongside senior policymakers like Howe, Brooke Claxton, and his old football coach L.B. "Mike" Pearson. More friction lay with the military, and is explored in chapter 10. Solandt's only major confrontation was with his natural rival, C.J. Mackenzie, Canada's senior wartime science advisor. Mackenzie, who supported the DRB and helped select Solandt for the job, attempted to circumvent Solandt's authority over personnel selection at the DRB. This could have resulted in a never-ending Machiavellian intrigue within the government-science community, but Solandt's handling of the crisis avoided this worst-case scenario and also secured his position as the nation's leader in defence research.

The official history of the DRB, written by Captain D.J. Goodspeed, was the first to explain the complex origins and organization of the DRB. This current chapter does not refute or revise Goodspeed's major arguments, but enriches that account of the DRB's origins with the view of Solandt, A.M. Fordyce, and others, on the difficult first year of the DRB's origins. As demonstrated in chapter 7, Solandt had clear ideas about the future of defence research, and they largely meshed with those of his Ottawa counterparts, especially Goforth. Goodspeed failed to discuss in any detail the role, function, and influence of Solandt as the DRB's chairman. Chapters 10 to 14 redress this weakness. First, however, the DRB's origin story and Solandt's role in it must be established.

Canadian Defence Research in the Twentieth Century

The scientific and technical demands of the First World War acted as the catalyst for Canada's premier state-science organization, the National Research Council. Created in 1916, the NRC's original mandate was to support Canada's war effort. After 1918, NRC focused largely on fundamental science and research until General A.G.L. McNaughton became president of the NRC from 1935 to 1939. McNaughton slowly shifted the organization's focus to applied scientific endeavours with military applications.[1]

When war began, McNaughton left the NRC to take command of the Canadian forces overseas, but retained the title of president. Acting president Doctor C.J. "Jack" Mackenzie would lead the NRC through the war. A decorated Great War veteran, Harvard graduate, and former dean of engineering at Saskatchewan University, Mackenzie was a shrewd and successful operator in Ottawa. While his war record with the NRC was not without mistakes and failures,[2] he successfully led his organization through a period of exceptional growth and important achievements. While not popular with prime minister W.L. Mackenzie King, Jack Mackenzie had Cabinet support from his former professor at Dalhousie, the American-born minister of munitions and supply, C.D. Howe. Howe was Canada's most powerful wartime minister. His Department of Munitions and Supply was tasked with both procuring munitions and organizing the nation's industrial potential for war. Howe's contacts in industry, academia, and government, as well as his vast capacity for work made him one of the pillars of Canada's war effort. That Canada ended the war as an industrial power of note is, in no small measure, the result of Howe's leadership and convictions.[3]

Over the course of six years, Canada employed or created a series of defence research assets through the NRC, which included the council's own divisions (biology, chemistry, aeronautics, etc.) as well as new establishments such as the Chemical Warfare Experimental Station in Suffield, Alberta; the War Disease Control Establishment (biological warfare) at Grosse Île., Quebec; the Armament Defence and Research Establishment at Valcartier, Quebec; and the Chemical Warfare Laboratory in Ottawa.[4] The services also developed their own research establishments.[5] Canada's wartime defence research efforts touched on nearly every key component of the scientific war waged against the Axis, including radar, jet propulsion, and atomic research.[6]

The Goforth Group

Many in both the military and civil service wished to see this momentum continue into peacetime. NRC president C.J. Mackenzie refused the job for his organization, though NRC would maintain a desire for some defence research work.[7] Still, Mackenzie wanted the NRC to return to fundamental research and be divested of nearly all its applied military research. Scientists who had served the government during the war now wanted to return to their respective careers in research. The fact that the NRC contained spies for the USSR within its ranks might have also suggested to Mackenzie that top-secret research should no longer be the NRC's mandate.[8] Each service also created schemes for future defence research. Some of the earliest work was done on behalf of Chief of the General Staff (CGS) Lieutenant-General J.C. Murchie in the summer of 1945. This group was led by Colonel Wally Goforth, an economist from McGill who was head of the Department of Staff Duties (Weapons) in Ottawa. His group consisted of Lieutenant-Colonel A.M. Fordyce, future comptroller of the DRB, and Lieutenant-Colonel Morton Mendel, an economist, lawyer, and future first secretary of the Royal Bank.[9]

The "Goforth" group, as it was known, produced a series of papers on future defence research organization during August 1945 that received apathetic acceptance from the Navy and distaste with the Air Force; the RCAF in particular wanted to control their own defence research agenda. Goforth refined his vision of post-war defence research from

August to September with his paper "The Future of Canadian Army Research in Development."[10]

It was a novel and innovative presentation. Defence research was to be regarded as a single concept. All service research and development organizations should be reorganized with appropriate civilian scientific representation into a new, fourth sub-department of National Defence. For this structure to work, the head of this organization had to hold status equivalent to that of a chief of staff and hold financial authority equivalent to a deputy minister. Four possible mandates for defence research organization were proposed, the final and most important being "to regard research for defence as a single concept and to reorganize research and development's goal."[11] Cooperation and compression of research was the aim, and central coordination appeared to offer the greatest chance for success. Speed to cement this proposal was vital in the face of the onslaught of peacetime demobilization.[12] Goforth's proposals were innovative and unique. Since they would deny the services their own research empires, however, they required staunch military and political backing to survive. Mackenzie backed the concept, as did Murchie's replacement, the new chief of the general staff, Lieutenant-General Charles Foulkes.

Foulkes's support was instrumental to the DRB's creation, though he did not originate the concept.[13] He became CGS on 20 August 1945, much to the astonishment and, in some cases, disgust of his contemporaries or subordinates. Foulkes remains a most enigmatic general in Canadian military history. While debate continues on his mercurial personality, salacious habits, and relative merits as an operational commander, Charles Foulkes was the most successful and influential political general Canada produced during the Cold War.[14] Like his contemporaries Guy Simonds and E.L.M. Burns, he had maintained a long professional interest in military technology,[15] and his war experience convinced him that technical advances would define any future conflict, one that would include advanced rockets and atomic payloads. Canada could no longer afford the time to harness its scientific capital, labour, and material for war. It required these capabilities "in being."[16] Foulkes believed, as did many British and American colleagues, that there was a five-year gap of peace before the Soviet Union would contemplate the use of war to obtain its objectives. This period would provide breathing room for the West to develop its post-war assets in a constrained fiscal environment.

In late August 1945, after further revisions, Foulkes accepted a modified version of the Goforth Group's proposal and sent it to the other services.[17] If the Navy and Air Force could agree in principle to an integrated defence research plan, he would forward the proposal to the Defence Committee of the Cabinet. What Foulkes actually sent was a version of the Goforth proposal that he himself condensed. Like Goforth's, it argued for an interim defence research organization to be created immediately, followed by the creation of a sub-department in January 1947.

He added a final step of creating a Department of Research, with a division concerned with defence research. He asked the services' chiefs to support the first two phases immediately.[18] Both the RCN and RCAF had schemes for controlling their own defence research after the war and contested Foulkes's arguments. Minister of defence Douglas Abbot settled the disputes at a meeting of the Defence Council 26 October 1945. An interim organization with a mandate to survey the current defence research needs would be created, to be led by a director-general: a civilian with rank equivalent to the chiefs of staff, who would coordinate the research effort, provide scientific advice on defence matters, and keep close liaison with the NRC. The services would maintain their own scientific advisors and operational research staff. No mention was made of a separate government department as the new organization's ultimate goal.[19]

Committee on Research for Defence

The Defence Council agreed to the terms of the draft memorandum and forwarded it to the Cabinet Committee on Research for Defence.[20] Formed on 10 August 1945, the committee was chiefly advisory and investigative. It met once, 4 December 1945, to consider the memorandum the Defence Council approved. Here it was agreed that all defence research for the services would be coordinated under a director-general of defence research, who would be a civilian with scientific training and would hold equivalent rank to the chiefs of staff. Initial concerns would harmonize research within the services, coordinating research done by outside agencies for the services, and managing the financial burden of focusing defence research efforts as a whole within Canada. After the meeting, the selection of the first director-general was discussed by Howe, Mackenzie, and Foulkes. According to Mackenzie,

> A group of us met in CD [Howe]'s office to decide whether or not to separate military and normal NRC activities. The Cabinet Ministers agreed

to such a decision and the next question arose as to who would head the organization. Charles Foulkes and I were appointed to a Committee to make recommendations. As we left CD's office and went down the corridor Charles said "The only Canadian that I know who has any connection with defence research and who would likely be available is a man called Omond Solandt." I immediately said "That is interesting because I have had him recommended to me by many people whose judgment I respect and I had intended to suggest the same name." We stopped immediately and said why don't we go back at once and make our recommendations, which we did. I doubt if any recommending Committee ever acted so soon after their appointment and I don't think any recommendation ever proved a better one.[21]

Cambridge or Ottawa?

Choosing Solandt as director-general for defence research preceded the meeting with him by some months.[22] While Solandt had made a good impression with certain Canadian soldiers and diplomats, he was still a relative unknown in Ottawa. What is clear is that there were no serious rivals.[23] Indeed, his unique experience, reputation, contacts, and accomplishments made him the most qualified Canadian for the job, next to Howe and Mackenzie (neither of whom wanted the job). After the 4 December meeting, Foulkes contacted Charles Best about his old pupil Omond Solandt. After the phone chat, Best wrote Foulkes a formal reply. "I have always felt that [Solandt] has a finer mind and broader outlook than any other research student of my acquaintance."[24]

Arriving in Canada in early December from Japan, Solandt met with Foulkes and Goforth in Ottawa to discuss the work of a director-general. Solandt was still unsure if he wanted the job or a return to Cambridge, so he had a plan. First, if his wife found a nice house in Cambridge to live, he would return to physiology. Second, Solandt would present a list of requirements he saw as absolute and necessary for him to accept the Canadian offer. If they could not be met, he would turn down the job. Most of his criteria were based on his evaluation of AORG and his thoughts regarding post-war defence research organizations for the Army (see chapter 7). He had heard serious complaints about the Canadian government's bureaucracy and decrepit management style. Remembering his own fights with the Ministry of Supply and in the War Office, he insisted that a director-general report directly to the minister. The organization also had to be independent of the Civil Service Commission. It would be hard enough to convince scientists to continue military research in peace, so Solandt required a free hand to hire the best of the best without reservation or interference. Also, he required a position equivalent to deputy minister and to have substantial military rank to participate in the chiefs of staff committee with influence. Finally, he wanted a salary of $10,000 a year, a sum he felt was large but not outrageous. He was convinced at the time that his demands would not be met.[25]

Unbeknownst to him, Solandt's demands were in step with the Goforth report. He never told anyone his criteria and was floored when Foulkes mentioned all of the major points (except a separate personnel system, though this would eventually be accepted). Solandt, astounded, contacted his wife, who had not found a house. The decision was clear: he took the job. Privately, Solandt told himself that he would accept it for five years, his goal being to create and run a defence research organization that could stand on its own two feet, after which a successor could take over to lead into the next phase of development as military technology progressed. Solandt could then follow his interests and opportunities to different work. If he was successful, such opportunities would be vast.[26] The Privy Council confirmed his appointment on 28 December 1945. He enjoyed a short leave, was discharged from the Army on 15 February 1946, and began his career as Canada's director-general of defence research on 25 February 1946.[27]

The Road to the Defence Research Board, February 1946 to April 1947

Later in life, Solandt recalled the long and gruelling early years of the DRB. Getting the right people was critical. Solandt's initial staff consisted of Goforth, Fordyce, and a clerical staff that included a captain and warrant officer who had been transferred from the director of staff duty weapons (DSDW) of the Army. Nothing could be set up before the organization was a legal entity, so they spent the year planning, using Goforth's organizational ideas, with a target date for being up and running in September 1946.[28] Goforth then secured Marge Eves, a former major in the Canadian Women's Auxiliary, as Solandt's first personal secretary. Her fine organizational skills were complemented by her long-term association with the deputy minister of national defence, W.G. Mills; Eves had been his secretary during the war and knew how to get the best out of Mills. During that initial year, when so much seemed to be foundering and so much paperwork required the official stamp of the department, Eves went to Mills directly and secured his signature.[29] Solandt recalled Mills as an elderly, incompetent civil servant who had worked at Canadian Military Headquarters (CMHQ) in London during the war. Without Eves, Mills would have had the DRB at a standstill until his retirement.

Solandt never took credit for the origin or original vision of the DRB. That laurel he laid at Goforth's feet. For Solandt, Goforth had the duality of imagination to extrapolate the future of defence research as well as any science fiction writer, and a fiscal pragmatism to keep that vision grounded. They talked numerous times of the future of research, the need for recruiting the "young Rutherfords" of Canada, the growing importance of computers, and the feasibility of satellites and intercontinental missiles. This imagination was balanced by a fiercely rational and organizational streak and the fiscal realism of an economist. For Solandt, a wild imagination with a pragmatic intellect was the perfect combination for a right-hand man, and he soon made Goforth his deputy, a position he served during the formative year at the DRB. Among Goforth's more remarkable traits was his experience in balancing the Army budget and keeping the Treasury Board happy.[30] However, at the first meeting of the DRB on 16 April 1947, Solandt informed the board of Goforth's resignation. He was replaced by Welsh chemist and former Suffield Experimental Station director, E.l. Davies.[31]

Why Goforth left is contested. According to Solandt, the DRB's workload taxed Goforth and he was under doctor's orders to find less strenuous work.[32] Fordyce was less certain. Goforth had provided the DRB with its initial structure and vision, but it took Solandt to turn that vision into a working reality in government. Fordyce felt Goforth was reluctant to go, but knew that his lack of scientific background was a limit to what role he might have in the organization's future.[33] It is also possible Solandt did not want to sow the seeds for future confrontations with Goforth by maintaining one of the "fathers" of the DRB as a subordinate. Friction might naturally ensue.

Later in life, Solandt reflected that it was easier to create an organization as it was growing than to change an established organization. Expectations had not yet hardened on the DRB's purpose or shape, allowing some freedom of movement to make improvements on the fly. The same could not be said for organizations bound by tradition and experience. Broad guidelines were established for the DRB's goals and specialized fields of interest. The malleable structure of an embryonic organization allowed it to mould into areas of import as needs arose and policy changed emphasis.[34]

Solandt's interim duties as director general were:

1. to coordinate research activities within the Armed Services with a view to economy of effort in fields of common interest;
2. to conduct such liaison between the Armed Services and the National Research Council and industrial and other research agencies as necessary to obtain economy of effort in matters of common concern;
3. to make a survey of research facilities and to recommend such organization as may be required for the further performance of his duties; and
4. to undertake such other duties of a scientific nature as the Minister of Defence shall require.[35]

Solandt saw his view for a successful post-war defence research organization coming into fruition. His ideas had meshed with Goforth's, and the emerging organization was backed by Foulkes. One million dollars was included to the War Appropriations for 1946–7 to be dedicated to defence research. Initial proposals on a defence research policy were vetted by the chiefs of staff on 17 April 1946, with some friction.[36] But Solandt managed to persuade them of the general principles he felt should guide the organization's future. Canada's defence research focus, he felt, should be predicated on the belief that any future war would be fought alongside the United States and Great Britain, so interchangeability and standardization of equipment was a concern. Designing peculiarly "Canadian" equipment not in step with the development and operational standards of their allies should be rare.[37] The organization should take the long view towards applied and fundamental research. Projects would be chosen where the design phase was not in the immediate future. Canadian research facilities would work best managing a relatively small number of projects well. Projects should make use of important Canadian resources such as skilled labour, natural resources, industrial strengths, distinct experimental establishments, and climactic and geographical features. Projects with direct or indirect commercial or industrial benefits should also be prioritized.

Nineteen research programs were listed for the new sub-department:

1. Atomic warfare
2. Bacteriological warfare
3. Chemical warfare
4. Guided missiles

  5. Rockets
  6. Supersonic aerodynamic gas turbines
  7. Jet propulsion
  8. Electronics
  9. Underwater research
 10. Conventional weapons
 11. Motor vehicles
 12. Operational research
 13. Air transport
 14. Biological research
 15. Service engineering
 16. Civilian defence
 17. Survey
 18. Meteorology
 19. Metallurgy[38]

At the request of the chief of the air staff, the services would maintain control of operational research establishments, but the new sub-department would act as a coordinating body that could provide scientists and other resources to the three services. As can be seen, Solandt played to his strengths. In all of these fields of research he had either direct or indirect research experience or contacts.[39]

By August 1946, with some of the financial and organizational structures in place, Solandt focused the team on surveying work. Indices of scientific personnel for future employment and emergency mobilization were created. Doctor Otto Maas, the German chemist from McGill who had been critical in the successful operation of the Suffield Experimental Station during the war, headed a Personnel Selection Committee. Responsibility for CARDE, the Chemical Warfare Establishment (Ottawa), and Suffield Experimental Station was ceded from their respective authorities to the director-general. Solandt and his staff soon participated on critical committees dealing with their future responsibilities such as atomic warfare, electronics research and development, and radio wave propagation. The nucleus of a scientific intelligence staff and security section was created. The director-general was also mandated to administer the newly formed Joint Intelligence Bureau. Solandt chaired the Sub Committee on Winter Warfare Research, and his staff were a critical component in the planning, development, and organization of the adjoined testing and training station being created at Fort Churchill.[40] Solandt's organization was taking shape.

Creating the Board

On 28 September 1946 Solandt drafted a memorandum for a "Defence Research Board," the executive policymaking body on defence research for all three services. The board would represent the interests of the services, industry, and academia on defence research. The chairman of this board should be the director-general of research. Membership would include the service chiefs, the president of the NRC, and four civilian members representing industry and academic scientists. Solandt included a list of prominent Canadian scientists and science backers for the four civilian members.[41] While alternative structures were considered, none were popular, and on 17 October an Order in Council was signed establishing a "Defence Research Board" in line with Solandt's proposal. The board would be interim until legislation could be passed to establish the DRB as a permanent body within the Defence Department. On 28 November a ministerial directive appointed the four original civilian members of the board: Doctors Charles Best, Otto Maas, and P.E. Gagnon,[42] and Colonel (retired) R.D. Harkness.[43]

Clearly, Solandt modelled the board after his experiences in the war. Here, in one body, were many of the "big shots" of industry, academia, the military, and government. In many ways, the board mirrored the MRC's Sub Committee on AFVs, where Solandt had cut his teeth in government research discussions, and where all the major players attended. The difference now was that he was no longer a junior scientist beholden to Mellanby, Carmichael, or anyone else. As chairman of the DRB, Solandt was now the boss and would personally deal with all the operators involved in defence research.

The interim DRB's duties were

1. to advise on scientific and technical problems relating to defence which may be referred to by the Minister of the Director General of Defence Research;
2. to consider and recommend new proposals for research and development in fields of defence research in relation to the most recent advances in scientific knowledge;
3. to make recommendations regarding the most effective employment of scientific personnel for defence research and development as may be in the interests of national security;

4. to create such advisory committees and panels as may be
   expedient; and
5. to report to the Minister of National Defence.[44]

On 26 December 1946, Abbot resigned and Brooke Claxton became minister of defence. A special meeting of the Interim DRB was held for his behalf on 20 January 1947. The meeting concluded with the board's agreement that defence research was critical to Canadian security needs, and with the end of the National Emergency Transitional Powers Act it was required to have a defence research organization established on a permanent basis.[45] By an Order-in-Council on 3 April 1947, the Defence Research Board was created. On 7 June 1950 Parliament passed Bill 133, and the portion regarding defence research was similar to that of the 1947 legislation. However, now there were provisions for a vice-chairman, as well as minor constitutional changes.[46] In April 1947, the Defence Research Board came into legal existence.[47]

The DRB was a significant example of state-building in peacetime. During the war, Canada had grown its own defence research capabilities largely through Crown corporations and government science establishments. But as the catalyst of conflict ended with the defeat of the Axis, these facets of the state began to recede as normative peacetime conditions returned.[48] And yet the momentum of this state growth did not vanish completely. It carried itself into the technologically defined Cold War. While the military suffered under severe budget and personnel cuts, at the same time a state-run defence research organization, acting as a fourth service of the military, was birthed and began the long journey to a position of respect and authority within the government. Solandt led this organization for ten years, building its strength and stamping it with his own personality and vision. The result was a significant addition to the Canadian state, one unlike any other.

American and British Counterpoints

The DRB was a unique creation. American and British post-war defence research efforts were categorically different. The United States disbanded the Office of Scientific Research and Development (OSRD) at the end of the war, which had centralized wartime defence research. Post-war attempts to create an effective organization were mired in service rivalry, Congressional wrangling, and OSRD president Vannevar Bush's ambitious plans to create his ideal research organization, the National Science Foundation. The result was division. Before the war ended, secretary of the Navy James Forrestall backed plans for a naval research organization, the Office of Naval Research (ONR). With funds and authority to create research contracts, the ONR was a major defence research force in the United States for over a decade. Bush fostered the creation of an interim organization, the Joint Research and Development Board (JRDB) in the Pentagon in 1946. Its role was to coordinate all defence research for the services through committees and panels. Bush served as its first president. It was renamed the Research and Development Board (RDB) in 1947. Most U.S. historians have seen the RDB as an important coordinating body but a relatively ineffectual organization when its defence research assets, including internal organization and external think tanks, are compared with those of the ONR and the other services.[49] The lion's share of defence research funds were vested in the Atomic Energy Commission and organizations concerned with developing atomic power and weapons.[50]

As we have seen, British defence research work during the war had been dispersed among many ad hoc committees,[51] and centralization was desired by the new Labour government of prime minister Clement Attlee. Under the guidance and efforts of Sir Henry Tizard, P.M.S. Blackett, and Solly Zuckerman, Britain created the Defence Research Policy Committee (DRPC) within the Ministry of Defence. DRPC was the gatekeeper for military research advice. Tizard served as its first chairman until 1952, when he was replaced by Sir John Cockcroft. DRPC's influence on centralizing defence research matters has sparked debate, especially since they had few funds for directing research. Current investigation indicates they were strong and effective proponents of chemical warfare (CW) and biological warfare (BW) research, but largely locked out of atomic affairs, which had been the sole purview of the Ministry of Supply.[52] Indeed, Michael Dockrill has argued they had minimal influence on government policy.[53] After Sir Henry Tizard's noted "Science and the Services" lecture in August 1946, Lord Maurice Hankey, the influential wartime minister without portfolio who supported wartime CW and BW endeavours,[54] said Britain should look to Canada for a model of defence research.

I might mention that I heard yesterday – I think it is correct, although I tried to check it up but was unable to – that a gentleman called Ormond [sic] Solandt has recently been appointed as Director of Scientific Works for the Canadian Services, for the Canadian Ministry of Defence, or whatever the right organization is, and that he has been graded on a level with the Chiefs of Staff. That is what I want to see here, a permanent scientific section, which must be sufficiently broadly founded to know a great deal about science and the activities of the Committee of Imperial Defence, and on the other hand, with the march of science in all its branches – and I do not believe we should leave anything out: physics, chemistry, civil, mechanical and electrical engineering, metallurgy, biology, nutrition, medicine, agriculture and the separate research conducted by the Government departments for their own needs, which must, of course, continue.[55]

This was high praise, but the United Kingdom charted its own course and did not follow the innovative Canadian example, and that was unfortunate. DRPC, as an advisory body only, never had the same influence on government policy that the DRB did (see chapter 13).

Solandt and the Canadian Establishment

Solandt arrived in Ottawa in 1945 as he had in Sutton in 1940, Lulworth in 1941, and Ibstock Place in 1943: as an outsider. While that was difficult, it was also an advantage. Solandt had made more admirers than enemies during the war, and his few adversaries were in the United Kingdom and not in positions of influence. The same could not be said of his contemporaries in the United States and United Kingdom. Both Vannevar Bush and Sir Henry Tizard had made powerful enemies in both the government and military during the war,[56] but Solandt arrived in Ottawa with a stellar reputation and a clean slate.

As chairman, Solandt sat on many critical government bodies outside the DRB. He was a full member of the Chiefs of Staff Committee with an equivalent rank of lieutenant-general, and as such he attended meetings of the Defence Council and the Cabinet Defence Committee. Solandt was also an official member of the Cabinet Panel on Economic Aspects of Defence, a body concerned with the fiscal management of defence that included key deputy ministers and representatives from the Bank of Canada.[57] His role on these bodies is considered in chapter 10. Here, we establish his place within the context of Canada's political elite.

The Prime Minister's View

That a new, innovative, and costly addition to the Department of National Defence was created in the post-war era says much of its value and how that value was appreciated by the government. Prime minister W.L. Mackenzie King had no interest in large standing armies, a navy with aircraft carriers, or massive bomber and fighter forces. Demobilization and peacetime strength reductions would be the main goals of the respective ministers of national defence Douglas Abbot and then Brooke Claxton.[58] Yet defence research, as Goforth envisioned it, was a new capability that would require funding in the budget-slashed Canadian military. Mackenzie King harboured distrust, suspicion, and dislike of all things military, and he hoped peace would rid him of having to concern himself too much with their needs.[59]

However, what we know of his opinion of Solandt and the DRB was generally positive. Here was a chief of staff who dressed not like a soldier, but like any other of King's civil service mandarins, one of high intellectual calibre, quiet disposition, and no military bearing. While never one of Mackenzie King's inner circle, Solandt was a known quantity to the prime minister. On 14 November 1946, Solandt delivered a lecture on Canada's embryonic defence research posture for the whole Cabinet. For the Prime Minister, it was a "remarkably good presentation."[60] Two months later, Solandt reviewed proposed Army research at Valcartier for the Cabinet. While Mackenzie King found most of the chiefs of staff's service plans too expensive and in need of further cutting, he noted in his diary, "I feel in the long run research will do much more than aught else."[61] Solandt cultivated a better relationship with King's successor, Louis St Laurent, but the DRB would not have survived its early formation if Mackenzie King had found defence research extravagant or without merit. But by 1946, Canada had established a solid reputation as a nation of intellectual calibre within the field of applied science, and Mackenzie King had an interest instead of apathy for the prestige of science compared to other aspects of the war.[62]

Senior Ministers

Solandt was not an Ottawa Liberal civil service mandarin, but he cultivated good relationships with almost all of his colleagues. His initial boss was minister of defence Doug Abbot, who held Solandt and DRB in high regard, even after he left the Defence Department to become minister of finance in December 1946 (see chapter 10). Solandt's boss for most of his tenure at the DRB was Brooke Claxton, minister of national defence, 1946–54. Claxton enjoyed touting the fact that Solandt "knew more British secrets than the Americans, and more American secrets than the British."[63] Solandt would be one of the pillars of his department. He was proud of Solandt and the DRB in large part because their work and connections translated into credit with Canada's allies. As Claxton wrote in his memoirs in the late 1950s,

> [The DRB] today is admittedly greatly superior to that in any other country. The Board managed to achieve the greatest possible integration with the research activities of other government services, of the armed forces, of the universities and of industry. Particularly through Dr Solandt's own personal leadership, and with the high standards of the scientists he recruited, DRB was established with such high scientific and administrative standards that it won the respect of other agencies in the field, not only in Canada but in allied countries. Substantial contributions were made to the pool of allied knowledge. Altogether DRB won for us a place in the councils and work and secrets of our allies. Canadian defence research probably had closer relations both with Britain and the United States than these countries had with each other.[64]

Solandt was critical in this regard. He had learned from the British that access to high station and secret information meant having chips to bargain with. "The Chiefs and Dr Solandt persuaded me that this was a big league," Claxton noted, "and that in order to obtain the advantages of membership, including particularly the exchange of information, it was necessary that Canada should make a proper contribution. In other words we should have some secrets to trade."[65] The DRB's research in CW, BW, Arctic affairs, anti-submarine technology, short- and long-range wireless, and the McGill Fence radar-defence installations, as well as increasing knowledge in armament and propellants, contributed to Canada's ability to become "inside partners" with its allies. This success was built in part upon the excellent relations the DRB fostered with academia and industry.[66] "For these and other reasons," Claxton felt, "we were a respected member of the club."[67] Solandt was his trusted scientific advisor who, on occasion, reviewed the minister's speeches and filled in on his addresses.[68] The view of Ralph Campney, who replaced Claxton in 1954, and of Solandt's role in government, is covered in chapter 15.[69]

In Ottawa Solandt was reunited with his old football coach, L.B. "Mike" Pearson, whom he had worked with during the war and was now rising to senior diplomatic posts within External Affairs. While he both liked and respected Pearson, Solandt lamented that he had failed to persuade either Pearson or Norman Robertson, Canada's high commissioner in London, to accept scientists and engineers as part of the diplomatic corps. For Pearson and Robertson, this was historian territory, and Solandt found this outlook myopic.[70]

Solandt also cultivated a positive relationship with General A.G.L. "Andy" McNaughton, Canada's premier soldier-scientist. They had met during the war when McNaughton was in command of the Canadian Army in England and Solandt was at Lulworth. McNaughton resigned as minister of defence in August 1945, became head of the Canadian section of the Permanent Joint Board of Defence (PJBD), and was Canada's representative at the UN. In 1946, McNaughton provided unofficial assistance to Solandt to stop internal rot at the Canadian Armament Research and Development Establishment (CARDE), an establishment at Valcartier he was fond of.[71] He told Solandt to make a list of the materials he would need for CARDE and offered to acquire a few men and trucks and get the equipment at surplus. According to Solandt, all the general had to do was send a soldier with a note to the quarter master general and it would be done. Solandt credited McNaughton for giving CARDE a running start that would have taken six months to a year through normal channels.[72] Once again, the value of knowing the key "big shots" in the government was paying dividends.

Omond Solandt and C.J. Mackenzie

The only serious challenge to Solandt's authority that first year came from NRC and later Atomic Energy Board of Canada president C.J. Mackenzie. "Jack" Mackenzie had been the former wartime tsar of defence research in Canada. He was instrumental in selecting Solandt to lead defence research and was a member of the DRB and an integral contributor to it. Before Solandt's arrival, Mackenzie's opinion on matters of defence research and production were second only to those of his former teacher and boss C.D. Howe. It is possible that Solandt's youth (he was only thirty-seven in 1946; Mackenzie was fifty-eight) and cooperative demeanour convinced Mackenzie he was malleable and perhaps naive in the new environment of Ottawa and thus could be manipulated. If so, he was in for a shock.

During the first official meeting of the DRB on 16 April 1947, Solandt presented the board with a series of administrative and other appointments. Mackenzie participated in the discussion and, along with the board, gave his approval.[73] The very next day, Solandt met with Claxton to discuss the appointments, and Claxton told Solandt, frankly, that Mackenzie had suggested he not sign them because Solandt had shown poor judgment in his selections.[74]

Forty years after the event, Solandt recalled his temper shooting "through the stratosphere" over Mackenzie's attempt to undermine his authority after his first board meeting. Knowing full well Mackenzie's reputation and connections in government, Solandt nonetheless immediately confronted him at NRC headquarters. He repeated Claxton's speech word for word. Years later, Solandt surmised that Mackenzie likely miscalculated how frank and open the new minister would be with his new DRB chairman.

Mackenzie apologized, but Solandt needed to make a point about contesting his authority. He asked Mackenzie if he should inform the council about his conduct. Mackenzie suggested against it. Solandt agreed, with the stipulation that Mackenzie admit to Claxton that he had been "talking out of his hat" regarding Solandt and the appointment list. Pursuing this kind of Machiavellian relationship instead of a cooperative one would be firmly resisted. Mackenzie agreed.[75]

Solandt made it clear that the DRB was his organization and that his leadership would not be undermined. Still, Solandt needed Mackenzie as his gateway to C.D. Howe's office. Neither man held a grudge. By the next DRB meeting Mackenzie made clear his support for Solandt's leadership of the DRB,[76] and an axe was buried in favour of cooperation.[77]

While Solandt had Mackenzie's support at the official level, the NRC was less cooperative at the working level. Despite protests of wanting to be free of military affairs, free of industrial projects, and desiring a return to fundamental research, the NRC dragged its feet while handing over some wartime responsibilities, especially radar and aeronautics, to the DRB. Indeed, the DRB never achieved full control of aeronautical research, despite the NRC having no reason to maintain this applied research except for military purposes. Solandt lamented that even on areas of mutual control, such as the National Aeronautical Establishment, the NRC was a difficult partner, and after he retired from the board NAE was effectively in the hands of NRC because his successors lacked strength in resisting NRC influence. Successful cooperation was achieved in supporting research for subjects such as computing at the University of Toronto, but the DRB and NRC were two towers of government research, and a certain amount of friction over jurisdiction was to be expected.[78] Mackenzie's account of these events has yet to be found.

Conclusion

Solandt had translated his wartime experience into a job that, in 1939, he could never have conceived of. For the better part of a decade, he pursued the defence research arm of Canada's Cold War strategy while simultaneously building the organization responsible for this task. His leadership in state science in this period had numerous dimensions that are considered in the following chapters. First, he was an active and valuable member of the Chiefs of Staff Committee, holding his own when fighting for the DRB within the highest echelons of Canada's strategic debates in the Cabinet Defence Committee. Second, he was a strong and effective leader within the DRB itself, working with the military, industry, and academia to provide the Canadian Armed Forces the best scientific and technical support Canada could offer. Third, he held and maintained the DRB's course against domestic and international pressures and allowed the organization's establishments to become respected in the fields of chemical and biological warfare defence, as well as rocketry and guided missiles. Fourth, he established himself as Canada's premier defence research diplomat, working with senior allies to gain access to their research in exchange for entry to Canada's work. Finally, for a decade, he was the chief advisor and defence research leader on atomic warfare. The legacy of Omond Solandt took root during the Second World War, but manifested in each of these arenas of Canada's Cold War defence research policy.

10 A Doctor among Soldiers: Solandt, Canada's First Scientfic Chief of Staff, 1946–1956

For Solandt, being chairman of the Defence Research Board comprised two jobs. First, he was the scientific member of the chiefs of staff, responsible for representing all major scientific and technical knowledge relevant to defence. Second, he was chief executive of a large and complex applied research organization.[1] The following chapter examines his role as Canada's first scientific chief of staff.

In this role, Solandt contributed primarily to defence policy on three government bodies: the Chiefs of Staff Committee (COSC), the Cabinet Defence Committee (CDC), and the Privy Council Panel on Economic Aspects of Defence Questions (PEADQ).[2] He kept each group abreast of the DRB's efforts and capabilities, and brought his intellect to bear on a host of defence matters. Between 1946 and 1950, his most significant contributions were to Canada's emerging continental defence policy, and, to a much lesser extent, the formation of NATO. At the same time, he kept the COSC and the CDC aware of the new DRB's existence, efforts, and capacity to fulfil its mandate.

From 1950 to 1955, as the Korean War and NATO responsibilities ushered in Canadian rearmament, Solandt found himself contending with the pressing demands of active warfare and the DRB's goal of meeting future service needs. The DRB provided resources and scientists for the Korean effort while holding onto the long view of defence research. In the CDC, he defended the DRB's expansion against such adversaries as C.D. Howe, minister of the Department of Defence Production and the most powerful operator in the government. When the dust settled from this confrontation, Solandt was the victor, and from 1952 to 1955 he focused his efforts on the COSC and CDC towards continental defence and scientific intelligence. This included contributing his own thoughts and resources to the development of the CF-105 interceptor, also known as the A.V. Roe Arrow.

As a Canadian chief of staff, Solandt achieved a level of influence and connection with the senior "operators" in Canadian government that he never achieved in Britain during the war. His ten years as Canada's first scientific chief of staff offers a unique window and perspective on Canada's post-war defence policy. Through intellect, presence, and political savvy, he found the apex of his influence in state science at the DRB.

Canada's Strategic Outlook during the Solandt Era, 1946–1956

Solandt's tenure as DRB chairman coincided with two distinct phases of Canada's post-war strategic vision. The first is aligned with the emerging Cold War and the subsequent strategies Canada developed, on its own and alongside the United States and Great Britain, against the Soviet Union. These efforts resulted in embryonic continental defence agreements between Ottawa and Washington, and the creation of the North Atlantic Treaty Organization (NATO) in 1949. The second phase began with the Korean War in 1950 and the change in American strategy associated with President Eisenhower's "New Look Strategy" of 1953, which focused on rearmament and then increased reliance on a nuclear deterrent, as both the United States and the USSR increased their nuclear and then thermonuclear arsenals. While Solandt was affected by these changes in international affairs and technological development, he kept a cool hand on such hot topics. His strong and rational mind was a helpful salve against periods of uncertainty in the thermonuclear age.

The Soviet Threat

With the common German enemy defeated, the wartime alliance began to splinter soon after VJ day. All the major powers contributed to this dissolution,[3] but for Canada the most alarming conduct came from the Soviet Union. While war ravaged, the Soviet Union maintained the world's largest army and massive industrial strength. Its scientific capital was bolstered by massive espionage efforts against its Western allies. The defection of Soviet cipher clerk Igor Gouzenko in Ottawa, September 1945, revealed this last point to the Canadian government in fine detail.[4] From the Gouzenko revelations until 1949, the Soviet Union revealed itself to be, in the eyes of most Canadian and Western observers, a hostile and aggressive nation.[5] Obstinate conduct in the

UN over the future of Germany, and the crushing of free elections in Poland were soon complemented by initiatives beyond the Iron Curtain. Moscow's failure to remove its troops from Iran, support for communist guerrillas in Greece, and aggressive ultimatums to Turkey over control of the straits to the Black Sea raised fears of a new war between 1946 and 1947.[6] The 1948 communist coup in Czechoslovakia was largely believed to have been initiated and supported by Moscow, a belief later confirmed.[7] Josef Stalin's attempt to force the issue of German unification in his favour by blockading Berlin in the spring of 1948 made it clear that the Soviet dictator was willing to risk international instability to obtain his insatiable security needs.[8] Stalin's desires were checked by the West's unwillingness to appease him, as witnessed by the American-led Berlin airlift to relieve the city. Stalin's public statement of not being frightened by atomic weapons did not prevent him from driving his country's scientists to create their own as fast as possible. By 1949, years before most in the West thought it possible, the USSR had detonated their first atomic weapon.[9] With communist victory in China that year, the West confronted a multi-front communist adversary with massive conventional forces and emerging atomic stockpiles. Many U.S. policy strategists felt that by 1954 the Soviet Union would have recovered sufficiently from the war to consider the use of military force instead of propaganda and subterfuge to achieve its expansionist goals, making 1954 the "year of maximum danger."[10]

Canada's contribution to deterring the Soviet threat spanned many levels (economic, ideological, domestic, etc.), but the two paramount fronts were continental defence for North America and collective security in Europe. By geographic necessity and historical tradition, Canada had always worked with Britain and the United States to guarantee its own security interests. The Cold War would not be different. The means of this security, however, would change.

Canada's Strategic Efforts 1945–1950

Prime minister Mackenzie King had hoped 1945 could see a return to the isolationist policies of the interwar years and acted accordingly. Minister of national defence Douglas Abbot was tasked with a single goal: demobilization. Brooke Claxton, his successor in 1946, had reduction and centralization of the Armed Forces as his lodestar. These efforts rested on hopes that the wartime allies could successfully maintain international peace and security through the newly formed United

Nations. For Mackenzie King, Canada would avoid anything resembling alliances or imperial entanglements.[11] But by 1948, the Soviet Union was acknowledged as the greatest threat to Canada and the West. In November, the aging King was replaced by Louis St Laurent, who desired a more international role for Canada.[12] Initial post-war strategic thinking on this emerging threat was muddled and unfocused until Claxton became minister of national defence in 1946.[13] By 1947, Claxton told Parliament what Canada's military purpose was in peacetime:

1. to defend Canada against aggression;
2. to assist the civil power in maintaining law and order within the country; and
3. to carry out undertakings that, by our own voluntary act, we may assume in cooperation with friendly nations or under any effective plan of collective action under the United Nations.[14]

For the Army, this meant continental defence as part of the airborne-brigade group Mobile Striking Force (MSF). This innovative unit was to meet any enemy incursion on Canadian soil in the northwest of the country while the nation mobilized.[15] For the RCN, it meant an anti-submarine role off Canada's three oceans.[16] For the RCAF, it meant continental defence and early warning against Russian bomber threats.[17] For the Defence Research Board, it was to support all of these goals within the realm of defence research and technical/scientific advice. This required Solandt's active membership in the COSC.[18]

The Chiefs of Staff Committee

This was Solandt's unique and most difficult job, in part because he was a complete and utter outsider. Since 1939, the COSC was the highest non-command echelon for soldiers from all three services to coordinate and execute the nation's military strategy. When Solandt arrived in the new position of chief of staff for defence research, the service chiefs were soldiers of distinction with proud war records and years of service in the profession of arms. Lieutenant-General Charles Foulkes, chief of the general staff (CGS, Army), rose rapidly through senior appointments in the Canadian Army and commanded the 2nd Division of the 1st Canadian Army during the Normandy campaign before replacing Major-General E.L.M. Burns to command the 1st Canadian Corps in Italy in 1944. He and the corps returned to Europe in 1945, where

Foulkes would receive the surrender of German Army in the Nether-lands.[19] Vice-Admiral Harold Grant was a decorated naval officer who served with the Royal Navy as captain of HMS *Enterprise* through a series of difficult victories in the Battle of the Atlantic, and later con-tributed to the Normandy landing. In 1947, he took the position of chief of the naval staff after the previous two appointments were retired in quick succession.[20] Scottish-born Air Marshal Robert Leckie had a dis-tinguished career as a fighter pilot with the Royal Naval Air Service during the First World War in Britain, and came to Canada in 1940 to take charge of the British Commonwealth Air Training Plan (BCATP), one of Canada's chief strategic accomplishments of the war. His success managing this program landed him a promotion to Canada's wartime chief of air staff in 1944, a position he held until 1947.[21]

Into this inner sanctum of veteran soldiers came the relatively unknown Doctor Omond Solandt from the British Army Operational Group, wearing a suit and tie and with the equivalent rank of a lieu-tenant-general; this no doubt raised the ire of any professional soldier. Worse was Solandt's relative young age: he was only thirty-seven. His presence on the COSC was backed by Claxton and Foulkes, but ini-tially resented by the other service chiefs as one more part of Brooke Claxton's efforts to civilianize the COSC. But while the COSC hosted many unofficial members such as deputy ministers and members from other government departments, Solandt was an official chief of staff.[22] Solandt's impressive war career could not compare with the military record of the other service chiefs. He was also the most educated man in the room. All of these differences (age, profession, academic creden-tials) should have minimized his influence on the COSC.

Impressively, Solandt was the second-longest-serving member of the COSC. Only Charles Foulkes served longer, though in two positions (CGS of the Army, 1945–50, and chairman, Chiefs of Staff Committee, 1951–60). Solandt would work with three different chiefs of the general staff (CGS), four different chiefs of the naval staff (CNS), and three dif-ferent chiefs of the air staff (CASs), giving him, along with Foulkes, an unprecedented position of consistent influence in the COSC until his resignation in 1956.[23]

Relationship with the Services

Solandt was an active chief of staff. He had long ago learned to get along with soldiers, but never as an equal among them. Now he had the influence in government denied him during the war, but he did not abuse this influence, nor did he relish power for power's sake. During the war he had cultivated a dislike of "empire builders" in government science. Solandt's approach was cooperative and aggressive as he tried to convince the chiefs that he was there for them, not against them.

He had a good working relationship with most of the service chiefs, especially Foulkes, who had picked Solandt and supported the DRB.[24] Solandt respected Foulkes as a cagey operator in the corridors of power in Ottawa, though they were not friends. Foulkes, never personable, was a solitary creature. They made an odd pair, the cunning general and the dour doctor, and they had disagreements,[25] but they generally supported each other's positions. Foulkes backed the DRB, and Solandt acknowledged Foulkes's primacy of place in the higher strategic decisions of the COSC. From 1949 onward, Foulkes regularly briefed the DRB on the strategic situation, and Solandt maintained the chiefs of staff as official members of the Board.

Lieutenant-General Guy Simonds, Canada's most praised and innovative commander of the Second World War, was harder to work with than Foulkes when he became CGS in 1950. After Claxton resigned in 1954, Simonds became a vocal critic of defence research as a diverter of funds from more important defence needs.[26] A disciple of Montgomery, Simonds likely shared his mentor's disdain for scientists telling the military how to do their job. Schonland had initially encountered this resistance at 21st Army Group. But, unlike Schonland, who had served Montgomery as a scientific advisor of junior rank, Solandt met Simonds on a more even playing field. He had no problem telling Simonds when his ideas were weak, especially regarding civil defence (see chapter 14). Still, Solandt tried to placate the decorated but prickly Simonds's flights of fancy on defence research, including his universally panned idea of a "flying tank," a heavily armoured but short-range transportation device, for the services.[27]

For Solandt, the Army was generally positive towards the DRB. They supported and requested Arctic research at Defence Research Northern Laboratory at Fort Churchill, chemical and biological warfare field trials at Suffield, and armament research at the Canadian Armament Research and Development Establishment at Valcartier, though they resisted the handover of full radar work from the Signals Corps.[28] While his opinion on the naval chiefs of the era is not recorded, Solandt had strong praise for the working relationship established with the RCN, whose collaborative approach to defence research with the NRC during the war made them natural partners with the DRB during peacetime.[29]

The RCAF, however, had both the greatest involvement and the most difficult relationship with the DRB. As James Eayrs has noted, the RCAF wanted a dominant role in the Cold War, one they believed had been denied them in the Second World War. That included control of all Air Force aspects of defence research.[30] Solandt recalled Air Marshal Leckie's confrontational attitude towards the DRB from the start. On 16 April 1947, Leckie was quick to press for increased RCAF membership to fill two vacant seats on the board.[31] Solandt and the board dismissed the notion because of conflict of interest and jurisdiction, and later approved academic and industrial members instead.[32] Leckie and Air Vice Marshal Wilf Curtis, along with their member for technical services, Art James,[33] were fiercely resistant to the DRB's initiatives and attempted to maintain as much control over aviation medicine and aeronautics research as possible. These manoeuvrings were largely behind the scenes. Solandt fought a rearguard action for years against the RCAF at the official level to maintain the DRB's place in Air Force research. But over the years, at the working level, the DRB's funds and scientists became critical to RCAF continental defence needs. The RCAF staff enjoyed working with the DRB scientists and technicians because of their skills, and their access to support from the DRB, industry, and academia.[34] Solandt also found cooperative measures on joint DRB-RCAF research.[35] For Solandt, Curtis never fully understood how well the DRB worked with his service. But despite this friction, Solandt had high regard for the RCAF in planning its post-war role in continental defence, and in generating the future requirements for it.[36]

Continental Defence, 1946–1950

Continental defence required close cooperation with the United States. With this cooperation came fears over Canadian sovereignty, making it a highly political defence issue. Since the Second World War, the United States had many troops and operations nestled in Canada, including the Alaska Highway and joint work on winter research at Churchill, Manitoba. In 1949 Newfoundland joined Confederation, and after much discussion Canada allowed the U.S. Strategic Air Command (SAC) to maintain a launch site at Goose Bay. This was the only North American launch site from which bombers could reach the USSR and return without refuelling, making it "the most important all-round strategic air base in the Western Hemisphere" at that time.[37]

The military plans for coordinated continental defence were managed post-1945 through the Permanent Joint Board of Defence (PJBD), which was created in 1940 as a cooperative organization for mutual Canadian and American defence needs, and it continued after the war. The Canadian section was headed by General McNaughton, the American by former mayor of New York City Fiorello LaGuardia.[38] The most important planning body of the PJBD was its Military Cooperation Committee (MCC), which reported to senior Cabinet officials and the Chiefs of Staff Committee and generated the earliest post-war strategic plans to counter Soviet incursions of North America. The MCC believed that by 1950 the war-ravaged Soviet Union would have the capability to launch strikes on North American targets. Their Basic Security Plan (BSP) of 1946 identified this threat and generated plans to counter it, which included the immediate construction of an early warning defence system that consisted of layers of radar defence lines stretching from the Arctic to the northern United States, anti-aircraft assets, and an all-weather interceptor to attack any Soviet penetration of Canadian airspace.[39]

The CDC found the BSP expensive, alarmist, and technically and nationally unfeasible in November 1946. In December, discussions with George Kennan, U.S. diplomat and author of the U.S. "containment" strategy, emphasized the need to counter Soviet intentions rather than the immediacy of a Soviet attack. This would be over a variety of non-military fields, including the economic strategy imbued within the Marshall Plan for European recovery, formally announced 5 June 1947.[40] So Canada focused its continental defence efforts on areas of research on the Arctic and the feasibility of the Arctic radar stations called for in the BSP. These studies involved the DRB and the RCAF and were completed by 1947. They found that Arctic stations alone would provide negligible defence, given their distance from air bases, and complete radar coverage of Canada was too expensive.[41]

Yet the Canadian government remained interested in the technical possibilities of continental security. As such, a series of radar screens that would protect vital areas instead of complete coverage was envisioned as the most feasible option. The first of these was the Pine Tree Line, started in 1951 and completed in 1954. "It consisted of more than thirty radar stations following a route running northeast from Vancouver Island into the Peace River district of Alberta, down through the northern states of the American prairie, up again into Ontario and Quebec, and ending at the Atlantic Coasts of Newfoundland."[42] They included electronic apparatus to direct fighters towards incoming targets. The project was manned and paid for by the United States and Canada, the former paying two-thirds of the $430 million cost.[43]

Solandt contributed to most discussions on the BSP and continental defence from the start. His critical eye no doubt made him unpopular with Leckie and his successors. On 4 February 1947, Solandt criticized the controversial air defence appendix of the BSP and its suggested performance requirements for air defence. He found the plan inefficient and suggested that the program be two phases: the first involved optimum development of present equipment, instead of future needs; the second examined what equipment would fall short of the requirement and what further R&D was needed. Of course, he offered the DRB's aid to the Joint Planning Committee to fix these problems, which were solved.[44]

Solandt also critiqued the RCAF plans for air defence discussed in September 1947. The plan called for immediate construction of a complete and functioning radar system. This was desirable, Solandt argued, if war was in the immediate future, but the system could not be made for years to come. As such, the system should be designed to be operated and integrated concurrently, and he argued against the RCAF desire to construct runways first and radar stations last. These should be planned to be created as a single operational fighting unit.[45] Solandt also suggested the RCAF review the trunk lines communication system used by RAF Fighter Command in defence of Great Britain, which was more extensive than all those in civilian use in the United Kingdom before 1939. The RCAF, however, showed little interest.[46]

Solandt also challenged the RCAF's choice of radar. On 4 February 1947, the COSC reviewed a RCAF memorandum with PJBD recommendations for installation during 1947–8 of six low-frequency Long Range Navigation (LORAN) stations in the Arctic: three in Canada, two in Greenland, and one in Norway. The memorandum emphasized the value to military and commercial air needs, which could be converted to early warning use at a later date. The proposal originated with the U.S. section of the PJBD and offered three alternatives for apportioning responsibility for installation in Canada. The cost was $1,160,000, with an annual recurring cost of $300,000.[47]

Solandt disputed the technical feasibility of converting the LORAN equipment to early warning use. LORAN was also a questionable navigation tool, if it was desirable to set this up as a purely military project without thinking of civilian system integration. "From a military point of view, however," Solandt said, "it would be desirable to restrict the use of these facilities to our own aircraft, and therefore to code the signal. This was difficult, though quite possible from an engineering standpoint." But he said this was a slight objection, and left it to the RCAF to decide.[48]

The RCAF went ahead with its LORAN proposal, despite Solandt's misgiving. In 1952, as Canada and the United States reconsidered continental defence needs, Air Marshal Curtis warned against "the undesirability of embarking on experiments which had not previously been adequately evaluated," and used the RCAF's experience with LORAN radar stations as an example. While the details of LORAN's failure were not mentioned, Solandt's warning had proved right, though he made no comment.[49]

Solandt also acted as a link to continental defence issues in Britain and the United States. In June 1948, he reported to the COSC and the CDC on the RAF Exercise Pandora. Pandora was a dual-purpose exercise examining air defence against conventional and atomic bombs, and Britain's defence research needs to meet these threats. The British outlook was pessimistic because most government research funds were diverted towards industry. Short-term rather than long-term weapons system projects were now being considered. But for Solandt, "the conclusion of the Royal Air Force on Exercise Pandora had appeared to him to be unduly pessimistic. Among the weapons that they considered would not be available by 1955 were supersonic aircraft, guided missiles (ground to air and air to air), and improved blind-flying bombing radar."[50] Solandt believed these systems could be available sooner than predicted. He also shared the RAF's view of operational methods for delivering atomic bombs.[51]

As well, Solandt maintained connections with U.S. counterparts in defence research aspects of continental defence. Claxton told the CDC in August 1948 of the nature of U.S./Canada defence collaboration and BSP 1. "In the important field of defence research, close liaison arrangements have been established between Dr Solandt and Dr Bush [president of the Joint Research and Defence Board], and between the interested agencies of both countries."[52] The full nature of Solandt's defence research diplomacy is considered in chapter 13.

The USSR detonated its first atomic device in late August 1949. As Canada came to grips with this event, Solandt suggested that the Joint

Intelligence Committee of the COSC focus their increasing attention to consider "the USSR plan for the strategic bombing of Canada," and work with the RCAF on exercises to appreciate RCAF needs and training requirements to meet this threat, though he lamented the current lack of information being released by the Atomic Energy Commission in the United States.[53] Solandt's efforts on atomic affairs is considered in chapter 14.

Arctic Research and Intelligence

Lack of information was not solely the result of tight-lipped atomic scientists in Washington. The RCAF's continental defence plans required comprehensive studies on Arctic geography and cartography, the effect of the magnetic pole on navigation, and the impact of sub-zero temperatures on weapons, equipment, and people.[54]

The Arctic was the DRB's responsibility. The organization had created the DRNL for most of this work, but Solandt had also agreed to administer and select personnel for the Joint Intelligence Bureau (JIB) of the COSC's Joint Intelligence Committee (JIC). JIB was to be an intelligence-gathering body, including topographical and geographic intelligence for the Arctic.[55]

The United States desired Canadian-U.S. Arctic research and exercises at Fort Churchill as early as April 1946, and Solandt argued for DRB scientist participation as soon as the organization was formed.[56] Solandt also initiated reorganization and streamlining of COSC subcommittees on winter warfare, contributing to the creation of the Inter Service Committee on Winter Warfare, as well as a Sub Committee on Arctic Research. By 1 October 1946, the British chief of combined operations wanted cooperation in winter warfare research, and soon active American, British, and Canadian cooperation on winter warfare exercises and research became normalized at the Army facilities at Fort Churchill and, later, DRNL.[57] The Winter Warfare Committee was then streamlined into the Arctic Research Committee under H.L. Keenleyside, deputy minister of mines, though it maintained service representation.[58] By November 1948 the CDC agreed to the expansion of the DRB's joint work with U.S. scientists on ionospheric research, which was believed to be a vital part of "the international Arctic chain, and their purpose was to supply technical information which was essential preliminary to the development of reliable radio communications in the north."[59]

Solandt also worked to centralize government Arctic research, civilian and military, at Fort Churchill. He coordinated all geophysical research.[60] Solandt also argued for restraint against the RCAF desire to build airstrips in the Arctic without U.S. support, as "it would not be possible to do so as the ships and other equipment required were not presently available."[61]

These joint Arctic efforts were not without friction and delay. The problem of the JIB provides one case. In June 1947, the COSC had received word that the United States was worried about lack of strategic information about the north. The COSC agreed it would be embarrassing for Canada to let the United States lead on this issue. Secretary to the Privy Council A.D.P. Heeney suggested that the Bureau of Mines be tasked with this if JIB was not yet ready, because JIC had not yet found a suitable director for JIB. Foulkes tasked Solandt to find a suitable director, with JIC approval.[62]

Solandt consulted with the deputy minister of mines to coordinate efforts as he searched for a director and organized the Topographical section of JIB, while the Bureau of Mines Geographic Section began work on Arctic topography. In 1949, British scientist Ivor Bowen was finally selected as the director of JIB. Air Marshal Curtis worried that since Bowen had been in Canada only since 1948, he might not qualify for the residency requirements for government service. Solandt reminded Curtis that the DRB, free of the Civil Service Commission, had no residency requirements, and Bowen was selected and began work in 1950.[63]

NATO

While continental security in the Arctic devoured the RCAF's attention, fear of Russian machinations in Europe and the need for collective security were also paramount to Canada's defence. Canada took an active and important role in the creation of NATO in April 1949, a twelve-nation alliance that many in Ottawa hoped could act as a counterbalance to U.S. influence, including in North America.[64] However, NATO's requirement for ninety divisions from member nations remained wishful thinking. Most nations, the United States included, had quickly demobilized after the war. Canada had prepared statements of proposed contributions to the alliance for planning purposes only.[65]

In the COSC, NATO was largely Foulkes's domain.[66] Solandt attended every meeting regarding NATO's formation, structures, and goals, though his role in the alliance's origins was limited to precise criticism and suggestions. In June 1949, he noted that the military and political concepts of NATO's organization, while individually sound, were not compatible.[67] The military concept had NATO responsible for global strategy. This was not politically possible and was likely in contravention of Article 51 of the UN charter. If North Atlantic countries were to plan world strategy, Solandt felt "they would in fact be substantiating the Soviet claim that the North Atlantic Treaty was not a regional defence pact within the meaning of article 51."[68] The complex subject of NATO's origins and future force structure cannot be dealt with here. Suffice to say, others shared Solandt's views. The Regional Planning Groups that eventually formed to provide the initial embryonic structure for NATO members was a result of these concerns.[69]

Maintaining the DRB's Presence

Throughout this period, Solandt kept the COSC abreast of all major DRB developments as well as his own efforts in defence research diplomacy (see chapters 11, 12, and 13). Most of these projects were the result of service needs, suggestions, and recommendations. Indeed, by 1949, Solandt and the DRB were a known entity.[70]

Becoming a known entity was no small feat, as "DRB amnesia" often occurred. In March 1947, a joint appreciation and plan for the Canadian Armed Forces excluded the DRB from their review. Solandt noticed, and the appendix was revised.[71] On 28 April 1947, Solandt challenged the assumption of the Joint Equipment Policy Committee, whose work on standardization could be read to include defence research.[72] In 1949, the terms of reference for the Canadian Joint Staff in Washington initially had no mention of the DRB liaison member. Solandt stated the DRB member would not participate in NATO discussions since this was beyond his mandate, but he demanded that his liaison officer in Washington be a member of the joint staff so the DRB would remain in the loop.[73]

The Korean War and Canadian Rearmament 1950–1955

But Solandt's efforts on all these fronts changed as a result of a war in Asia. Canada's peacetime disarmament reversed gears when North Korea invaded South Korea on 25 June 1950. With UN authorization, a military force under U.S. command was used to repel the North and their Chinese allies. By the end of the fighting on 27 July 1953, Canada mustered over 26,000 troops for active operations in all services and sustained 1,600 casualties.[74] Most Canadian and Western powers believed that Stalin had instigated the war in Korea as a quagmire for UN forces so that he could make aggressive moves in Western Europe.[75] The fear of a new world war was palpable between 1950 and 1953. NATO Supreme Allied Commander Europe (SACEUR) General Dwight D. Eisenhower called for members to make good on their contribution to the alliance.[76] As such, Canada raised and sent the 27th Brigade group to Germany in 1951, with two additional brigade groups raised in Canada should they be required (together known as the 1st Canadian Division), as well as an air division of twelve squadrons, and ships for Supreme Allied Commander Atlantic (SACLANT). For the first time in its history, Canada would maintain armed force in peacetime Europe as a deterrent to prevent a foreign aggressor from starting a war.[77]

The active fighing of the Korean War had ended, in part, because of changes in the U.S. political landscape. Eisenhower, no longer in uniform, was elected president of the United States in November 1952. His strategic vision was established in NSC 162/2, known in public as the "New Look." Eisenhower wanted to stem the tide of the massive increase in the defence budget triggered by the Korean War and encapsulated in the Truman administration's strategic paper NSC 68. This comprehensive policy paper called for, among other things, massive conventional and nuclear capabilities as well as the development of the hydrogen bomb to provide the United States with maximum means of facing the Soviet Union.[78] Eisenhower feared such spending was playing into Soviet hands by destroying the American free market economy. NSC 162/2 sought to protect the economic strength of the United States and prevent the creation of a "garrison state" by an increased investment and reliance on nuclear weapons rather than expensive conventional forces.[79] A key aspect of the "New Look" was the administration's greater willingness to consider the use of nuclear weapons in a variety of theatres. By 1955, the year that Eisenhower's administration felt the USSR would have matched the United States in power projection, it would even consider their use in limited wars. The "New Look" would return the initiative to the United States with an asymmetric strategy that would force their adversaries to question the value of taking general or limited action and risk nuclear war, while cutting costs on large conventional forces.[80]

This increased value of nuclear weapons as a deterrent against Soviet aggression influenced Canada's view of continental defence and its NATO obligation. It was a nuclear era, defined as much by the advances in weapons technology as by changes in the international situation. The British Global Strategy Paper (GSP) of 1952 outlined the need for nuclear weapons in NATO to reduce its general weakness in conventional forces vis-à-vis the Red Army, and the GSP in turn influenced the generation of NATO Military Committee paper MC14/1, the heart of NATO's nuclear strategy in 1954.[81] By 1955, with the Soviets now defended by the Warsaw Pact alliance, Canada supported NATO's decision to arm and equip its forces with nuclear weapons.[82]

Nuclear weapons and nuclear nations increased. Britain had exploded its first atomic device in 1952, the same year the United States detonated its first hydrogen bomb.[83] The Soviets tested their first hydrogen weapon in 1953 and by 1955 had exploded a thermonuclear device.[84] The increasing megaton yield of thermonuclear weapons reduced the need for accuracy and increased the risk of their possible use in Canada in the event of an East/West confrontation. As well, the shifting means of delivery, from bombers to the emerging threat of intercontinental ballistic missiles, complicated the defence planning effort. Solandt's role in managing Canada's atomic efforts without atomic weapons is discussed in chapter 14, but the his roles in continental defence and NATO from 1950 to 1955 are addressed here.

The DRB and the Korean War

Solandt took a limited role in the COSC and CDC deliberations on the Korean War, reserving his judgment for specific issues of research merit. War was the senior service chiefs' business, but the Korean War was an unexpected divergence from what Solandt and others felt was the "front line" in Europe. Korea and its resulting demands for immediate defence research needs almost derailed the DRB's main objective of taking the long view towards defence research. Yet Solandt managed to find a balance between immediate war needs and maintaining Canada's strength in science and technology as one pillar of Canada's deterrent contribution against the Soviets.

These were mostly Army projects. Among the less glorious aspects of the DRB's contribution to the war effort was straightening out the Army's recruitment and mobilization schemes. The rush to fill the ranks had led to embarrassing situations where elderly and crippled Canadians managed to pass muster for the original deployments to Korea. The DRB's Personnel OR teams corrected this mishap.[85]

Solandt went to South Korea as part of Brooke Claxton's visit to the region 31 December 1951, just before the Canadian Army despatched it first post-1945 operational research team, known as the Canadian Army Operational Research Team (1-CAORT).[86] Not surprisingly, Solandt was largely responsible for the idea of 1-CAORG. He pushed Lieutenant-General Simonds hard to have the Canadian Army send an operational research team to Korea, the first active theatre of war in which to do research on modern infantry weapons. Simonds reluctantly agreed, and the two-man 1-CAORT was despatched to Canada's 25 Brigade Group in Korea in early 1952. The DRB even provided one of 1-CAORT's two scientists: veteran and future DRB strategist Doctor Robert J. Sutherland.[87] Simonds, however, demanded Sutherland rejoin the militia and be officially "hired" through the Canadian Army Operational Research Establishment instead of the DRB.[88] Despite some communication snafus, the DRB received 1-CAORT's reports on battle and casualty analysis, equipment studies, and body armour and helmet research. The DRB establishments also provided 1-CAORT with research subjects for field study.[89] 1-CAORT's efforts have been largely forgotten or dismissed, though current scholarship shows they did substantial and interesting work in a short period of time, especially on casualty analysis, a field Solandt had pioneered with Laurie Chute a decade before.[90]

The DRB's Expansion 1950–1953

But the major impact of the Korean War for Solandt and DRB was the increased risk of a war in Europe and possible attacks on North America. Foulkes told the CDC in July 1950 that even if the Soviets had not instigated the Korean War, they might take advantage of it when their atomic stockpiles increased over the next year.[91] Canada soon began rearmament to field its commitments to the UN forces in Korea, to NATO, and to re-examine its continental defence requirements. The DRB's staff was increased to meet the needs of the Armed Forces.[92] By December, Claxton had asked for a $900-million increase in the defence budget, including for the DRB, whose projects would be increased.[93]

As the war dragged on for another year, Solandt felt the pinch in his organization (see chapter 12) and requested more funds for research. Howe, now minister of the Department of Defence Production, challenged Solandt's request. He believed Canada was spending too much on research and not enough on the production of critical equipment like the CF-100 and the Orenda engine. DRB's fiscal management is demonstrated in chapter 11.

The CF-100 project was critical to Canada's air defence strategy. To meet the Russian Tu-4 intercontinental bomber, the RCAF had required an all-weather jet fighter. No nation was creating an aircraft with these specifications, so the RCAF suggested the airframe and engine be made in Canada. Howe and Claxton championed the RCAF desire for Canada to produce the air frame and engine. Although Foulkes and Solandt were both sceptical, arguing that the talent was there but the support needed to make it work would outweigh its value, their criticisms were less persuasive. Eventually, A.V. Roe Canada was given the task, and succeeded at high cost. The result was the CF-100, powered by the Orenda engine.[94]

At the CDC meeting in September 1951, Howe suggested that Solandt and the DRB reduce their research agenda so money could be spent producing the CF-100 and Orenda engines. As well, Canada should buy what advanced technology Britain had created and have it produced in Canada, instead of increasing its research mandate.[95] Solandt was now being tested by the most powerful minister in the government on an issue of great strategic import. He did not shrink from the challenge, but countered it. The production problems of A.V. Roe, Solandt said, were a result of poor management, the implication being that the DRB should not pay the price for A.V. Roe's inefficiency. Howe may have taken this cue, since he placed one of his deputies, Crawford Gordon, as manager of A.V. Roe Canada in 1952. Solandt continued, "The company had been good in the development field but poor in organizing production. Last year the development shop at Avro [sic] had been turned over to production. Most of the airframes and engine development contemplated in the DRB programme was not related to the CF-100 and the Orenda but looked to types to be produced a few years hence. Development of the CF-100 and Orenda did not entail a duplication of work in other countries since a special type of interceptor had been required for Canadian purposes."[96] Most of the DRB's research and development plans for the following year, including in fields such as guided missiles (see chapter 12), were done as part of tripartite agreements with Britain and the United States and thus were not being duplicated by those countries.[97] The CDC asked Solandt to make a full report on the DRB's development program for the following year, which he did.

When Solandt tabled the further expansion of the DRB budget to $42 million at the CDC meeting on 19 May 1952, Howe again challenged him, believing that the DRB's expansion was worrisome, and Howe was backed by minister of finance Doug Abbot, who noted that, despite the DRB being the one investment in the defence budget that paid dividends, he saw little value in increasing its mandate during rearmament.[98]

Solandt was prepared; Claxton and deputy minister of national defence C.M. "Bud" Drury were in his corner. He reiterated the DRB's alliance obligations. The staff increase was substantially smaller than the defence research and development budget of their allies. Of the $42 million, $19 million was going to the services' individual development fund, which the DRB had no control over, leaving roughly $24 million for research. Throughout the Korean War and before, the DRB had spent time, money, and energy in coordinating government science departments to manage the defence needs of the Armed Forces efficiently and effectively. This included coordinating with the Ministry of Health, the Ministry of Agriculture, and the NRC, among others. All of this was done in the name of fiscal efficiency and to avoid increasing the budget for as long as possible. As his budget made clear, the DRB was not taking on new work, but trying to work more efficiently. Claxton concurred and so did Drury, who noted the DRB had been critical to efficient weapons procurement projects.[99]

Prime Minister Louis St Laurent had the final word. The DRB had become a valuable government asset, and had done great service in getting highly qualified scientists into government employment. The staff and budget increase was justified. After further wrangling, the CDC agreed to the increases over the next two years.[100] Solandt likely prepared long and hard for this confrontation, consulting with Claxton and perhaps the prime minister before he decided to get in the ring with Howe again. Indeed, few men could walk away from confronting Howe and Abbot on their own turf with victory under their belt. But few men were like Omond Solandt.

Continental Defence 1950–1955

The Korean War and Canadian and American rearmament reopened the discussion on Canadian and U.S. efforts on continental defence. Between 1951 and 1952 a series of summer study groups in the United States, most notably Project Lincoln and Project Charles at the Massachusetts Institute of Technology (MIT), re-evaluated continental air defence needs. The DRB was represented at both by nuclear physicist and OR scientist Doctor George Lindsey, who had first worked for Solandt at AORG. While several proposals and variations on continental air defence were made, the final policy for both governments was the development of complementary radar systems in the Arctic in 1954: the Canadian-developed Mid-Canada Line along the 55th parallel and the U.S.-developed Distant Early Warning (DEW) in the northern Arctic from Alaska to Baffin Island.[101] Both used automated radar station technology, decreasing their operating cost. The DRB and its partners in academia and industry designed and produced the McGill Fence radar stations that used an automated audible alarm system for the Mid-Canada Line, which is discussed in chapter 11. The United States developed the Semi-Automatic Ground Environment (SAGE) network of control and radar stations, which became the "electronic nervous system for both the Canadian and US air defence efforts."[102] Both projects involved funding, support, and resources from government, academic, and outside sources.[103]

Solandt and the DRB remained active on these developments and discussions between 1952 and 1954. In early February 1952, the COSC agreed that the RCAF and the DRB's Operational Research Group (ORG) should continue to study "the limitations and capabilities of air defence weapons with a view to making specific recommendations concerning the Canadian air defence system to the Chiefs of Staff as soon as possible." The operational research staff of the DRB[104] now made supporting the RCAF study a "top priority project."[105] A joint Army-RCAF-DRB paper on air defence was tabled in 11 June 1952. By this time, the DRB had begun research and development in guided missiles (see chapter 12). For the COSC, Solandt discussed the T33 fire-control system employed for certain kinds of guided missiles in the United States, and warned that more effective control systems were being planned for the NIKE missile; early developments would likely be replaced by more efficient systems. The DRB needed another six months to study the guided missile problem and trend of future control systems, so the air defence paper was to be redrafted.[106]

By February 1953, the United States had approached the Canadian government for cooperation on experimental early warning radar sites for continental defence. The COSC was tasked with creating the agency to coordinate these efforts, known as Project Counterchange. The COSC and CDC both discussed the issue, and Solandt led the CDC discussion. If these experiments were successful, Solandt noted, the United States would want a portion of new permanent stations on Canadian soil.

The experimental stage would have three stations in the north, two in Canada. The United States was fully ready to staff, finance, and work on the sites alone but would welcome Canadian participation.[107]

Here, Solandt pitched the radar defence system being created through the DRB support at McGill.

> Defence scientists in both Canada and the United States had cast some doubts on the efficacy of the proposed new radar chain. Although communications might be difficult, [Solandt] was confident that this problem would be solved. One of the main questions was the value of the information which would be received from the stations so far in advance of likely targets and fighter aircraft defences. He reported on successful work which had been done by Canadian scientists on a new type of radar fence [the McGill Fence]. A pilot station had already been established with excellent results. It was hoped that this experiment would enable the establishment of unattended stations whose equipment might cost as little as $10,000 per unit and which would give coverage from 60,000 feet almost to the ground with absolute reliability. The Canadian equipment had the additional advantage that it was difficult, if not impossible, for enemy aircraft to detect the position of stations until the enemy aircraft themselves had been detected.[108]

Foulkes said the chairman of U.S. Chief of Staff had misgivings about the radar fence, that it was being considered largely because of pressure from civil defence officials. But militarily there was no reason to oppose the establishment of three radar stations.[109] The CDC now agreed to full participation of Counterchange.[110]

While Solandt won support for the project, he lost the chance to lead the Canadian end of Counterchange. There were two main concerns with the project: technical and scientific investigation of the feasibility of operating radar stations in the far north, and the study of continental air defence. Solandt wanted these problems separated. The first was entirely experimental and the DRB was the most appropriate organization to handle it. The USAF was the lead agency for the project in the United States, but the Air Defence Research and Development Command of USAF (ADRNDC) ran it. The only Canadian equivalent of the ADRNDC was the DRB. The COSC, however, maintained that the RCAF should deal with the United States regarding requirements. So the RCAF took the helm, though the DRB provided critical support on Arctic as well as OR studies of the growing radar system in Canada.[111] Solandt disagreed, but let the matter rest and got on with the job.

The Military Study Group

To act as a centralizing force on all the groups involved in early warning radar, the COSC created the Military Study Group, first proposed on 6 October 1953. The Canadian section worked primarily on continental air defence, early warning line problems, and machinery for joint planning. Solandt felt that the first step should be a report on immediate actions needed while future studies were carried out. As it stood, it was too early to report to the CDC on the MSG, since they had no concrete conclusions from the U.S. and Canadian sections, and would generate questions that the COSC could not yet answer. Foulkes disagreed. He feared senior U.S. officials would likely be making decisions before the immediate report was done, so a preliminary report was better than no report. For Solandt, who studied under Charles Best, Foulkes's suggestion, regardless of its practicality, must have been abhorrent.[112] But such were the political realities of government science.

Construction of the Pine Tree Line, the first major radar defence system for North America, began in 1951 along the 49th parallel. During the early construction period (1951–3), there were tangled communications among the various agencies, despite the MSG's creation. The COSC was not to advise on the U.S.-constructed DEW Line until the survey of the radar stations was complete. Solandt warned against too many agencies having divergent visions of continental defence and emphasized the need to see all early warning as part of the same system. If planning were done by different agencies, "the system would be difficult to manage as an integrated unit."[113] By 1954, construction of both the Mid-Canada Line and the DEW Line had begun. Solandt left government service before they became operational in 1957, but was one of the guiding hands for their creation.[114]

The CF-105

The RCAF had been adamant about developing Canadian air power to meet the continental defence needs in the unique Canadian environment. They pursued, with DRB cooperation, the creation of the CF-100 all-weather fighter in partnership with A.V. Roe, supported by research from the DRB, the NRC, and the joint RCAF-NRC-DRB National Aeronautical Establishment (NAE).[115] Solandt had a mixed opinion of Canada's industrial capacity for varied aeronautics work, and about A.V. Roe itself (see above). At the 18–19 January 1951 meeting of the

Panel on Economic Aspects of Defence, Solandt argued against Canada's increasing the volume of types of aircraft to meet its defence needs. "There was a danger that Canada was trying to build too many types of aircraft and engines ... the emphasis should be on increasing the production of the CF-100, which could be bartered to meet other Canadian requirements. He felt that the same was true in the electronics field. The aim should be large production runs in the interests of economical programmes."[116] Development work on the CF-100 and Orenda was done and production was soon commenced. But the RCAF was more interested in creating a new all-weather fighter to meet changing priorities than modifying the CF-100.

The Korean War raised new fears of Soviet long-range high-speed bombers carrying nuclear warheads with megaton yields that dwarfed the kiloton carnage of Hiroshima and Nagasaki. By 1953, Air Marshal Curtis had laid the requirements for Canada to have a new all-weather supersonic interceptor to replace the CF-100, designated the CF-105 "Arrow," and due for service in 1957. While the weapons and navigations systems would be bought, the airframe and engine were to be made in Canada by A.V. Roe Canada, Ltd, with research support provided by the DRB, NRC, and NAE.[117] In October, the Treasury Board put a ceiling for aircraft development at $1 million, and $200,000 for a design study. A.V. Roe had completed the design study on the CF-105. The COSC wanted the development of the aircraft to continue, but A.V. Roe would be left idle if funds were not allocated. The Treasury Board stipulated that more money could be requested if projects went beyond design study phase.

Simonds was cautious about the project and the aircraft industry in general. They were notorious for poor estimates for government funds, and the discrepancies had gotten large and needed to be watched. Solandt, recalling his tangle with Howe and knowing Crawford Gordon, one of Howe's cronies, was now managing the Arrow project, suggested that the Department of Defence Production be responsible for accepting the costs of such projects. He also noted that A.V. Roe Canada, Ltd, was aware of DND's concern over the discrepancies in the estimates and was strengthening their costing department. With a push from Curtis and Solandt, the COSC agreed to continue backing the CF-105 while maintaining a watchful eye.[118]

By February 1955, the cost of the Arrow project was still on the rise. But new Chief of the Air Staff Air Marshal C.R. Slemon was pursuing procurement of the CF-105, trying to convince the COSC of its continuing value. For Simonds, it was a hard sell. Guided missiles and ICBMs would likely make manned aircraft obsolete, so the entire procurement program was flawed. Foulkes disagreed. The CF-105 was intended to stop the Soviet T37 bomber, and it was unclear if this threat actually vanished just because ICBMs were developed. There was no effective defence against the existing T37, and if a suitable weapon to counter it could be produced fast, it should be pursued. Simonds was not convinced, and, surprisingly, looked to defence research as a way of reducing the CF-105's value and getting Solandt's backing. Clearly, Simonds reasoned, anti-ballistic missiles research had more value. It was dangerous to underestimate Russian missile development, as had happened with atomic and other weapons.[119]

Solandt was not goaded and kept to the facts. The future threat to Canada's continental security could be considered in four stages:

1. T37 bomber;
2. supersonic bomber;
3. efficient cruise missiles like the Navajo supersonic cruise missile; and
4. the ballistic missile.[120]

The CF-105 was suitable against T37 and supersonic bombers and could be capable of "marginal defence against the Navajo guided missile." But there was no defence in sight against ballistic missiles. "Progress in the guided and ballistic missile fields generally has been disappointing, and while it must be considered that the Russians were doing at least as well, these weapons should still be looked upon only as future possibilities. It was better to invest in defence against weapons we were certain would be available than against a future possibility."[121] Simonds lost this round. The COSC agreed to tell CDC to accelerate the program for CF-105 procurement and proving, stressing the threats Solandt had emphasized as the T37 and supersonic bomber.[122]

According to Solandt, any research done on or to the CF-100, CF-105, and Orenda or Iroquois engine had close DRB cooperation.[123] The DRB worked with A.V. Roe unofficially, and not on behalf of the Air Force, on these projects, though the extent of its involvement is currently unclear.[124] At the official level, Crawford Gordon, director of A.V. Roe Canada Ltd, pressed Solandt to tell his old boss Howe that the plant in Malton, despite its production difficulties, was still in good order. Solandt's response is not known, though it is doubtful he told Howe anything but the truth that even Howe realized: Gordon, and his own personal failings, was now part of the production problem.[125] Solandt was also involved in the little-documented controversy regarding the arming of the Arrow, dealt with in chapter 12, but he had left government service before the debacles regarding the Arrow reached their apex in the Conservative government of John Diefenbaker. Solandt was both shocked and appalled at the way the weapons program was cancelled and the A.V. Roe plant in Malton, Ontario, closed.[126] Regardless of the Arrow's soaring costs and A.V. Roe's poor management, Canada had invested millions of dollars in the technical excellence of the teams at A.V. Roe, much of which was lost due to short-sighted cost-cutting policies of the Diefenbaker government.[127]

NATO and Strategy, 1950–1955

As noted, Solandt was not a key contributor to Canada's approach to NATO force development or strategy. However, the most significant strategic debate he participated in was the 16 April 1953 COSC meeting with Field Marshal Bernard Montgomery of Alamein, deputy commander of NATO. It was a rare occasion, and Solandt, recalling Schonland's pride at working with "Monty," must have relished the opportunity to be the famous soldier's scientific advisor for a day. The discussion focused on future Canadian defence efforts and the effects of an atomic attack on the size of Canadian and Allied forces. Solandt led the latter discussion. He did not feel that a large and wholly trained reserve force was as important as a force-in-being that could act as shield to hold off a Russian "onslaught" in Europe until strategic and atomic attacks on the enemy had their impacts. "The effect of new weapons would be so great that the mobilization or reserves and industry could be done at comparative leisure." Montgomery disagreed. Allied powers tended to overlook the possibility that Russia would also have atomic and strategic weapons, and the wishful thinking by NATO governments that the introduction of tactical nuclear weapons would reduce the staffing requirements in Europe – a coy jab at Eisenhower's "New Look" strategy. A SACEUR study of force requirements based on the use of new weapons had not been completed, but there were indications that the need for staffing would not be lessened. According to Montgomery, "The only way to win the next war would be to have forces in place and their effectiveness increased by atomic and strategic attack. If a concept of leisurely mobilization based on the theory that the effect of our strategic and atomic attacks would provide necessary breathing space were accepted, the West would be in serious danger of losing Asia and other vital areas in the global defence system."[128]

Solandt agreed that the shield, taking the brunt of the attack, should be appropriate. However, the tendency during the next ten years would be to decrease the reserve and build up a larger shield. Montgomery countered that the trouble today was "that nobody had any reserves worth mentioning." The scale of a Russian thrust in Europe could only be guessed at. Thus, the ratio of shield to reserves was a matter of conjecture. Every nation should be prepared for maximum effort. "If for reasons such as communications it was not possible for a nation to mobilize rapidly, that nation should concentrate its efforts on forces in being."[129] Even though the encounter had friction, Solandt no doubt enjoyed the tête-à-tête with the man he wanted to work for in 1945.

Intelligence

Intelligence matters were also critical during the deepening of the Cold War 1950–5. The DRB did not have foreign intelligence assets but used those of the senior services. In 1952, RCAF Group-Captain Doyle, the current service attaché in communist Czechoslovakia, was returning. Should his position be withdrawn? The RCAF and the DRB wanted to maintain the attaché. Solandt said JIB and the DRB's Directorate of Scientific Intelligence (DSI) thought Doyle was doing an excellent work. For JIB, "Over the last 12 months, 49 per cent, or 100 out of 203 items published, of the original factual intelligence disseminated to other Canadians, UK and US Intelligence Agencies had been based on the work of the Air Attaché, Prague. If we withdrew a Canadian representative from Czechoslovakia and expelled the Czech attaché from Canada we would in fact be doing little harm to the communist's information network but would ourselves be losing an important source of information." COSC agreed it should continue."[130] Solandt had told the Pentagon in 1947 that the DRB did not want to conduct field research in technical and scientific intelligence. Clearly, there was no need when the RCAF was providing it for them. The attaché was maintained.[131]

In January 1952, Solandt also challenged how the JIC was analysing Soviet capabilities for intelligence estimates. Regarding the JIC report on SG 176 "Report by the Intelligence Committee to the

Standing Group on Intelligence Guidance for the Standing Group," written in 17 December 1951, Solandt noted that most intelligence reports on enemy potential assessed capabilities rather than possibilities in weapon development. In short, estimates were made on the basis of an assumed uniform progress in each field, but such assumptions were wrong. "The enemy frequently selected particular fields and made great strides in them. An example of this was the progress Russia had made with the atomic bomb which had been completed much sooner than had been estimated. It would be advisable to bear this in mind in future intelligence estimates of enemy capabilities."[132] This was a valuable correction for realistic strategic analysis, coming from a man who had witnessed the contrast of the Soviet Union's grand advances in science and its decrepit technical infrastructure and poverty in 1935.

Conclusion

For a decade, Solandt served as Canada's premier scientific chief of staff, and it would have been easy to fail at this job. Solandt was confronted by senior service chiefs who did not know him and had no reason to trust him or his embryonic organization. Far from failing, Solandt thrived in the highly contested arenas of senior defence policy in Canada, and rose to become an active, aggressive, and confident speaker on Canada's growing defence concerns during the Cold War. Continental defence, with its massive and diverse requirements, dominated most of his years as chief of staff, though he also added his voice to such critical issues as NATO's development, while proving, at each and every meeting of the COSC or CDC, that the DRB was active and willing to support Canada's defence requirements.

It was a hard ten years, but Solandt outlasted all the other service chiefs, with the exception of the wily Foulkes. By his very presence in the COSC, he made enemies and grudging partners of the chiefs of the air staff, and yet of all the services it was the RCAF that worked best with the DRB. That this happened is in large part thanks to Solandt's ability to make partners of enemies based on excellent research and resources, which was, despite being disliked or unappreciated, critical to this success. Following his late father's example, Solandt put service before self.

These tensions and pressures only increased as Canada entered the Korean War and prepared to arm itself in peacetime to deter any

Soviet threat in Europe or North America. Solandt relished the chance to be so close to the networks of power in Ottawa, the same ones that were out of his reach in Britain during the war. But he tempered his enthusiasm with cool rationality and the attitude that one could confront serious challenge from soldiers and ministers without burning bridges. His reputation on the COSC and the CDC rested on his ability to make good on the promise of the DRB's contribution to Canada's defence research needs. And such promises were forged within the board itself.

11 A Decade of Leadership: Omond Solandt, Chairman of the Defence Research Board, 1946–1956

For a decade, Omond Solandt shaped the unique and innovative Defence Research Board from idea to an established defence research organization. Much of this success rested on Solandt's leadership and belief in the DRB and its mission. As chairman, Solandt was strong in will, quiet in manner, and fierce in determination to achieve the DRB's goals.[1] For Archie Pennie, a DRB member at CARDE, Suffield Experimental Station, Defence Research Northern Laboratory, and secretary of the board in 1953, Solandt was the heart as well as the head of this unique organization.[2]

Indeed, before president Dwight D. Eisenhower coined the term in 1960, Solandt was directing research in a small and specialized military industrial complex (MIC) for the Canadian government.[3] As chairman, he fostered relationships with and provided guidance to leaders in industry and academia towards achieving the DRB's goals for the Canadian Armed Forces. In doing so, he developed his own management philosophy on development of weapons systems, which embodied his personal views of defence research relations as well as those he had garnered from the board. Solandt distilled these views into one and gave them a voice in government.

The value of this leadership approach in defence research was best witnessed with the development of the McGill Fence automated radar defence system in 1951. It involved translating military requirements into an effective and efficient continental defence system for the Mid-Canada Line. Solandt and the DRB stimulated research interests and served as a nexus point that allowed the McGill Fence to exist.[4]

Solandt also demonstrated leadership by avoiding research of questionable value. In the wake of the Korean War, fear of communist brainwashing experiments led to often dubious and unethical medical and psychiatric research. Doctor Ewen Cameron's "de-patterning" experiments on his patients at the Allen Memorial Hospital at McGill University, secretly funded by the Central Intelligence Agency, were among the most shocking and tragic results of this era of fear-driven research.[5] The story of Cameron's failure to secure the DRB funds is a testament to Solandt's respect for medical ethics of conduct regarding patients.

As chairman, Solandt also grappled with three perennial problems at the DRB: staff, secrecy, and finance. Each was a constant bugbear to the organization, and Solandt developed strategies to manage them while maintaining his vision of the DRB's mission, which included the DRB annual symposium, hiring British experts and avoiding German scientists, and fighting to gain control of the service's development budget. These tasks required a strong, stable leader who could provide critical insight across a broad spectrum of scientific and technical matters and communicate with civil servants, generals, and industrialists with equal ease. And it was in this role, as chairman of the DRB, that Solandt excelled.

Despite the mounting stress of acting as maestro to this unique orchestra, Solandt maintained the cool, rational approach to conducting the DRB's business that had served him so well during the Second World War. Archie Pennie, when secretary of the board, believed there was no member of the board – military, civilian, industrial, or academic – who would have had a negative thing to say about Solandt. They all respected him for getting the DRB off the ground and providing it with the momentum for success.[6]

Chairman of the Board

While he was fascinated with the DRB's research, Solandt's chief occupation as chairman was to formulate sound defence research policy. Specialization and excellence were his watchwords. These would not only produce the best research, but that research would then become capital to use to influence allies.[7] Policy was formed by consultation with the official board members, though Solandt had the final say on proposals. The board met three to four times a year. Once formulated, policy was translated into action by the establishments, by the DRB committees and panels, or through grants and contracts provided to universities and industry through the Standing Committee on

Extra-Mural Grants. The deputy director (after 1950, vice-chairman) was responsible for transmitting policy from the DRB headquarters in Ottawa to the appropriate panels, committees, universities, or industrial contacts, or the DRB's own establishments. Most of the panels and committees were at DRB HQ in Ottawa and were chaired by experts from all walks of life (medical, academic, industrial, etc.).

Solandt, as chairman of the Standing Committee on Extra-Mural Grants, maintained a decisive influence on the DRB's connections with industry and academia. No longer would he be denied the chance to sequester and work with industrial and academic "big shots" outside of government, as had been his experience during the war. Now, he could guide defence research with a lever of influence previously denied him. As chairman, he was now a "big shot" himself.

Industry

Solandt always maintained military, academic, and industrial membership of the board to facilitate its goals. He was concerned about the need to maintain close and strong ties to Canadian industry. His experiences at Lulworth, where he was kept off committees with key members of the Ministry of Supply (the gatekeeper to British industry), were no doubt still fresh in his memory. After the war, most Canadian industries had retracted from military work and returned to peacetime commercial enterprise. The Crown corporations that the Department of Munitions and Supply had created to deal with optics, electronics, and computing were largely closed. When he first arrived in Ottawa, Solandt was struck by how quickly industry had disentangled itself from weapons production. As such, all of the people in key production positions during the war were gone.[8]

Solandt leaned heavily on industry leaders to maintain active interest in projects of defence research value by providing grants and access to the DRB's staff. Colonel (retired) R.D. Harkness, president of Northern Electric, was the founding DRB member for industry and a leading Canadian industrialist with strong ties to the military.[9] He advised the Board on contracting research from industry and served on grant and contract committees, though he abstained from voting when his company was considered for a contract. The board also maintained a separate industrial advisor for consultation, starting with Mister R.W. Diamond, president and general manager of Consolidated Mining and Smelting, Ltd, in December 1948.[10] The board actively worked with and provided contracts for both companies, though no one claimed conflict of interest. This was likely the result of Solandt's reputation for being a straight-shooter. There is no indication that he received any personal gain from his industrial contacts other than professional respect.

From 1946 to 1950, much of the DRB's industrial focus was on the problems of future industrial mobilization in war. The Industrial Defence Board (IDB) was created in April 1948 to focus solely on this issue and existed only until its responsibilities were transferred to the Department of Defence Production in 1951.[11] However, the DRB was the main body for coordinating the services' defence development plans. Clearing lines of responsibility between these bodies was critical. At the June 1948 meeting of the DRB, in the middle of the Berlin Crisis and its subsequent war scare, Solandt asked the board's views on development coordination.[12] "[Solandt] pointed out that one of the reasons the problem was becoming acute was that in order to be prepared industrially for war, it would be necessary to work out a comprehensive development programme based on the needs of the Services and also on the needs of industry. This entails a sound development scheme as a support for industry in time of peace. The Canadian development programme must, therefore, be a positive instrument of policy."[13]

At the time, the DRB had money for development, but the services picked the projects, a policy Solandt disliked and fought throughout his tenure as chairman (see below). All the service chiefs emphasized the need of standardization. For Foulkes, industry was to back the mantra of standardization and not specific Canadian developments, and he warned Solandt that the DRB should avoid getting too involved in service development, given the specialized knowledge required to prioritize it.

Solandt disagreed. Knowing the military's stance, he specifically asked the academic and industrial members for their input and support. Harkness suggested the services set out their own development needs. If the DRB was to be involved in any of them, however, they required a voice on what organizations should be used, and IDB be consulted on the industries able to carry out the work from the development to production phases. For Harkness, the services, the DRB, and the IDB had distinct and non-transferable responsibilities. IDB should analyse the nation's industrial potential, but if industry was to take on any development work, the DRB needed a voice. Industry also required the capacity to produce necessary equipment. Key areas like aircraft and electronics production had suffered in peacetime and were not self-sustaining without government support. The board agreed that there was a serious gap in the cultivation and coordination of development plans in government and rectified it with greater representation and discussion with IDB.[14]

If war came, Solandt knew, it would be sudden and devastating, with no "phoney war" period to allow for unthreatened mobilization. Canada and its allies would require the "industrial front" to be strong from the beginning. As he told a public audience in 1948, "War will be launched entirely from the homeland so that preparations for aggression will be very hard to detect." Destruction of vast enemy resources in industry or human resources would be achieved in days. If two nations of similar size prepare for this kind of war, "victory will go to the one that first destroys the means of retaliation in the other. The whole war would be over before there was time to begin mobilization plans for the armed forces and industry. Only those resources that are continually ready for immediate action will be of any value." This was the face of total war, and preparation for it was the DRB's job.[15]

Solandt, like Foulkes, believed at least five years would pass before the Russians would contemplate the use of military force to achieve their goals. The DRB provided voluminous contracts and grants to industry and academia to sow seeds for future development of forecasted industrial war needs in computing, aeronautics, and food research. Among the more prominent contracts between 1946 and 1950 were those focused on armament and electronics. Consolidated Mining and Smelting received DRB funds to conduct experiments on producing picrite from natural gas as a cheap and alternative armament propellant. Solandt also feared the effects from the rapid demilitarization of the electronics industry. This was a grave matter because advanced electronics were at the heart of radar, weapons fuses, and other advanced weapons and weapons systems. Solandt maintained the electronics industry's interest in Canadian defence needs with contracts and funds to Smith and Stone Ltd., Northern Electric Co., Canadian Marconi Co., and RCA Victor Co., all of whom had little interest in the military application of electronics or electronics components after 1945.[16] By the end of 1948, the board also agreed to increase industrial membership, given the importance of industry to the DRB's own goals.[17]

A Management Philosophy for Research and Development

Many of the ideas and policies on industry from the DRB's first two years became the grist for Solandt's management philosophy on research and development. Solandt, echoing the old Best approach of using narratives to explain research, used the analogy of a chain to discuss the development of new weapons and equipment. The chain consisted of research, development, design, modification, production, inspection, and use.[18]

The DRB was on the research section of the chain. Given Ottawa's post-war disentanglement from war production, the weakest link in the immediate post-war period was development. Development had three interested parties: armed services, the users, who know what they want; production people, who will have to make it and know how it can be made; and the applied research or "research and development" people (including DRB scientists, OR specialists, and technicians), "who have the new ideas which should be applied to the new weapons." These parties had to cooperate if success was to be obtained, and he hoped IDB would serve this function.[19]

Next in the chain were design and modification, leading to production. Solandt believed development both preceded and followed engineering design for production, a contention that was forged in his experience on tank design at Lulworth.

> When a new idea has been brought, by the applied research worker, to the point where a working model can be made, then development begins. At this stage development consists in applying all that is known in related fields of science and technology to perfecting of the design of the weapon. As the designer goes ahead with his task he may well ask the development engineer to test out various elements of the design and may possibly send back to the research laboratory for some new ideas. At this stage the production expert must be a full member of the team so that the finished design will be suitable for production. The user must also take a hand to ensure that the equipment will be a usable one when complete. The scientists must also watch to see that the essentials of his new idea are retained. Finally the inspector must begin to show an interest so that he will know which features are essential and which are merely desirable, and will have new gauges or other special equipment ready in time.[20]

At another stage of development, Solandt said, bugs were ironed out. Here the party responsible for production, normally industry, has the greatest interest, but the user, normally the military, must maintain close contact during user and technical trials and their conversion of results into a final design. OR scientists should be present at the formation of original requirement and appear for the final stage of user trials for evaluation. Development was the most complex part of the chain, given the varied interests involved, and lack of coordination or appreciation for the roles of each part of the chain would lead to delays, soaring costs, and obsolescence.[21]

The DRB's goal during the development stage was to ensure that new ideas were fully exploited and obsolete ones avoided. Industry ensured that the developed project could, in fact, be produced so that time, money, materiel, and human resources were not wasted. The Armed Forces had two interests in development: first, to ensure that equipment and weapons produced were indeed better than what was being replaced, and that they fit into tactical and strategic plans; second, that new equipment did not interfere with standardization. Solandt lamented that the services handled their own development but still required extensive consultation among all interested parties. The special characteristics of Canadian industry were not to be forgotten. Even during the war, Canadian industry had not produced a complete range of military equipment. Canadian troops used only a small fraction of Canadian goods produced. Unlike in the United Kingdom or United States, industry could not fully provide for Canada's defensive qualitative and quantitative production needs.[22]

The Department of Munitions and Supply under C.D. Howe had been responsible for war production until 1945. IDB was attempting to fill the gap that DMS left in its wake, facilitating organization between government and industry, and between key government and industrial bodies.[23] Solandt hoped that IDB would successfully organize and mobilize Canada's war industry, but all DRB members were adamant that something similar to the Department of Munitions and Supply was needed to fulfil this task.[24] IDB was disbanded in 1951 after the creation of the Department of Defence Production, which took over its role under C.D. Howe, whose deputies became members of the DRB.[25]

Academy

Academic members of the board were also critical to the DRB's success, maintaining and generating interest on the DRB's objectives throughout Canadian universities. Physicists J.H.L. Johnston,[26] Dalhousie; Gordon Shrum,[27] University of British Columbia; and G.S. Field,[28] University of Alberta; as well as Otto Maas, had all done important defence research work during the war and were progressive and active members both on the board and on the Standing Committee for Extra-Mural Grants. They provided a breadth of contacts in academia that allowed the DRB to effectively promote areas of interest to the military and also act as the focal point of defence research relations between industry and academia.[29]

Still, Solandt, as chairman of the standing committee, was the steady hand on what project was supported. As with industry, Solandt supported many fields that were in danger of post-war malaise – such as radar research at the electrical engineering department at McGill[30] – or were under-appreciated by the military, such as oceanographic research off the Pacific coast for the RCN at the University of British Columbia.[31]

The DRB supported research in almost every conceivable field of interest to the military, from medical to psychological to entomological, but Solandt's insight on the military value of multiple fields of science was not shared by some academic board members. Pennie recalled that a chemist (likely NRC chemist Ned Steacie, who joined the board in 1952) resisted the appointment of a psychologist to lead a DRB program, as he saw no merit in the field of psychology. Solandt quickly and clearly spelled out the nature of the program (the Army used psychologists to good effect for personnel selection, among other fields) and then asked, "Who do you think should run it? A chemist?" Publicly chided, the member said no, and the grant was passed.[32]

Among the many research endeavours supported between 1946 and 1950 were joint contracts with the NRC to support a computing centre and aeronautical research lab at the University of Toronto. Solandt was adamant that both computers and aerospace research would be critical to any future war effort Canada produced, and required early support.[33] But perhaps the best example of Solandt's eye for valuable research lay in electronics, the nervous system of continental defence.

The McGill Fence

The outbreak of the Korean War on 25 June 1950 and the fear of general war in Europe made DRB's relationship with industry and academia even more critical. In his annual chairman's report in September 1950, Solandt noted that defence policy had been predicated on choosing guns or butter during times of peace or war.

> There now seems to be every reason to believe that we are approaching
> a period of international affairs when there may be years of tension

and limited warfare, without total war. The best way for the country to weather a period such as this is to attempt by hard work and personal sacrifices to have both guns and butter. When this general policy is applied to research it suggests that we must try to increase the volume of effort put into research and development for the Armed Forces without diverting too much effort from the research in Universities, industry and federal agencies that does so much to build up the strength of our economy.[34]

The most strategically important project the DRB supported with funds, staff, and human resources to universities and industry between 1950 and 1955 was the development of the McGill Fence radar system. As noted in chapter 10, the RCAF had established clear priorities and requirements on continental defence that went through successive revisions and consultations with senior Canadian and U.S. defence authorities. By 1951, the United States and Canada had increased their efforts on technical solutions for the air defence of North America.

The DRB initiated and contributed significant research assets that fostered the development of the McGill Fence. In 1949, Solandt had provided Doctor G.A. Wooton with $25,000 to assist and equip an instrument-maker's shop for the Eaton Electronics Lab set up at McGill University. Wooton was then selected as chairman of the DRB's Electronics Committee to consider electronic dimensions and requirements of radar.[35] By 1950, the RCAF was sufficiently interested in Wooton's work that they asked the DRB to provide him funds to work on radar-scattering measurements.[36] These efforts, as well as general grants to support basic research, were the early seeds of the McGill Fence project.[37]

The research end of the program was soon led by Doctor J. Rennie Whitehead, a British physicist and radar expert. During the war, he had directed and produced the IFF equipment used to "identify friend and foe" on radar screens.[38] Whitehead joined McGill's Physics Department in 1951.[39] However, according to Whitehead, the central idea of the McGill Fence project originated with physicist Wilfred Bennet Lewis. A disciple of Cockcroft and the last wartime director of the British Telecommunications Research Establishment (TRE), Lewis came to Canada after the war to run the atomic research laboratories at Chalk River. Solandt had met Lewis during the war and always believed Lewis was one of the two best physicist-engineers to ever work in Canada (second only to Cockcroft). Lewis's incisive mind and high capacity for work were matched by a critical insight to forecast future needs, as was witnessed in the McGill Fence.[40] Solandt later employed Lewis at the DRB as chairman of the Electronics Components Development Committee.[41] Before Lewis became vice-president of research and development at Atomic Energy Canada in 1952, he proposed to the DRB and Whitehead (one of Lewis's star employees from TRE) the idea of "using a double-Doppler radar technique to meet a new air defence operational requirement for a 'cheap' trip-wire line of radar defence along the 50th parallel, across Canada." He urged Whitehead to take up the work at the Eton lab at McGill with Wooton. Whitehead agreed.[42] According to the DRB official history, the DRB had initiated the project, though it seems clear they did so at Lewis's suggestion.[43]

Doctor Field, Solandt's point-man for academic relations on the board and responsible for telecommunications research, was assigned responsibility for the project, which required managing all four groups that the board represented: the DRB, the military, academia, and industry. Whitehead conducted his research at McGill on behalf of the DRB and maintained close liaison with the board. His project was supported by many DRB scientists and establishments, including Doctor J.W. Scott and the DRB's Radio Physics Laboratory,[44] Mister L.G. Eon, a radar expert and senior telecommunications staff officer at the DRB HQ,[45] and DRB OR scientists Harold Larnder and George Lindsey, who were working for the RCAF's Air Defence Command (ADC) in Montreal.[46] The RCAF provided several technicians as well as logistics support for the DRB-funded project,[47] and the NRC supported the project with staff from their Radio and Electrical Engineering Division.[48] The DRB acted as the link between Whitehead and RCA Victor, the key industrial partner of the project, whom, as noted before, the DRB had supported with grants for electronics research.[49] Solandt took a personal interest in the project and developed a friendship with Whitehead, who was given an office at the DRB HQ in Ottawa, so there was a direct line from the lab to the chairman.[50]

Solandt reported to the board in October 1953 that draft plans for an early warning system in the north were going ahead. War gaming exercises had been conducted at Colorado "in order to find out how [the USAF] Air Defence Command Headquarters dealt with situations and information reported to them." The results were a tentative recommendation to the COSC that "an early warning line should be placed along the 55th parallel and that a McGill Fence type of equipment should be employed. This should be backed up by a set of scanning radars at points of anticipated high traffic density. The study also indicated the need for early warning systems out over the sea, particularly over the Pacific and out from Alaska."[51]

Development of the McGill Fence was promising and would likely be integrated into the future early warning systems. Solandt cautioned, as the U.S. experts had also surmised in 1952, that the McGill Fence alone was not a "complete" solution, but was a significant contribution.[52] The Arctic testing of Whitehead's radar system would take a year, but Solandt was confident enough in the McGill Fence to present its attributes to the Cabinet Defence Committee, stating that it "showed high promise of making an important contribution to radar defence."[53] A series of experiments in the Arctic on equipment performance and the effectiveness of the automated nature of the proposed system were conducted that year, codenamed Spiderweb,[54] as well as tests of the McGill Fence's automated system, known as Lonewolf.[55]

By 1954, the McGill Fence was a hit. The test results were successful, the equipment was cheap, and Canada could create this efficient and effective radar defence system without U.S. support, which a critical political advantage of the system. In November, minister of national defence Ralph Campney announced to the CDC that Canada would construct the Mid-Canada Line with the McGill Fence.[56] It would be integrated with the U.S.-funded and constructed Distant Early Warning (DEW) line in the far Arctic to provide increased radar coverage. It went into service in 1958.

The McGill Fence stands as a testament to Solandt's early strategy of harnessing academic and industrial potential to Canada's critical defence needs, and as an example of effective use of the relationships inherent in a small military-industrial complex. Solandt's strategy of seeding industry and academia with resources that could be used for defence came to a successful end with the McGill Fence.

The Ewen Cameron Controversy

Solandt also navigated the DRB away from one of the most controversial and tragic defence research projects of the early Cold War. During the Korean War, several Western POWs appeared to have been "brainwashed" by the communists. They spouted propaganda and professed "communistic" and anti-capitalist beliefs that they had not held before. Fearing the Soviets' success in this field, American, British, and Canadian officials discussed collaboration on psychological research as a means to understand what the Soviets might be capable of.

This meeting occurred at the Ritz-Carlton Hotel in Montreal, 1 June 1951. The Canadians included Solandt, psychologist and the DRB OR superintendent Doctor N.W. Morton, and psychiatrist O. Hebb. The United States was represented by Doctor Tyhurst, Doctor Carl Haskins of the Carnegie Institute, and Commander Williams of the CIA. Sir Henry Tizard represented the British.[57] They decided that it would be of value to conduct basic research on the means that "might be employed by totalitarian governments for effecting fundamental changes in the attitudes of persons from free nations who temporarily fall within their control. It was thought that if we knew more about what was possible, any remedy or safeguard which we should use would suggest itself more readily."[58] Hebb's work at McGill University on the effects of isolation and boredom on subjects listening to propaganda had been supported by the board for three years. Strict precautions were made regarding the subjects. They were not pressed into the research, they were paid for their participation, they were free to leave the experiment at any time, and their physical comfort was maintained. Later in life, Solandt maintained that Hebb's research was all the work the DRB supported in this field, though other work on totalitarian brainwashing techniques was being done elsewhere.[59]

Indeed, the most ethically appalling "research" was also being done in Montreal, at the Allen Memorial Centre by Scottish psychiatrist and past president of the World, American, and Canadian Psychiatric Association, Doctor D. Ewen Cameron. Cameron had created a theory called psychic driving, or "de-patterning." His patients, who suffered from mental diseases ranging from mild depression to schizophrenia, would have their damaged personalities destroyed and rebuilt into healthy ones through the unorthodox use of degradation, LSD, electroshock, sleep deprivation, and conditioned-response therapy that included repetitive tapes that played personality-rebuilding suggestions.[60]

Early on in his research, Cameron had secured funds from various Canadian government agencies. As Anne Collins has argued in her convincing account of Cameron's experiments, Cameron also attempted to persuade the DRB to fund his research, but Solandt was not impressed. As a trained physician himself, Solandt saw Cameron as a personal empire-builder and questionable physician. Cameron was not a Canadian citizen, and Allen Memorial had more than enough funds from the government to support him if they chose.[61] But there was a more personal reason for Solandt's distaste of Cameron and his methods.

Doris Davies, the wife of E.Ll. Davies, Solandt's vice-chairman for most of his DRB days, had been admitted to Allen Memorial to receive treatment for mental problems in the late 1940s, where she became one of Cameron's patients. When she arrived home after treatment, her problems remained and she seemed worse. Cameron told E. Ll. Davies that he had done everything possible, nothing more could be done for his wife, and Mrs Davies was not to return for treatment. A very upset Davies, himself not an emotional man, turned to Solandt for medical advice.

Listening to Davies's worries and other professional views of Cameron, Solandt formed the opinion that Cameron's work was bunk, and his manners as a physician were tactless and callow.[62] Cameron would never be supported by the DRB. The experience was tragic enough that Davies returned home to the United Kingdom with his family, resigning as vice-chairman of the DRB in 1954 and taking over the position as the DRB liaison officer in London.[63] This shift had unforeseen consequences. Solandt lost the best deputy he would ever have, thanks to Cameron. His replacement would be Hartley Zimmerman, who eventually succeeded Solandt. While it is not known if Davies would have been Solandt's equal, it is clear from current evidence that he likely would have done a better job than Zimmerman. That possible future was denied, thanks to the unethical work of Ewen Cameron.

The CIA had none of Solandt's hesitation. Through a front organization called the Society for the Investigation of Human Ecology, based in New York City, they funded Cameron's research as part of their own brainwashing research program, designated MKULTRA, until 1964.[64] Cameron died in 1967, but the survivors of his "research" filed a lawsuit against the CIA in the late 1980s. Allegations and concerns were raised about Canada's involvement with Cameron. After explaining the DRB's involvement with Hebb and not Cameron, Solandt worked with the lawyer in charge of the case for the survivors, James C. Turner. Rosemary Davies Hofberg, daughter of E. Ll. and Doris Davies, also contributed, writing to Solandt of her childhood memories of her mother's illness and Cameron's treatment. As Solandt told Turner in 1988, "It is indeed a sad story. Much worse than I thought."[65] Solandt provided a deposition and was prepared to testify on behalf of the survivors, but the case was settled out of court. Solandt always maintained that the DRB never supported Cameron's work, and this argument was bolstered by Anne Collins. No evidence has been found during the course of this research to contest Solandt's or Collins's arguments.

Perennial Problems: Staffing, Secrecy, and Finance

Solandt's most constant battles as chairman of the DRB concerned the more pedestrian needs for staff, secrecy, and funds. The success of the DRB rested on its ability to attract the best scientists for work in defence research, and Solandt and the board did it by viewing the entire scientific community (government, university, industrial, medical, etc.) from the stance of national defence needs. Goforth and Solandt both believed they needed to get "the young Rutherfords" interested in defence research.[66] The primary focus for recruiting was academia and industry, because the Canadian government had an "unblemished record" of forgetting about defence research between the wars.[67] In 1946 the DRB selected Otto Maas to chair the new Selection Committee on personnel selection and salary analysis.

From 1948 to 1949, staffing problems increased. The DRB's proposed efforts to create a first-class supersonic aerodynamics research centre at the University of Toronto were threatened when the lead researcher, Doctor Patterson, was approached by the United States with more lucrative work. The issue was critical enough that Solandt himself raised it at the CDC in April 1948 so that prompt action would be taken. Howe worried that Canada could not support such a complex field for research. Undersecretary of state Lester "Mike" Pearson, however, backed Solandt, his old football player, and suggested that allowing the United States to scoop up Patterson would leave the impression that Canada had no interest in this work, which would have even more negative impact on allied cooperation.[68] Solandt and Mackenzie were instructed to work out the issue of joint NRC and the DRB support for the work,[69] and Patterson was retained.[70]

Solandt also turned to his connections in the United Kingdom to bolster the DRB's staffing strength. Free of the Civil Service Commission, he had the power to hire the best and the brightest from overseas, even for senior positions, in a matter of days instead of months or years.[71] In 1947, E. Ll. Davies, then working at Suffield Experimental Station (SES), was due to return to Britain to become deputy to Brigadier Owen Wansbrough-Jones (SA to the Army Council) when Solandt stole him for himself to replace Goforth.[72] Davies's replacement, H.M. Barrett, was British, as was his replacement, Doctor E.A. Perrin. Doctor D.C. Rose left CARDE in 1947 and was replaced by former Woolwich Arsenal employee Doctor W.B. Littler, on loan from Britain.[73] Solandt also sought former AORG members such as Hey and G.D. Kaye to join the organization and even secured radar inventor Robert Watson-Watt as a DRB advisor.[74]

In 1949, lack of staff was delaying work in electrical research, as well as Arctic and bacteriological warfare,[75] so the DRB initiated annual recruiting drives. These were largely successful, as were summer programs for university students at DRB establishments, in minimizing the strain on scientific human resources. Solandt told the board in late 1949, "The success that DRB has achieved in recruiting competent scientists is the best possible evidence of the wisdom of controlling research in one agency large enough to give good opportunity and to preserve a scientific outlook."[76]

The Korean War multiplied staffing problems. The services demanded more from the DRB staff, OR teams, and resources. Funds and staffing limits were increased from 1950 to 1951, but by 1952 the Treasury Board declared a ceiling on both. After talks with the CDC and the Privy Council's Advisory Panel on Scientific Policy, the ceiling was set at 582 professional scientific staff, and a fiscal limit of $22.8 million. Solandt lamented, "It will no longer be possible to undertake new projects regardless of the work already in hand; indeed if new projects are undertaken a corresponding number at present under investigation will have to be dropped." However, he thought it was "foolish to allow the limitations imposed upon the Board's activities to become the excuse for rigidity in its structure and attitude to demands upon its resources."[77]

The staffing crunch during the Korean War also raised the thorny issue of employing expatriate German scientists at the DRB. Unlike Britain, the United States, and Russia, Canada never recruited large numbers of German war scientists at the end of the war.[78] The DRB maintained a policy of not hiring German scientists to work at the DRB establishments, but supported their candidacy for work in university and industry so long as they were not a security risk.[79] However, in September 1951, the Department of Citizenship contacted Solandt about German scientist Rolf Engel, who wanted to work on guided missile research in Canada. He was working for the French government and had been "a senior scientist on the German [guided missile] program at Peenemunde and elsewhere." Engel was a colleague of Werner Von Braun, the chief mind behind the V1 and V2 rockets Solandt had faced during the war and who now worked for the U.S. Army's missile program. The board reiterated their previous policy of not hiring German scientists, but wondered if it should be changed. While this was being deliberated, they instructed the liaison office in London to interview Engel, and his candidacy would rest on merit.[80] Nothing came of it. Perhaps the Liaison Office concluded what historian Michael Neufeld discovered, that Engel may have been an early rocket pioneer, but he was also a compulsive liar, self-aggrandizer, and former police informant and officer in the German SS. The DRB never hired him.[81]

Solandt also had to face Canadian industry siphoning his staff. In May 1951, he informed Claxton that Sorel Industries Ltd in Quebec was inducing technicians at CARDE to leave for higher pay. This was hurting CARDE projects, including guided missile development and the Heller anti-tank weapons system. Claxton raised the issue at the CDC, and Howe, as minister of defence production, agreed to take up matters with the company, and that apparently settled the issue.[82]

Like staffing, secrecy was a persistent problem for Solandt. It silenced the dialogue that kept science vibrant. As Solandt told a public audience in 1947, "It is universally agreed that scientific research thrives and is most productive when there is the freest possible interchange of ideas and results between all workers. Interference with this interchange, whether on grounds of national security or for commercial or personal gain, always delays the progress of science. The old adage that 'when you lock the laboratory door, you lock out more than you lock in' is still true."[83] Solandt and the board devised a strategy to harness the best and minimize the worst effects of secrecy on research. First, the DRB would limit the amount of secret work done at university labs and instead focus on research that was unclassified. Secret research could be done in the DRB labs with university members working for the board. Second, key researchers were hired to head the DRB committees and panels. These included some of Solandt's best friends and family, including Charles Best, Don Solandt, and Laurie Chute. The committees and panels served as main links between the DRB and the universities and industry.

The greatest concern was how to attract researchers to this kind of work for steady as well as part-time DRB employment. Classified research could not be used for professional accreditation in peer-reviewed journals or conferences to build the employee's career, and the DRB could not remove the "lock" on their research. So Solandt devised an academic-styled fix that would allow the maximum amount of academic freedom possible in the highly secret world of defence research.[84]

At the June 1948 DRB meeting, Solandt proposed the creation of a DRB symposium held in December each year, where scientists and other DRB staff could present papers and exchange ideas of import, and members from U.S. and British establishments were invited. The inaugural symposium was held on 15–17 December 1948 and became an annual affair on which Solandt maintained a keen and critical eye throughout his tenure. He pushed for fewer attendees and higher security clearance requirements. This led to more valuable research being presented and contested. Solandt wanted the symposiums to be a critical forum for scientific discussions on defence research, and so there were long question periods.[85]

These "discussion" periods were colloquially known as the "Murder Board." Solandt pushed for vigorous, critical, and thorough debate. Cecil Law presented a symposium paper during the fifties while working for the DRB's ORG and recalled the excitement, but "you damn well had better know your research," he said. Solandt often participated in these sessions with acumen, no matter the subject. He would not only ask probing questions of the presenters, but challenge them on points of fact, research, and scientific merit. While not cruel, Solandt demanded excellence, and few walked away from his sessions of the "Murder Board" untested.[86] Solandt always recalled the symposiums with affection, and he was glad to report in 1950 that most papers were of high scientific merit.[87] The symposiums also served as a chance to have all DRB establishment chiefs in Ottawa to meet with Solandt and review their progress.[88]

While the greater part of Solandt's time was spent dealing with the "big shots," he never forgot the importance of the DRB's staff. He held an annual address to the DRB staff only, presented at a theatre across the street from National Defence Headquarters. Every year, Solandt lectured on the importance of the DRB's work to international affairs and global goals. It was important to him that all in the organization, even those working at menial jobs, understood that their contribution was helping maintain the security of Canada and its allies. Sadly, the texts of those speeches have yet to be found.[89]

Budget 1946–1950

Financing the new DRB was difficult. Solandt relied heavily on Alec Fordyce, the DRB's chief comptroller for the era and member of the original Goforth group. Fordyce considered his job more difficult than managing a board of directors. Much of his time was spent trying to reconcile and manage the needs of the government and the conduct of science with the realities of government finance. Fordyce regularly confronted the "Treasury Board commissars" who attempted to apply standard government rationales to a scientific establishment. Rightly, they were responsible for proper use of public funds, but expected everything to be capable of being audited, but such precision of accountability was almost impossible in a scientific organization. Eureka moments do not happen according to deadlines. Treasury Board officials were rarely competent in scientific matters and thus not qualified to evaluate the DRB's work, but if the work was explained clearly and the money accounted for, they were generally satisfied. Fordyce worked hard and maintained good relations with DND comptroller Owen Faire.[90]

Solandt, as chairman, had signing authority. For work exceeding $50,000, however, approval was require from the minister of trade and commerce, C.D. Howe,[91] as well as the Treasury Board, was required. This was another reason why Solandt required C.J. Mackenzie as a DRB supporter: he was the gateway to Howe's office. Solandt had tussled with Howe in the past, but if Mackenzie backed Solandt's plans, they stood a better chance of being approved by Howe. The Department of Defence Production (DDP), created in 1951 under Howe, became responsible for large defence contracts. As Fordyce put it, you could not escape Howe and the DDP any more than you could escape the Treasury Board.[92]

The DRB's initial estimate of expenditures for fiscal year 1948–9 was $14,008,627.[93] Claxton asked the DRB, like every other service, to reduce their estimates. But as a growing operation, it was hard to measure DRB's budget needs. As Solandt put it,

The Chairman explained that last year an attempt was made to outline roughly the size of the organization which was based on a scientific strength of 300 with all the ancillaries. Using this as a basis it is essential that sufficient funds be provided to run an organization of this size efficiently, but if sufficient funds are not going to be made available to run the organization on this scale then it would be advisable to reduce the size of the over-all organization so that it would run effectively with the funds that can be made available. The Chairman stated that he considered that an organization should level off at a figure of about 7,000,000.00 a year for the Defence Research portion of the vote for research and development. This figure allows for the taking over and operation of the Naval Research Establishment this year.[94]

This estimate was reduced from $7 million to $5 million.[95]

As previously noted, service development estimates, which were included in the money voted to the DRB, were not in fact under DRB control. They were turned over to the individual services to be expended at their discretion. In 1948 the RCAF had the largest cut, primarily for contract work with A.V. Roe on the CF-100, and they had asked for an increase.[96]

The separate service funds for R&D brought headaches for Solandt. Initially, the services had no obligation to tell the DRB what they were doing. Duplication of effort, which the DRB always struggled against, went undetected. Solandt told the board in December 1948 that while the Army and Navy had done well in communicating their individual R&D plans, "very little was known of the Air Force Programme." The NRC, whose Division of Mechanical engineering worked with the RCAF, had better relations with the RCAF than the DRB. However, C.J. Mackenzie's comments earlier in the meeting on the need for a senior body to control development made Solandt question even this coordination. CAS Air Marshal Curtis claimed ignorance of any problem and suggested Solandt and Mackenzie work on a coordination body for RCAF development programs.[97]

By 1949, Solandt no longer wanted the defence research appropriations lumped with the main defence estimates of the three services. Instead, the amount of money the federal government spent on defence research should be "related to the total federal budget for research and development" and reflect the percentage of the total federal research effort allocated that was to be allocated to defence rather than "considered on the basis of how much of the defence estimates should be allocated to research development."[98] He hoped presenting the DRB's budget in this wider, non-military context would generate support for his desire to control service development funds. It fell on deaf ears.

In September 1949, Solandt tabled his first annual report as chairman. As the DRB left its formative phase of development and began tackling major defence research problems, his financial estimate for the fiscal year 1949–50 was $6.94 million, and for 1950–1, $7.20 million, excluding the flight research establishment that DRB was helping to support.[99] In December, he tabled the DRB's first Five-Year Plan, and the board agreed with his estimates:[100] for 1950–1, $25 million; for 1951–2, $26 million; and for each of 1952–3, 1953–4, and 1954–5, $27 million.[101]

Total estimates for the DRB, excluding Service Development and Signals Research (termed "Y")[102] for 1949–50, was $8.5 million, and the total allotment for the board for 1950–1 was \$10.7 million. By 31 August 1950, \$8.2 million had been encumbered, and \$362,000 had been allotted to the RCAF development vote.[103] These were large increases from the initial start-up budget, but as the DRB moved from planning to research and development, costs grew in step. Support for large capital projects like Velvet Glove, as well as the resources for continental defence radar work, grew as research was translated into results. These expenditures were compounded as the DRB tried to meet the growing and immediate needs of the Armed Forces during the Korean War.

Budget 1950–1955

The outbreak of the Korean War in the summer of 1950 increased the Canadian defence budget from \$425 million to one billion: \$200 million was to equip European allies and \$800 million was assigned to the Armed Forces, while \$40 million went to defence research, and of this amount roughly \$16 million went to the DRB and \$25 million to service development. Yet the DRB's tasks were still growing. "Whatever the final amount," Solandt said, "it is clear that new developments can only be undertaken at the expense to some element of an existing program. The adjustment of priorities for this purpose will be the task of the Committee on Development."[104] By September 1951, the largest single expenditure for the services' R&D vote was work on the RCAF's CF-100 fighter aircraft and Orenda engine. The DRB's largest expenditures were in chemical and bacteriological warfare, with armament coming second.[105] The DRB estimates were approved for 1951 at \$11.8 million in round figures, and most of this had already been spent. The estimated DRB budget needed to maintain its growth and efforts for 1952–3 was \$24.0 million, as opposed to \$16.6 million in the previous year.[106] By 1952, after Solandt fought Howe over the validity of his budget (see chapter 10), the Treasury Board agreed to the increase and then placed ceilings on the DRB's staff and budget. The cash estimate approved for 1952–3 was \$42.0 million, of which \$22.8 million was for research and \$19.2 for development. Air Force development remained the most expensive at \$14.1 million.[107]

By 1952, the board was responsible for administration and financial control of development funds for Armed Services for reasons of efficiency and coordination. Solandt had finally convinced the services that handing over this control meant their needs would still be met. It is unclear from current documents how this was actualized, but it meant an increase in funds for the DRB, as the estimate for 1952–3 of $59 million was approved. Allocations for research were $35 million, and development $23 million, for a total of $56,250,493.[108] 1953 saw a modest increase in the estimated budget to a total of $60 million,[109] and the DRB's budget from 1954 onward remains classified.

Conclusion

In ten years, Solandt had successfully led the DRB from idea to established defence research organization. By working with government, military, industry, and academic members concerned with Canadian defence needs, he was able to set the stage for the research and development of critical technologies for Canada's defence, including the McGill Fence radar defence system. He also kept the DRB out of questionable activities brought on by brainwashing hysteria, while at the same time keeping his eye on the reality of this threat. All of this occurred while Solandt devised strategies on and fought battles against the perpetual problems of staffing, secrecy, and finance for an organization that had no track record of success and was led by a man who, in 1946, was a relative unknown in the corridors of power in Ottawa.

It was a gruelling decade that often tapped Solandt of his own dry and dark sense of humour. When former AORG scientists and the DRB OR troubleshooter George Lindsey's first child was born, the DRB secretary Archie Pennie went into Solandt's office, excited at the news, and informed the chairman that Doctor Lindsey's wife just had a boy. Solandt looked up from the work on his massive desk and asked, with no enthusiasm, "Well, is that what they wanted?" Pennie left with "my tail between my legs."[110] Such was Solandt's mind, focused on priorities and efficiency, service before personal interests. For ten years, this mindset guided the policy and plans for the DRB's own establishments to make effective contributions to the needs of the Canadian Armed Forces. The DRB staff and funds Solandt had fought for were funnelled into critical research projects conducted by the chief internal arms of the DRB: the establishments.

12 Rockets, Germs, and UFOs: Solandt, the Establishments, and Public Perceptions of the DRB, 1946–1956

As chairman, Solandt was responsible for research establishments even more diverse than those at AORG, conducting research on behalf of not one service but the entire Canadian Armed Forces. He had abandoned his own research in 1943 and embraced administration and management of research for the British Army. From 1946 until 1956, he embraced the extraordinary task of transforming himself into a Renaissance man in the science of war. Anything else would be irresponsible, as he said in 1953.

> No narrow specialist can possibly make intelligent judgments concerning future courses of action in a system as complex as the modern armed forces. Such judgments can only be made by a person who has attempted to assess and to understand all the factors involved. I therefore felt that it is the job of the Chairman of the Defence research Board to struggle with the obviously impossible task of comprehending all the developments of modern science and their possible impact on the future of war. This means that the Chairman has to try to pose at one moment as an expert in electronics, at another as a nuclear physicist, at another as a chemist, a physiologist or a mathematician. He can probably do this job pretty effectively if he recognizes his own limitations and retains his sense of humour.[1]

While many lamented Solandt's hard sense of humour, most admired his few limitations as overseer of Canada's major defence research projects.

This chapter considers Solandt's role in directing research from Ottawa to the Establishments. From the start, he maintained a cool and steady grip on the DRB's commitment to long-term research, balancing it with immediate service needs. While an overview of the key developments of each establishment is provided, the focus here is on Solandt's influence and support of two major and controversial the DRB research programs: armament development, and chemical and biological warfare (CW and BW) research.[2] Each is a useful case study of Solandt's multifaceted view of how Canada could best conduct defence research. In armament, Solandt supported the development of conventional weapons systems for the Army, and the development of Canada's first air-to-air guided missile for the RCAF, the Velvet Glove. In CW and BW, Solandt promoted and maintained Canada's leading role in conducting field trials of these deadly weapons to study their effects and provide clues on defence. These trials focused on the effects of nerve agents such as sarin (GB) and biological viruses like rinderpest. Both cases provide insight into Solandt's vision of defence research. Each was an important contribution to the armed forces, and also served Canada's alliance commitment to avoid duplicating research being done by Britain or the United States.

By 1952, after six years of public addresses on the DRB and war and technology, Solandt appeared in the national limelight.[3] Solandt became the public face for the DRB and did his best to weather the distortions and accusations that cropped up about him, his organization, and its fields of research. The Soviets and their allies portrayed him as a cold and villainous tool of Western imperialism, whose machinations ran from supporting the creation of a robot army, to his glee at the prospect of biological warfare (discussed below). At the same time, Solandt was confronted by public fascination in the UFO hysteria of the early 1950s. With a cool hand and manner, Solandt navigated these storms of controversy and got back to business, keeping Canada's defence research agenda on course.

Headquarters

While Solandt dealt with "big shots" in the military, government, academia, and industry, his vision was executed primarily through Vice-Chairman Davies at the DRB headquarters (HQ) and its administrative structure in Ottawa. The DRB HQ was the organization's information hub, which maintained contact with industry and academia through a series of panels, committees, and subcommittees on defence research concerns. By 1949, HQ also housed the DRB's Operational Research Group (ORG). The services had had responsibility for their own OR sections, but "Service operational research sections had ... been discontinued" after the war. Solandt suggested OR sections for each service and within the DRB.[4] HQ also contained the DRB's scientific intelligence assets. The DRB was responsible for the administration of the Joint Intelligence Bureau (JIB) of the Joint Intelligence Committee (JIC) of the COSC.

Short- or Long-Term Planning, 1946–1956

The DRB was founded on long-term research goals with a productive peak expected in 1953. But international crises, starting with the Berlin Blockade and culminating with the Korean War, kept short-term projects as an issue at the DRB. Solandt had seen how peacetime preparation in civil defence, especially at the blood depots, had aided Britain's survival during the war. He also clearly understood that ad hoc management of short-term defence research problems could create structural problems for research, as had happened at AORG. Solandt told the board in 1948 that DRB would maintain the goal of long-term planning as long as the Armed Forces were not required to rearm substantially. Vice-Admiral Grant agreed but wanted short-term planning for emergencies maintained, especially on atomic, biological, and chemical weapons. The Army and RCAF wanted long-term planning continued.[5]

Solandt made the DRB's long-term commitment public in an October 1948 address. He asked the Montreal United Service Institute, "Should we switch to short-term objectives each time that international tension rises or should we keep our eyes continuously on more distant goals?" His answer was predictably pragmatic: continue to aim for long-term basic and applied research but with necessary diversions to solve urgent problems as they arise.[6]

The DRB's Master Plan

With the embryonic phase of his organization complete, Solandt took the first steps to long-term planning with a review of Canadian research and development policy in 1949.[7] After it was finished, the DRB constructed its first five-year "Master Plan." Solandt's annual report from the chairman, DRB, emphasized the need for specialization, of including the services on all major decisions because "the Services, having been told that the DRB could not undertake a research project, have made other arrangements with very mixed results." Solandt suggested the creation of a joint DRB-Services Project Control Centre "to supply the data needed for keeping the master plan up to date and for following the progress of projects," as well as a joint review panel.[8]

The board discussed the "Master Plan" on 17 September 1949. Solandt stressed the value of the Project Control Centre to keep the DRB informed of all R&D and let the services have access to information from the DRB, other services, and U.S. and U.K. development projects. This would eliminate duplication and keep everyone in the loop.[9] At the 1 December meeting the board approved the program.[10]

During the Korean War, Solandt attempted to balance the goals of the "Master Plan" with immediate service needs. Fear of possible CW and BW munitions use in Asia or Europe dominated most discussions at the DRB, but the Armed Services had myriad chemical research needs. Early in the Korean War, the Army had no portable flamethrowers and could not find a commercial vendor to purchase Octal, a thickening agent for flamethrower fuel, so they looked to the DRB for assistance.[11] Solandt shifted some of Suffield Experimental Station's (SES) work and most of Defence Research Chemical Laboratory's (DRCL) research agenda to answer this need. Prior to the war, DRCL's most important work had been developing Arctic respirators as part of tripartite agreements on the division of labour among allies conducting respirator research.[12] Now it switched to work on flamethrowers and fuels. Between 1950 and 1952, SES had finished development of the Iroquois flamethrower, and the DRCL had discovered silica gel as an effective stabilizer for flame fuels thickened with Octal.[13] Previously, commercial Octal gels were of low quality, but by 1952, DRCL's chemical pilot plant, running twenty-four hours a day and seven days a week, supplied the 25th Canadian Infantry Brigade Group with its operational requirements of Octal, and had generated commercial interest in their process.[14] Yet most of the DRB's chemical work remained in chemical warfare research (see below).

Leading the Establishments

In previous chapters, we have seen Solandt establish the DRB's policies through the COSC and the board, and, with the latter, through industry and academia. The DRB also maintained nine defence research establishments, scattered across Canada, to conduct research and development. Contact with the establishments was facilitated largely by Vice-Chairman Davies. Solandt, however, would not be tied to Ottawa as he had been in Roehampton. The DRB meetings were held at each of the establishments so that board members could deal personally and directly with the superintendents and staff. Doctor Cecil Law, a Second World War veteran who worked for the DRB at Suffield, Fort Churchill, and with the ORG in Ottawa, recalled how Solandt himself often asked sharp and penetrating questions of the working staff and offered advice on contacts while touring the establishments. As he had at AORG, Solandt maintained an active interest in the work of all the establishments and visited them as often as he was able. For Law, this maintenance of interest and genuine concern helped those in the distant establishments feel that their work was important and not forgotten by Solandt and the DRB HQ.[15]

Unless Davies or the establishments raised a concern, Solandt did not interfere with the running of the establishments. Davies kept tabs, the superintendents provided annual reports, and communication was handled through committees and panels on their individual research areas. Here, Solandt maintained the "maestro" approach he had learned at MRC and employed at AORG. So long as the best professionals were hired and were following board policy, he assumed that individual superintendents knew best how to run their operation. As George Lindsey recalled, "As Chairman, [Solandt] did not attempt to conduct or supervise research himself. What he did was select, attract and support the kind of scientists that were needed, and provided them with the atmosphere and facilities in which effective defence research could be accomplished."[16]

Solandt was adamant that the DRB's research agenda remain specialized. Scarcity of resources, unique expertise, and domestic conditions provided Canada an edge if it focused its efforts where it could excel, rather than spreading itself too thin. This expertise translated into better results for the Armed Services and aided the DRB's continuing efforts at allied coordination of defence research (considered in chapter 13). The priorities established at the DRB's origins remained the same. Research on Arctic affairs; aeronautics; atomic, biological, and chemical warfare; rocketry; telecommunications' and sonar work were the mainstays of Solandt's term as chairman.

The Chief Programs of the Establishments in the Solandt Era, 1946–1956.

The DRB inherited or created nine applied science establishments during Solandt's chairmanship.[17] Brief mention of the chief success of each establishment is considered before focusing on Solandt's distinct role in the senior research and development projects at the DRB during his tenure as chairman: armament research at Canadian Armament Research and Development Establishment (CARDE) at Valcartier, Quebec; and chemical biological warfare research, largely undertaken at Suffield Experimental Station (SES) in Suffield, Alberta.

Each of the nine establishments did tremendous work. The Naval Research Lab (NRE) had been a highly successful wartime naval research establishment and continued to work in anti-submarine research, most notably variable-depth sonar on Project Dunker, and early work with the RCN and Ferranti Canada on the digital automated tracking and resolving (DATAR) battlefield information system.[18] Its sister station, the Pacific Naval Laboratory (PNL), conducted research similar to that of NRE on sonar and corrosion particular to the Pacific.[19] The Defence Research Telecommunications Establishment (DTRE) consisted of two entities: the Radio Propagation Laboratory (RPL) and the Electronics Laboratory (EL), which were concerned primarily with the physics and mechanics of radar research, and, as seen in chapter 11, contributed to the development of the McGill Fence. The Operational Research Group (ORG) was headquartered in Ottawa, but its teams and individual scientists contributed to the services on myriad research projects throughout Canada, especially the RCAF on air defence in the Arctic.[20] Defence Research Medical Laboratory (DRML) was stationed with the RCAF base in Downsview, Ontario, and focused on occupational medicine.[21] Defence Research Northern Laboratory (DRNL) was situated at the army base at Fort Churchill, Manitoba, and conducted applied and basic research on a range of Arctic affairs, including topographical and navigational studies, weapons and equipment tests, military exercises, and physiological and entomological studies. With RCAF Squadron Commander Keith Greenway, DRNL supported, along with the DRB Electronics Laboratory, research on the "Twilight Computer" for Arctic navigation.[22] With the exception of the NRE and PNL, each establishment conducted research for each service and alongside Canada's chief allies, Britain and the United States.[23]

Solandt was fond of every establishment, with a particular interest in the valuable navigation and acclimatization work done at DRNL and ORG.[24] But armament and chemical and biological warfare research provide the most centralized, as opposed to dispersed, research programs with which to assess Solandt's influence as chairman.

Armament Research

The Canadian Armament Research and Development Establishment (CARDE) at Valcartier dealt with armament, propulsion, and explosives research. CARDE was responsible for the most demonstrable evidence of the DRB's efforts in research and development: the pot-sabot propellant, the anti-tank rocket known as the "Heller," and the guided-missile project known as "Velvet Glove." In September 1948, the DRB held a special meeting on the future of Canada's armaments projects at CARDE,[25] which included continued work on the development of picrite propellant (see chapter 11), the 17 lbs. pot-sabot projectile, and an infantry anti-tank rocket, the Heller.

The improved Second World War–era seventeen-pounder pot-sabot[26] projectile anti-tank weapon was an early success for CARDE. By 1952, it had passed the acceptance trials for use in Canada and the United Kingdom and was accepted as standard inventory by Canada, Britain, and the United States. The Canadian Army put in large orders for it during the Korean War. Solandt proudly noted in 1951 that this project was CARDE's first completed research and design project and stood as an example of the value of CARDE's research approach and the value of standardization.[27]

The "Heller" 3.2 inch anti-tank rocket was modelled on the German "Hammer" rocket launcher of similar size. In 1948, Solandt claimed it had been more of an "education project" for the CARDE engineers than anything else, though he felt it could be the best rocket in the field if supported.[28] The Heller anti-tank rocket garnered interest not only in Canada but from allies as well. Davies reported to the board in 1949 that both Britain and the United States were interested in the weapon and its possible development as "interim air-to-air and air-to-ground weapon."[29] When Solandt announced to the board in June 1950 that CARDE would now be researching air-to-air guided-missile work, Foulkes applauded the choice of missile development but worried that such a difficult task would drain CARDE's resources from projects nearing completion, including the Heller. Davies assured him that the Heller was almost finished, and the new guided-missile program was not a threat. The board agreed, but made it clear that, at the moment, the Heller was CARDE's "top priority."[30]

Canada's rearmament program during the Korean War accelerated work on the Heller. In 1952 Solandt suggested that "big" Heller anti-aircraft weapons should also be considered, though nothing appears to have come of this proposal.[31] By March 1952, Solandt informed the board of the commencement of the pilot production of fifty Heller launchers and 5,000 rounds of ammunition.[32] In October 1953, the Department of Defence Production announced that a $3 million order had been placed for the Heller, marking "the first successful transition of a major project from research and development to production and use."[33] Solandt always maintained that the Heller was, bar none, the best anti-tank weapon in the West at the time. The U.S.-made bazooka did not compare with its shoulder-fired comfort or efficient recoilless system,[34] but CARDE's most notable program was the creation of Canada's first guided missile.

Solandt initiated preliminary guided-missile research in 1946, with a Guided Missile Advisory Committee.[35] By 1947, he was convinced that the Canadian Armed Forces needed to keep pace with guided-missile development and be prepared to contend with these weapons. By 1949, the RCAF had developed requirements for an air-to-air missile for the CF-100's continental defence role.[36] However, given the scientific challenges and large resources required to excel in this field, Solandt initially focused CARDE's efforts towards research and training of staff. In 1949, Solandt established a small working group on guided missiles to assess if Canada could make an efficient alliance contribution to the field, one that avoided duplication of effort with allies. The DRB consulted with the tripartite Guided Missile Working Party 15 (WP 15) on the issue and discovered the American missile programs were based on future needs and advanced R&D, while the British were creating immediate-need weapons. The committee suggested Canada should pursue a missile development program that "would use a relatively simple concept, employ known technology and know-how, and incorporate the largest possible number of existing components."

Solandt backed the proposal and took it to the CDC in October 1950. As the committee discussed Canada's response to the Korean War and NATO commitments, CAS Air Marshal Curtis noted that the DRB wanted to invest in guided missile work since, if general war broke out, these weapons would be required. Solandt agreed, though he tempered the CAS's point by saying that guided missile work was "a normal development of the armament research programme." Such a program would not only benefit the Canadian armed forces, but act as a catalyst for development in industry in a range of fields, including the miniaturization of electronics. The scale of Canadian expenditure would be controlled, and it was considered that the cost of the project would be approximately "$1,650,000, per annum, of which $650,000 would cover

salaries and overhead already allowed for in the Defence Research estimates." The CDC approved the initiation of Canada's first guided missile
program, which became known as Velvet Glove.[37]

Over the next four years CARDE designed and tested prototypes for
the Velvet Glove and its supporting systems. It coordinated its research
with other the DRB and NRC establishments, including the joint DRB/
NRC/RCAF National Aeronautical Establishment in Ottawa and Aerospace Institute at the University of Toronto. By 1955, the Velvet Glove
program was virtually developed and was nearing full production,
with 300 missiles already in existence, produced by Canadair. By this
time, however, the RCAF had set priority on developing a supersonic
fighter, the CF-105, or Avro Arrow. While Velvet Glove could be used
on the CF-105, the RCAF chose to go with the U.S.-designed Sparrow
series of air-to-air missiles. At a cost of $24 million, Velvet Glove was the
largest DRB establishment project during the Solandt era, generating
roughly 400 guided-missile experts in Canada, and CARDE worked
hard to retain that expertise.[38]

It is unfortunate that the minutes for the DRB meetings for 1955 and
onward have been lost. Solandt, however, recalled a serious confrontation with the Air Force about arming the CF-105 after the cancellation of
Velvet Glove, one not well documented. The RCAF was adamant about
having a new weapons system for the CF-105, and not using Velvet
Glove. According to Solandt, the DRB, the RAF, and the USAF said it
was unwise to create a new weapons system and a new airplane, and
it made more sense to buy and adapt an existing weapon system. The
RCAF wanted the Sparrow II, a U.S. Navy missile, but the USN had
informed the DRB that the Sparrow would never be used. The RCAF
asked the USAF to ask the USN again, and this time they said they
would be proceeding with the Sparrow. At the working level, USN told
the DRB they never intended it to be finished, and Solandt tried to convince the RCAF to maintain the agreed armament for the Mark II Arrow,
the USAF Falcon. In 1956, the Sparrow II missile project was cancelled
in the United States. Canadair took on the difficult task of developing
the Sparrow and its guidance system, but both were cancelled in 1958.[39]
So, according to Solandt, the Arrow never had an armament.[40] Velvet
Glove was a near miss for CARDE and the DRB, though the experience they gained in the process would be valuable. Still, it was a bitter
truth that the project, much like Solandt's pioneering tank OR work,
was beaten to the punch by competing U.S. interest– hard reality in the
tough world of working with a emerging super power. Nonetheless,

Solandt was proud of everything CARDE achieved and was glad to see that their expertise in rocketry, navigation, and guidance would eventually allow the DRB to play a role in Canada's emerging space program after his chairmanship.[41]

Chemical and Biological Warfare Research

The DRB had four establishments focused on the threat of chemical and biological warfare. In part, this reflected a general perception among the Western allies that between 1945 and 1949 the USSR, while working on an atomic weapon, might resort to using chemical or biological weapons as a deterrent in lieu of their lack of atomic weapons.[42] This threat was made more pronounced when the allies discovered in 1946 that Russian forces occupying Germany had gained access to large stockpiles of advanced German nerve gas munitions and production facilities, and had reconstituted some of the facilities that had been destroyed during the war.[43]

Ever since his Lulworth days, Solandt had been dealing with the deadly nature of chemical warfare and the difficulties of protecting soldiers against its use. The need to protect against its most deadly current form, including nerve agents, seemed clear, but he was less certain about biological agents. As he told the United Service Institute in 1948, bacteriological (later biological) warfare offered more possibilities than clear characteristics for future warfare. The potentials were as great as the unknowns, so "we cannot yet say whether it will be a major factor in winning a future war, or just another weapon." However, Canada could not afford to ignore it for either defence or economic reasons.[44] In any future war, dictators would likely focus on weapons that kept the enemy home intact, as nuclear weapons might destroy too much of the target nation's industrial strength and wealth of national resources. This required anti-personnel weapons that either kill or destroy the will to resist, and such an aggressor would likely use CW and BW, as well as propaganda. "With chemical or biological agents, he might kill quite a few of those that he attacked, but the survivors would be sufficient to help his own surplus population to operate the intact industrial machine of the vanquished country."[45]

Canada and its allies had to be prepared to face such threats. Four of the DRB's nine establishments considered issues regarding both BW and CW. Grosse Île Experimental Station (GIES) was a wartime biological warfare station that remained largely dormant until 1949, when the

DRB took part in tripartite work on hoof-and-mouth disease. Under director of projects Doctor C.A. Mitchell, the DRB's effort focused on vaccines of rinderpest, a deadly virus affecting cattle. Solandt argued that Canada's work on rinderpest vaccines would contribute enough to world food supplies to justify all of Canada's wartime expenditure on defence research.[46] Defence Research Chemical Lab (DRCL) in Ottawa was the oldest establishment and concerned itself primarily with defensive aspects of CW, including respirator research and production. Defence Research Kingston Laboratory (DRKL) was concerned with BW and psychological warfare.[47]

The Suffield Experimental Station (SES), a 1,000-acre property in northern Alberta, was the largest of all the establishments. It had been a highly respected joint British-Canadian chemical weapons research establishment during the war under the direction of Welsh scientist E. Ll. Davies, who became Solandt's deputy in 1947. SES maintained its status as the leading tripartite field trial establishment for Canada, Britain, and the United States during the Cold War, using a vast array of animals for tests.[48] It also began limited BW and atomic warfare research during Solandt's era.[49]

Knowledge and Field Tests, 1946–1950

At the request of Lieutenant-General Foulkes, the DRB held a special meeting at SES to discuss CW and BW warfare policy in June 1948, arranged by Doctor H.M. Barrett, chief superintendent of Suffield. Doctor G.B. Reed, director of projects at Kingston Laboratory, outlined work done in biological warfare with reference to methods of protection; Doctor J.R. Dacey, superintendent of DRCL, Ottawa, outlined methods and facilities for the production of war gases and indicated the progress in development of protection devices. Solandt was impressed with SES as one of the finest CW establishments in the field, though he emphasized the need for secrecy on these matters, given the public's general negative perception of CW and BW research.[50] Field testing and some production of CW agents for research on defensive measures were clearly on the DRB's original CW and BW policy agenda.

At the DRB meeting the following September, Vice-Admiral Grant suggested the creation of a joint school to study the unique aspects of atomic, biological, and chemical warfare. Solandt agreed. He believed that the similarities between chemical, biological, and atomic warfare in protective equipment and decontamination required that they be grouped together for service purposes. He suggested that the DRB assist in the creation of joint service courses to study the nature and effects of what they referred to as "the New ABC of War": atomic, biological, and chemical warfare.[51]

The board agreed. Otto Maas, as SA to the CGS and member of the board, coordinated this project with the services.[52] The December meeting of the board dealt with Maas's efforts. "Dr Maas discussed the problem with the Army and a meeting was later held in Dr Solandt's office, where representatives of the Services and DRB agreed that, initially, the Defence Research Board would run a training course for a selected group of Service Officers. This group would be restricted to a maximum of ten people, and would be given a thorough background in ABC – what is available in Canada in detail – what is known of work being done in the United States and the United Kingdom."[53]

Foulkes wanted the DRB to be the authority on civilian and military protective measures on ABC warfare and reiterated that this new course did not exclude the board from its obligation. In fact, the course was the beginning of the work. "So little is known about the practical problems that it will require a good deal of study before the information at hand can be turned into actual training material."[54] Vice-Admiral Grant warned the board to give only policy direction on these affairs, since the services were responsible for working out protection against new weapons, but Foulkes rejoined that the public would hold the DRB responsible for defensive measures against ABC, regardless of official attitudes. Solandt surmised that "the problem resolved itself into determining at what level it became a Service problem. The DRB might run schools for top level instructors and leave the detailed training matters to the individual services." Solandt worked on a report for the next meeting.[55] In March 1949, Davies reported that arrangements were complete for the ABC school. The course would start April 7 and run to 21 May in Ottawa, Chalk River, and Suffield.[56] In September 1949 Solandt reported that eleven officers had taken these courses thus far.[57]

With education established, concerns now ran to modern research into defensive and offensive use of chemical and biological warfare. Solandt maintained an active presence in Canada's CW and BW arena, though Davies, as the board expert on Suffield's work, was the senior board member concerned with CW and BW policy.[58] The DRB committee panels on BW and CW had informed Solandt in December 1949 that DRB needed pilot plants for both CW and BW agents to continue field trials. The board agreed to the immediate construction of the CW pilot plant, but not a BW plant, since the United States and Britain were leading the way in the difficult arena of production. Instead the work at GIES, concerned with animal disease and vaccines, would continue.[59]

By the end of 1949, Solandt reported that SES had increased its staff and was still the main source of field trial data on CW in the "free world." The discovery and use of fumaryl chloride as a stimulant for GB nerve gas itself had "greatly extended our knowledge of this most promising agent."[60] At the time, SES was under the directorship of British applied chemist Doctor E.A. Perren as part of exchange with the British Chemical Warfare Establishment at Porton Down. Perren was famous in British science circles for "his extensive activities in the post-war intelligence exploitation of the German chemical warfare capability." The Germans had focused largely on nerve gas. As the Soviets were using these agents, nerve gas was the chief CW substance under investigation at SES.[61]

Chemical and Biological Warfare Policy, 1950–1955

CW and BW research took on new urgency after the Korean War. According to Doctor Cecil Law, there was a general fear at the DRB that not only could a war in Europe begin, but the war in Asia might have a CW and BW dimension that concerned those at the DRB. The Communist Chinese forces had inherited all of Japan's wartime chemical and biological warfare research facilities in mainland China.[62] With China's massive involvement alongside its North Korean allies, the threat of CW and BW attacks in Korea loomed large in scientific circles.[63]

In March 1950, three months before North Korea invaded the south, the DRB established a BW Review Committee under Solandt's mentor Doctor Charles Best to study the supply of bacteriologists and the state of the DRB programs.[64] Best's report in December 1950 focused largely on tripartite work on the outbreak of hoof-and-mouth disease in Britain (discussed above). Canada's focus was on researching, producing, and stockpiling a rinderpest vaccine (the R Project) for all tripartite countries in case of war. For Best, Canada's reputation as a BW power lay in rinderpest work. Given the difficulties of manufacture witnessed in the United States, he believed that Canada should not pursue creation of BW weapons. The staff and facilities of the U.K. Chemical Research Establishment, Porton, and the U.S. Army Chemical Corps, Camp Detrick, Maryland. were better suited, though some work could be done at the Physics section at SES. Still, there was a desperate need for bacteriologists because of the current threat of enemy use of bacterial or viral diseases. Creating a DRB bacteriological fellowship would strengthen the BW effort.[65] The board agreed and the fellowship was established.[66]

Regarding CW, emphasis at the DRB had remained on conducting field trials and designing preventive measures against "G" or nerve gases/agents. Solandt's annual report for 1951 reported an increased pace to existing projects at SES, where it was now possible to conduct, on the same day, trails of CW, BW, and radiological weapon (RW) materials, reflecting the new urgency of the research.[67]

In September 1951, the board held a special meeting with U.S. and British officials about the possible use of weapons of mass destruction and current defence research capabilities in this field. Participants included Sir John Cockroft, now chairman of the Defence Research Policy Committee, and Walter Whitman, chair of U.S. Research and Development Board (RDB).[68]

Major-General S.F. Clark, substituting for Foulkes, asked about the political implications of ABC warfare. While the primary concern was atomic weaponry (see chapter 14), Solandt raised the issue of CW, noting that CW had problems tied up with "moral objects on the part of the public and with military doubts. It was his opinion, however, that we should assume G gases would be used. This being the case, we are confronted with the fact that we have A bombs but no GB and that production of the latter is not anticipated before 1954. For this reason he felt that high priority should be given to accelerating production of GB. Canada's most effective contribution would be helping to supply intermediates to speed UK and USA production."[69]

Here, Solandt reveals two aspects of his view of CW and BW policy at the DRB. First, the DRB was not going to be involved in the difficult process of CW and BW production. Second, he maintained, as he had with the atomic bomb, a realist perspective. As a medical professional, he knew all too well what kind of havoc could be wrought by CW and BW. But Solandt also realized that the Soviet Union and Communist China had the capacity to wage chemical and biological warfare. The public could fear the weapons, the military could question their applicability, but it was Solandt and the DRB's responsibility to offer expert knowledge of their nature and to provide the best judgment on how to defend against them.

As fears of a general European war rose in 1951, chemical warfare became a NATO concern. Solandt was present when Claxton reported to the CDC in August 1952 that Supreme Allied Commander Europe (SACEUR) General Mathew Ridgeway had requested Canada's approval for his use of chemical weapons should they be employed by the enemy. After a clarification on the terms of use, in October the CDC agreed that the Canadian representative on the Military Representatives Committee should endorse the directive, on the understanding that the higher authority required for their use also included the approval from a representative of the Canadian government.[70]

By the end of 1953, CW and BW research had continued at a steady pace. Solandt reported to the board that Suffield was working mainly on the effects of temperature on performance of GB munitions, assessing BW munitions, developing and testing flame munitions, and "the study of irradiation from an intense radioactive source distributed in the open."[71]

Less is known about the DRB's specific work on BW agents and munitions as opposed to the rinderpest vaccine because the research is classified, though Solandt's involvement in Britain's anthrax BW program is discussed in chapter 13. The DRB did support allied tripartite work in anti-animal BW research. During the spring and summer of 1951, Solandt approved the decision to provide over 400 pigs from the SES farm to the U.S. Army Chemical Corps, who conducted successful field trials on hog cholera as an anti-animal weapon at Elgin Air Force Base, Florida.[72] Canada also participated in the U.S.- led anti-animal BW program East Africa Project 1001, which collected a series of viruses and developed vaccines for them, including for the DRB's rinderpest research. The United States left anti-animal research behind in 1954, and Canada, it appears, followed suit.[73]

But the majority of research on CW and BW during the Solandt era was on CW and BW munitions and defence against them. To facilitate this research, SES received Area E at Suffield for testing BW munitions. They created a new type of dynamic bursting chamber made from an aircraft hangar and three aircraft engines. This would neutralize the effect of weather on trials with BW munitions, which was ruining results.[74]

Solandt in the Spotlight

Despite the classified nature of all of the DRB's work, the Canadian and international press began to take a serious interest in Solandt, the DRB, and their research, starting in early 1950. As chairman, Solandt had never hidden from the public. He maintained an aggressive public speaking schedule throughout his tenure, addressing a variety of interested parties on the DRB, its mission, and his view of war and technology in general. His lectures were part of a DRB-approved strategy to keep select sectors of the public informed about developments in science, technology, and warfare.[75] It was hoped that Solandt's pragmatic and confident view of these matters, grounded in his expertise, would help mitigate the general fears about modern warfare and the myths about "indefensibility" against atomic, biological, and chemical war that had grown in the public mind since 1945. Foulkes always worried about the public's resolve to fight any future war in the atomic age.[76] Solandt's dozens of speeches, covering almost every aspect of war and technology he was involved with, rationally and coolly debunked myths but pulled no punches about the deadliness of modern warfare: many of his lectures were about his experiences at Hiroshima and Nagasaki. Still, Solandt remained rational and clear about how best to meet the threats Canada faced, as he told audiences in 1947.

> Until wars cease to occur we must be prepared to defend ourselves against all forms of attack, including atomic bombs. I do not subscribe to the views that there is no defence against the atomic bomb. At least for the next few years atomic bombs will be carried in ordinary bombers which can be intercepted and destroyed just as in the past. When the long range rocket carrying an atomic warhead becomes a reality, there is little doubt that there will be some means of defence against it. In addition to this active defence, which is the responsibility of the Armed Forces, much can be done in the field of Civil Defence. Here the important measures will include psychological preparation of the people, so that attack will not weaken our will to resist; careful planning of both the type and location of new buildings and industries of national importance and the preparation of plans for rescue and repair services and for the treatment of casualties.[77]

Solandt emphasized that Canada needed to retain the will to fight and accept casualties, but that it would best preserve peace by contributing to Western defence to deter any Soviet consideration of starting a war.[78]

As the DRB and its research caught the interest of both domestic and international audiences, however, Solandt and his organization came under scrutiny. In the Soviet press, Solandt was portrayed as a cold, inhuman tool of Western imperialism and an abuser of science. A May 1950 article in *Pravda* quoted Solandt as saying that robots were "cool headed and able to concentrate under fire." Such practical statements were taken by *Pravda* as evidence that the West was preparing to replace human soldiers with robots.[79]

While Soviet propaganda could be shrugged off, Canada's 1951 rearmament program increasingly put Solandt and the DRB into the media spotlight. By 1952, the DRB hired a veteran and former journalist for the *Ottawa Journal*, Charles Pope, to be its first public relations officer to help manage the increased interest.[80] The *Financial Post* cover story for 16 February 1952 was solely about the DRB, entitled "Science: Our Armed Forces' $35 Million Fourth Arm," with the subheading "The Job We're Doing in Defence Research; The Promise It Holds for Industry's Future." It featured photos of DRB scientists working on radar, aerospace, and naval research projects, and of a young Solandt, dark-eyed but with a firm and confident smile. Before interviewing Solandt, author Cyril Bassett provided a colourful picture of the DRB's activities:

> Men chasing radioactive insects over the northland with Geiger counters ... taking blocks of ice and snow apart ... spending their days getting sick under the impact of sudden and varying changes of altitude and temperature ... probing the mysteries of the inky deep of our coasts and our Arctic skies, night, and wastelands ... hunched over drawing boards, at laboratory benches, on firing ranges spread over 1,000 square miles of hush-hush wasteland in Alberta, even studying the soldier in the battleline [*sic*] at such simple tasks as washing his mess tin.[81]

In the article, Solandt presented the Canadian public with his vision of defence research: the focus was on excellence in a few fields, support for the Canadian Armed Forces and allies, and support for research in industry and university. The piece presented him as a cool and pragmatic leader and concluded that the DRB and its expanded budget was a valuable investment in Canadian military and industrial power, as well as a boon for research in universities.[82]

Less than two months later, Solandt's image was tarnished by Soviet allegations about his role in biological warfare. In April 1952, James G. Endicott, DD, a former Canadian missionary from Toronto who was in China during the Korean War, made public "speculations" about Solandt and the DRB's BW work. In an interview broadcast by Moscow, Endicott said Canada was manufacturing germs at Suffield to feed the alleged BW measures the United States was employing in the Korean

War. He then maligned Solandt by falsely quoting him as saying that the prospects for mass death through BW were "extremely heartening."[83]

John Diefenbaker, Conservative MP for Lake Centre, wanted to know more about Endicott's allegations.[84] Ron Kenyon of the *Toronto Telegram* interviewed Solandt at the time. "This was no fire-spouting dragon as Dr Endicott would have him. He was just not the type to describe the horrors of bacteriological warfare as 'heartening.' In fact he had assured me that he had said no such thing." The report concluded, correctly, that Suffield was a testing ground and not a manufacturing plant, and Endicott's statements were summed up by Solandt as a fishing expedition for propaganda purposes. The article absolved Solandt of the charge, and his quiet disposition likely aided in the assessment, though this incident might have been the spark of Solandt's distrust and dislike of future prime minister John Diefenbaker.[85]

That same month Solandt had to deal with the press over UFO hysteria. For three years he and C.J. Mackenzie had quietly ignored the rising tide of interest in "flying saucers" in Canada and the United States, but by 1952, so many reports had been made and so much interest had been generated that they were forced to respond. In Canada, the root cause was alleged UFO sightings in North Bay, Ontario, by civilians and soldiers in the region. Solandt told the press that he was "as mystified as anyone else" regarding the sightings of odd lights in the sky and said he and senior scientists were "keeping open minds on the questions."[86] This included the creation of the aptly titled DRB "Committee Set Up to Deal with 'Flying Saucers' Sightings" under the chairmanship of astrophysicist Doctor P.W. Millman of the Dominion Observatory, and would be given the name Project Magnet.[87] A full examination of this trivial but publicly popular part of the DRB's existence cannot be dealt with here.[88] Suffice to say that no evidence of UFOs was ever found.[89] One can only imagine the pragmatic Solandt politely dealing with this bizarre episode before focusing his mind on the serious matters of modern warfare. He likely enjoyed the diversion, perhaps wondering what Wally Goforth would have made of it.

Conclusion

Ten years of DRB research at the establishments had provided the Canadian Armed Forces with a vast array of tools, analysis, and support to prepare them for the nature of modern warfare, from hydrofoils to Arctic navigation computers to improved food rations. Omond

Solandt managed this multi-piece orchestra as few others could. His ability to not only retain knowledge but also decipher value across the spectrum of the DRB's research projects was impressive. Under his watch, Canada entered the arena of guided missiles and conventional weapons design with positive results such as Velvet Glove and the Heller anti-tank rocket. He also maintained Canada's leading role in CW and BW field test research, providing a body of knowledge on this type of warfare and contributing to allied strength in the field of research. He did so as his name and face became synonymous with all facets of Canada's defence research program, weathering attacks on his ethics and diversions into the borderlands of science fuelled by public fears of a technologically dangerous world. Through them all, his cool, rational, but positive countenance stilled the waters of controversy so that he could get back to the near-impossible task of being an expert on nearly every facet of the scientific dimensions of modern war. Solandt admitted this was possible only if one accepted one's limitations and maintained one's sense of humour. "It is probably harder to retain a sense of humour than to recognize your own limitations because in such a job these limitations are only too obvious and if one fails to notice them there are always friends to point them out. The outstanding fact that makes the job not only possible but extremely pleasant is the amazing cooperation received from everyone. I have never asked in vain for help from anyone in the Canadian services, in the Canadian government, in universities and industry throughout Canada or in the defence establishments of Britain or the United States."[90] While Solandt admired his organization's capabilities, these were generated largely as a result of a strategic appreciation of Canada's role in alliances with senior powers that guaranteed her security. As a result, Solandt took on with relish one of his least appreciated roles as chairman of the DRB: Canada's chief defence research diplomat. Solandt led the DRB to be a respected and valued partner with Great Britain and the United States, and worked with each ally towards mutual as well as Canadian advantage.

13 Canada's Defence Research Diplomat: Solandt and the DRB's International Connections, 1946–1956

For Solandt, Canada's success in defence research required cooperation and collaboration with senior allies. "The Board was established
primarily to meet the research needs of the Armed Forces. This is done
in part by the establishment of the Board; partly through extra mural
projects in universities and industry; and by following development
in other countries."[1] The CDC agreed. In May 1946, they noted that
while they would not commit Canada to "specific arrangements" on
intergovernmental research in peacetime, there was a mutual understanding that any future war would see Canada siding with Britain and
the United States. As such, "close cooperation" on defence research
between these nations was critical in peacetime.[2]

These informal peacetime relationships were collectively known as
the tripartite ABC (American/British/Canadian) agreements.[3] Each
nation worked with the others in areas of mutual strategic interest, contributing distinct research and sharing the results for mutual benefit.
Solandt's strategy for working with Britain and the United States was
multifaceted. While DRB maintained liaison offices in both London
and Washington, Solandt made defence research diplomacy a personal
obligation of the chairman, maintaining close contact with critical
defence research scientists in both countries.

Naturally, Solandt had much stronger ties with Britain's defence
research establishments. He was an active participant in joint British
and Canadian discussions on defence research through his personal
contacts in two distinct defence research bodies in Britain: the Ministry of National Defence's Defence Research Policy Committee (DRPC),
and the Commonwealth Advisory Committee on Defence Science
(CACDS). Solandt's personal relationships with key British scientists,

including Tizard, Cockcroft, and Owen Wansbrough-Jones, afforded Canada a valuable point of connection and collaboration with British science. The British perspective that emerges from this period shows Solandt and the DRB were held in high regard individually and as chief British allies.

Solandt was also cognizant of the need for even stronger connections with the United States. He established connections with the U.S. Research and Development Board (RDB) through its president Vannevar Bush and his successors, Karl Compton and W.G. Whitman, and personally introduced the DRB and its mission to Pentagon officials. All three nations worked hard on coordinating research on areas of mutual interest and concern. Solandt's role in tripartite work on biological weapons, a field that, as seen in chapter 12, provides an example of Solandt's personal interest and input into one of the DRB's main international concerns.

While the main goals of Solandt's efforts in defence research diplomacy were information and personnel exchanges between allies, they also served a domestic function. Solandt recalled how prime minister W.L. Mackenzie King had included himself in the famous "Tizard Mission" of 1940, when Sir Henry stopped by Ottawa before heading to Washington to show the United States the cavity magnetron radar.[4] King interposed and called Roosevelt to inform him of the importance of Tizard's mission, taking great satisfaction in this self-aggrandizing action in allied technology affairs. Solandt made sure that Bush, Tizard, Cockroft, and others visited Ottawa regularly, where they told King and his successor, Louis St Laurent, how valuable and important the DRB was. Such external validation made the DRB shine in Ottawa against any internal attacks about defence research in peacetime, and boosted Solandt's reputation as a peer to Nobel laureates.[5] Canada was lucky to have him.

The DRB Liaisons and Information Networks

Solandt established the DRB liaison members in London and Washington with full staffs in 1947. According to Solandt, all the services, and especially the RCAF, communicated to the United Kingdom and the United States on defence research problems through uniformed channels only. The DRB's Washington member, A.L. Wright, however, was attached to the Canadian ambassador's staff as well as the Canadian Joint Staff in Washington. Solandt personally introduced him to Canada's

U.S. ambassadors during his tenure, Hume Wrong and A.D.P. Heeney. He insisted to the diplomats that Wright and his successors were part of their staff and would cooperate in everything asked, including general science work the NRC members in Washington were loath to touch. There would be no secrets between him and the ambassador, though some things were too secret for the embassy staff.

Solandt held these staff members in high regard. If they did not have an answer for the ambassador, they knew how to get it. The "liaisons" also provided access through the embassy to key scientific and military research personnel that the other services could never reach through military channels alone. This included direct access to Carroll L. Wilson, the first general manager of the Atomic Energy Commission.[6] However, this information was never sequestered for DRB use alone and was always shared with the services. Solandt had no interest in creating his own "empire."[7]

Solandt had even greater appreciation for the structure of power in the U.K. defence research networks. On 20 August 1946, he told the COSC that the DRB liaison with the United Kingdom was critical and argued against the DRB and service personnel being attached to the high commissioner in London, as had been previously suggested. The high commissioner's office was not directly attached to the U.K. chiefs of staff and their important subcommittees, "and it was felt that the Services and Defence Research could not afford to be without such a direct link." The COSC agreed to this with certain conditions, and the DRB liaison HQ would be with the Canadian Joint Staff mission in London.[8]

The DRB's long-time liaison member in Britain was Colonel Roy Carrie, a Great War veteran and colleague of General McNaughton. Carrie represented DRB on committees and panels of mutual interest under the auspices of the DRPC. Owen Wansbrough-Jones, then scientific advisor for the British Army Council and an old colleague of Solandt's from Cambridge and the war, had high regard for Carrie, whose role as liaison was the most important in the Commonwealth. "His contribution to the general progress on the lines we all want has been outstanding."[9]

The British Connection

Solandt kept an active correspondence with Sir Henry Tizard, chairman of the Defence Research Policy Committee (DRPC), whom he had known during the war. These exchanges helped harmonize alliance cooperation and to avoid duplication of effort in fields such as aeronautics and petroleum research. They also allowed them to correct misunderstandings about defence research relations.[10] One such misreading regarded Canada's research and adoption of the 30 mm air-to-air combat gun for the CF-100 in late 1947, early 1948. Rumours had spread that Canada was simply inheriting a weapon the British had abandoned. This rumour was untrue and Solandt worked with Tizard to correct the misunderstanding.[11]

Tizard's first official visit to the DRB was in the fall of 1947. He was accompanied by senior DRPC members Ben Lockspeiser, SA to the Directorate of Scientific and Industrial Research; Doctor J.A. Carroll, deputy controller R&D (Admiralty); and Owen Wansbrough-Jones, SA to the Army Council. For the British it was a fact-finding mission on the DRB. Every SA in DRPC wanted information on Canada's defence research structure and resources, especially in aeronautics. Tizard also initiated discussions of combined initiatives, which included British concern over the effects of strategic use of atomic weapons and the need for dispersal of defence research capabilities, since Britain would be a battleground in any future war. They also discussed strategic aspects of biological warfare and atomic energy, collaboration between Canada and United States on Arctic warfare, enlarging OR work in Canada, and relations with the United States. The group came away suitably impressed and keen to see how Solandt would develop the organization.[12]

Tizard's next trip came in the summer of 1951. Vice-Admiral Sir Charles Daniel, controller of the Navy, accompanied Tizard and wrote a substantial report on both Canadian and U.S. defence research establishments. Daniel reported that the DRB, unlike the DRPC, had substantial funds at their own disposal and appeared to concentrate their efforts on projects of value. "I was most impressed in my talks with Dr Solandt and other members of the Board with the single-mindedness with which they are keeping their eyes on the main objective, and resisting all blandishments to take on projects which are already being adequately covered in the United Kingdom and the United States." Credit went to Solandt. Daniel warned Tizard that the Canadian senior services felt that the DRB focused on academic interest and not on pressing projects of vital importance, but he himself did not think this was the case.[13]

Daniel felt the DRB's naval projects were significant and had not duplicated British efforts. "This is a tribute to the Defence Research

Board's policy, though in the case of one project, DATAR,[14] I believe the work has been largely a naval affair and the Defence Research Board have only recently taken any interest."[15] Daniel was impressed by most of the DRB's naval research,[16] most notably the DATAR system, which he felt might be used by the RN.[17]

Every member of the Tizard group largely had praise and ideas for future collaboration. Tizard's aide wrote a paper on issues of common interest. Wansbrough-Jones wanted increased GB, or sarin, "nerve gas" research work, as well as increased attention to the development process of picrite. He found Canada's BW work was a valuable part of Britain's BW program, especially the field trials, and that the exchange of members at Porton Down to work at Suffield had been first rate.[18]

Solandt and The British Defence Research Policy Committee

Solandt, never one for sitting still, often visited U.K. establishments and attended military exercises,[19] but was most active at meetings of the DRPC. He first attended on 3 May 1949, but only for the discussion on the future of the Commonwealth Advisory Committee on Defence Science (CACDS). On the basis of his experience with the DRB symposiums, Solandt argued it would never be possible to keep the conference to general questions, and sooner or later top secret work would be discussed, though he acknowledged the inclusion of Pakistan and India in the Commonwealth made this difficult. Still, if top secret research was off the agenda, the conference would have limited value.[20]

During the Korean War, Solandt attended a more critical meeting of the DRPC on 4 September 1951,[21] to discuss each nation's senior defence research projects in almost all fields, and explore areas for cooperation. Solandt and his old colleague Tony Sargeaunt reported on successes in cold-weather testing in Canada and Britain. They agreed that operational work at Fort Churchill should not be reduced, but on purely technical matters, as much cold-chamber work as possible should be done in the United Kingdom. Solandt said "that it was now proposed to extend the establishment at Fort Churchill for Canada's own needs. It would be useful therefore to know what kind of tests on UK equipment would be required over a period of years." Solandt would receive a general policy statement to this effect. For Sargeaunt, Korea showed that insufficient attention had been paid to the medical dimension of extreme cold. Realism in field trials and exercises was now critical. Solandt concurred.[22]

The outbreak of foot-and-mouth disease in the United Kingdom in 1950 was also cause for concern. British, Canadian, and American research and development on this disease was harmonized through the tripartite agreements on BW. The United States covered crops research, Canada focused on rinderpest, and the United Kingdom addressed foot-and-mouth disease, "research on which was forbidden in the United States and Canada." Solandt said both defence and agricultural authorities in Canada supported the research program and that Canada's work in rinderpest was going strong.[23] They also discussed the development of manufacturing capacity for nerve gas. The DRPC Sub Committee on the subject was concerned about construction of facilities to produce fifty tons a week. Solandt said, "Canada had considered the possibility of undertaking nerve gas production, but had decided that, in view of the shortage both in the United Kingdom and the United States of the raw materials and chemical intermediates, it would be wisest to concentrate on increasing the capacity for these. He agreed to send a note to the JWPC on the Canadian proposals."[24]

Solandt attended his last DRPC meeting on 8 July 1952, a tripartite affair that focused on Solandt and U.S. RDB president W.G. Whitman.[25] Whitman reviewed the current U.S. guided-weapons program. He discussed the need to make decisions for the following year on which weapons should be pushed towards production, which weapons development should be slowed, and the relative lack of prioritizing that had occurred in RDB. The current yearly cost for R&D was $250 million and would be $400 million the following year.[26] The Weapons Systems Evaluation Group (WSEG) could not be wasted with operational analysis of all types of guided weapons, so RDB's chief job was now "selective emphasis" on key weapons. According to Whitman, priority assessment of weapons was conducted by supporting systems in the development phase that showed the most promise. One can imagine Solandt bristling at the seemingly inefficient approach to using vast funds for R&D. Whitman himself was adamant that it had to change.[27]

North American continental defence was also discussed. Whitman noted that coordination of U.S. services interest was done by the RDB's Guided Missile Committee, but "the study of the total defence system of the United States was really a combined US/Canadian problem. Project Charles had been followed up by Project Lincoln, which so far was mainly concerned with hardware but would also include operational evaluations of the air defence problem. There was as yet no real knowledge of how expensive a guided weapons defence would be."[28]

Solandt worried that the main preoccupations thus far were early warning radar and detection, "and that little thought had been given to the actual destruction of enemy aircraft, especially with future defence systems." In discussing the economics of continental defence, Solandt emphasized the cost of radar installations in Canada, including the ongoing expenditure for infrastructure, staff, and maintenance.[29] Compared to the United Kingdom, Canada had unique economic factors in air defence, and a very different strategic outlook on possible targets. "Targets in Canada were so far apart," Solandt noted, "that an increase of range from, say, 20 to 50 miles was of no real importance." He reported on the successful test flights of the Velvet Glove missile that would "fill a gap and to make use of the Hughes APG 40 radar, and [be] used by the CF-100." He also reported that fuse work was behind schedule and they had not yet obtained the required Sparrow motors from the United States.[30]

Solandt then reported on DRB successes. The board had stimulated industrial efforts in electronics research so that it now had a capacity several times greater than the wartime peak. The DRB had progressed on early warning and counter-bombardment radar. CARDE had also developed the Heller, which generated British interest. Canada had also completed her part on the British Blue Study project, an automatic blind-bombing system. Mister S.S.C. Mitchell of the Ministry of Supply noted, "The Canadian design for the beacon container had been adopted and was extremely successful. Sir Ralph Cochrange agreed, and said it was a very simple conception that could not be bettered."[31]

Both Solandt and Wansbrough-Jones felt that the DRPC under Tizard was an essential asset in doing good work for both U.K. and Canadian defence research.[32] Tizard retired in 1952, no doubt influenced by the return of a Churchill government, and with it Tizard's old colleague and nemesis, physicist Lord Cherwell. Tizard was replaced by Solandt's old boss from AORG, Sir John Cockcroft. DRPC's import and influence in the new Churchill government soon began to wane, despite Cockcroft's ability and reputation, both of which Solandt held in high regard.[33]

The Commonwealth Advisory Committee on Defence Science

Solandt also pursued Canada's participation in the Commonwealth Advisory Committee on Defence Science (CACDS). A British creation, CACDS served as a focal point on mutual interests in defence science for the United Kingdom and senior dominions. It consisted of general meetings, a working group in London, and conferences held each year in different countries. The first informal conference was held in London in 1946. Given his anti-imperial biases, Canadian prime minister W.L. Mackenzie King disliked the conferences and organization, but Solandt was adamant on its value to Canada.[34] Soon Canada became an integral member of the organization, hosting the conferences in 1951. Solandt led each Canadian delegation and was initially considered for the post of chairman. He declined, because he had an immense workload at the DRB, and the post went to Tizard – an indication of the esteem in which Solandt was held in Britain.[35] CACDS was to be a supportive mechanism in Commonwealth defence research relations, but, as all parties agreed, it was never the primary or sole means of cooperation, which would be bilateral.[36]

For Solandt, the conferences were of secondary importance because the papers were of limited security grade. However, the meetings of the CACDS bolstered coordination of effort and the sharing of research. At the 4 November 1947 meeting, Solandt discussed Canadian defence research efforts with equal skill, including cold-weather testing and its relation to guided missiles; soil analysis; reduction of bulk in essential foods; physiological studies on clothing and general stores; and more. Tizard again emphasized the need for defence research dispersal across the Commonwealth should Britain become a battleground for atomic or other strategic bombs.[37] Solandt and Tizard debated radar air defence needs and high-risk targets in each country, with Solandt emphasizing the need to keep the United States firmly involved in all Canadian continental defence schemes.[38] Solandt was also keen to discuss work on synthetic materials and civil defence, as part of defence preparation, should war occur.[39] The members were most impressed with Canada's work on psychological research problems of defence.[40]

Future meetings continued this multifaceted approach, though with a central research theme. In 1949, this was civil defence and how best to research its needs.[41] The meeting on 4 July 1950, in the heat of the outbreak of the Korean War, was intensive and extensive. The meeting discussed the purpose and principles of operational research,[42] air defence and Canada's collaborative efforts with the United States,[43] surveys of technical trained personnel,[44] physiological and psychological dimensions of equipment design and clothing,[45] scientific intelligence,[46] food research,[47] and oil research.[48] Solandt chaired the section on substitutes for strategic materials in short supply, where the DRB member, Doctor

G.S. Field, discussed Canada's need for light metals such as lithium and bauxite, which had to be imported, and the possible need to find alternatives. On this issue Solandt identified organizational weakness as the greatest problem and believed that they first needed to define what a strategic shortage meant.[49]

After the general meeting, Tizard held a special closed session for delegation heads alone on sharing the research and development load throughout the Commonwealth. Since the United Kingdom spent more per capita on research than any Commonwealth nation or even the United States, Tizard wanted better and more efficient cooperation from the dominions. Doctor Marsden said New Zealand agreed, if the United Kingdom took New Zealand into its confidence. The South African representative, Doctor Biesheuvel, said defence research in his country had never considered itself in these terms, since it was so new. Solandt said countries had to be of significant economic and industrial strength before they could effectively devote efforts to defence research. "It might well be that Canada was now reaching the stage of development where her effort on defence research would be considerably increased, and indeed during the last few years it had developed fairly rapidly. On the other hand Canada felt that research should be an indigenous and natural growth, stemming from the needs of the country concerned and should not be too much influenced by research tasks proposed from elsewhere. It was particularly important to encourage initiatives in the countries of the commonwealth."[50] For Solandt, Canada would not be a British laboratory, but a British ally.[51] Tizard agreed, but wanted defence research of the Commonwealth treated as a whole. Solandt felt Canadian scientists often "drifted" to the United States, and thought was needed on how best to attract them to the Commonwealth instead.[52]

In a letter to Tizard after this conference, Solandt said that he found that the meeting was most useful, especially as regarded India and Pakistan. He had talked about this with Claxton and he indicated willingness to assist India and Pakistan where they could.[53]

Solandt and Wansbrough-Jones

Solandt's other touchstone on U.K. defence research was his old friend from Trinity Hall, chemist Brigadier Owen Wansbrough-Jones, who served as SA to the Army Council (1945–51), principal director of scientific research (defence) (1951–3), then chief scientist (1953–9) at the

Ministry of Supply. Wansbrough-Jones was held in high regard by the Americans and Canadians (Foulkes was an admirer) for his war and post-war work on BW and CW. However, he himself found BW both frightening and disheartening because, as he told Solandt, a "certain amount of instinct suggests to me that it cannot quite have the potentials it seems to have, and yet I can find nothing in the evidence to suggest that they are any less." So he had to pursue all the work that was worth doing. Solandt shared this view.[54]

They maintained personal contact throughout Solandt's time at the DRB. Their discussions ranged across research interests of both nations, including the protection of tank crews against cold,[55] standardization of the British .280 small arm round,[56] and in the field of BW. Here, we find one of the actual controversial ends of Solandt's DRB work, albeit an unclear one.

In June 1951, Solandt had discussed with Wansbrough-Jones and "Sachs"[57] a project he wanted kept secret from the Americans that involved the transport of BW material to North America. According to Wansbrough-Jones, he and Sachs had found it difficult, if not impossible, to do so through the normal BW authorities, since they were obligated to share such information through the tripartite agreements. He also expressed alarm at being implicated in unauthorized transactions regarding these materials, whose importation to North America was forbidden by international agreement.[58] Solandt's original letter has not been found, and current research on Britain's BW program has not explained this incident, though it occurred during a general downturn in American interest in BW as the result of difficulties in production.[59] However, given the DRB's work on rinderpest, it is likely that Solandt wanted to increase his organization's work on foot-and-mouth disease, which was then illegal in North America.[60]

In 1956, on the eve of Solandt's retirement from the DRB, Wansbrough-Jones reflected that the DRB never looked back and was now "a major pillar of the research endeavour" between both nations. He also admitted that British support for the DRB was never altruistic. Helping the DRB meant helping the United Kingdom, as strains on the British economy weakened their defence research resources. He also mentioned that Britain considered taking over Canada's work on the troubled CF-105 A.V. Roe Arrow, which they hoped would bolster U.K. R&D resources. If this happened, he told Solandt that it "would be rather a fitting end to your own ten years occupancy of your chair."[61] Nothing came of this proposal.

Solandt and the Americans

Despite his extensive experience and relationships with British defence scientists, Solandt worked hard to cultivate positive relationships with Washington. He had met Vannevar Bush during his 1943 visit to the Pentagon as deputy superintendent of AORG and reintroduced himself as director-general of the DRB when the two met in Washington in April 1946.[62] Bush found Solandt a young but kindred spirit in the unique challenges of directing government science, and wanted to exchange ideas and information with him regularly. They included Tizard as part of this informal network. For Solandt, Bush's wartime experience as head of the Office of Scientific Research and Development (OSRD) was an invaluable source of guidance for his new position. During the DRB's initial year, Solandt canvassed Bush for his view on "the general problem of the balance between civilian and military control of research and also on the relationship between groups working on weapon research and the industry that will produce these weapons."[63] Bush facilitated Solandt's introduction to key members of his staff, like Doctor Lloyd Berkner, the RDB member concerned with continental defence. Bush indicated that Berkner was the key player to talk to in order to understand "the somewhat unusual way in which this affair is operating." Bush's displeasure with the RDB's lack of influence in this area is painfully clear.[64] Unfortunately, Allan A. Needell's otherwise fine biography of Berkner is almost devoid of Canada's participation in continental defence, and makes no mention of Solandt, the DRB, or the McGill Fence in relation to Berkner's work on the DEW Line.[65] But this relationship existed. In April and May 1947, Solandt and Bush discussed collaboration on R&D in early warning radar equipment.[66]

Solandt introduced himself and the DRB to the Pentagon's key defence research players on 3 October 1947. Solandt discussed the DRB's origins, purpose, and vision, and his own view of how defence research was best accomplished. The DRB "differs from anything you have in your organization because we have concentrated on service functions which here are separate and distinct. The Board has actual financial and executive control of the various research and experimental establishments." During the question period, Solandt discussed the importance of industry for the production phase of R&D. He also stated the DRB's only likely role in tactics or strategy would be through the use of operational research, but OR would be working for the services, not above them. "We feel you cannot separate operational research from the services. It is very much the application of the methods of science to immediate operational problems. We feel the great strength of your operational research and experience was due to the freshness of approach. We would like to avoid, therefore, a large permanent staff, but would like to get scientists with special qualifications in their particular field and borrow them for two or three years, returning them to their own fields, and then bring in others." He admitted that Canada's work on scientific intelligence (SI) was embryonic at best but growing, and made it clear that fieldwork in SI was not being pursued. "We feel we have no reason to go into foreign fields; that it is wiser for us to stay out of them."[67]

In March 1948, Solandt invited Bush to return the favour and visit the DRB to meet senior members of the Canadian government and services.[68] Bush was keen to visit Canada and the DRB "for the very important interrelationships lead me to desire very much to know better some of the men with whom you have close relations and to see something of what you are now doing."[69] On 19 March 1948 Bush attended the DRB meeting and gave a short but memorable lecture to selected employees and senior officers with secret or top secret clearance. Bush, normally a dynamic and excited speaker, proceeded with a mind-numbing speech about inconsequential research to the stunned crowd. He then leaned against the lectern and complained that his speech had been all that remained after being screened by four intelligence agencies in two countries. Now, though, he would talk off the record and if anyone mentioned what he said he would deny having said it. He then began an hour-long speech on current secret U.S. R&D work in defence, to everyone's edification and Solandt's relief.[70]

On Bush's invitation, Solandt visited Washington in July 1948, though what he worked on is unknown. Bush thanked him for coming, saying, "I feel that there is no substitute for these personal exchanges for achieving a clear insight into matters of common interest." Solandt agreed and attempted to maintain this personal touch with Bush's successors Karl Compton and Walter Whitman, though none were a match for Bush, the pre-eminent government science tsar in the United States.[71] Solandt had high regard for U.S. scientists, organizations, and efforts, especially in atomic affairs and OR. But he also felt that the United States ignored British developments and thus failed to concentrate their efforts on fewer fields, an opinion he believed Walter Whitman was starting to share. Whitman, Solandt believed, had increased RDB's effectiveness. In 1952 he was invited to join Solandt on a trip to England to improve tripartite relationships in defence research (see above). However, Solandt was embarrassed that Whitman expected Solandt to do all the arrangements for the suggested trip, since he did not want to be seen as running the U.K. show.[72]

Tripartite Work on Biological Warfare, 1945–1956

All three nations maintained extensive defence research liaisons through a series of tripartite agreements that had originated in wartime work on chemical and biological weapons, as well as propellant and armaments. With the exception of the Combined Policy Committee on tripartite atomic weapons and energy research, Solandt largely left Davies to represent Canada on tripartite panels and conference. However, he maintained an active interest and influence on most of the programs the DRB participated in with allies through the ABC agreements, including biological warfare.

By late 1948, General Foulkes wanted the board to work against public fears and discuss the problems inherent with WMDs.[73] Solandt agreed and took a more active involvement in BW policy and linked it to allied research in the field. In December, he informed the board that the CACDS meetings had made clear improvements in BW research. Coordination of tripartite research in CW was also facilitated by visits of key allied personnel, including Doctor F.J. Wilkins, principal director of scientific research (defence) for the Ministry of Supply, who toured Canadian and U.S. CW, BW, and armament facilities in Canada and the United States. Solandt participated in these tripartite meetings in October 1947.[74]

The British post-war BW program was closely integrated with the U.S. and Canadian programs and they hoped to hold field trials by 1948. The DRB sent University of British Columbia professor Doctor Alex Wood, an expert on animal research, to assist the U.K. efforts, and he earned high praise.[75] As such, Canada was invited to participate in the first tripartite sea trials of biological weapons, known as Operation Harness.[76] Wood was specifically requested, and Solandt attended the trial himself.[77]

Operation Harness was executed off the coast of Parnham Sound, Antigua, from November 1948 to January 1949. It was a costly, difficult, and highly secret affair that suffered from poor security and accidents that led to public interest and investigation.[78] In February 1949, Solandt reported that journalists in the United Kingdom had discovered the true purpose of the trials. Solandt concurred with the U.K. BW sub-committee on aborting the original cover plan of conducting CW trials and suggested "that a policy of refusing to answer press enquires be adopted." It was public knowledge that Canada was keeping abreast of developments in biological warfare and it would be of little value to attempt to conceal this fact. "However no information beyond this point should be made public." Foulkes agreed, fearing that any tripartite BW talks might prejudice his work in the North Atlantic pact discussions.[79]

Still the research was deemed of sufficient value to foster more field trials on land and sea on the effects of various forms of *Brucella*, tularaemia, and anthrax.[80] Solandt, however, was not impressed with Harness. The whole affair had been a costly and unsatisfactory way of carrying out the trial, and the same result could have been obtained by "a shore base using small ships." Doctor Wood's work on the operation, though, had earned high praise.[81]

Conclusion

Solandt always felt that international cooperation in research produced the best results. His experience as a Canadian scientist working for the British government during the Second World War validated this thesis. For six years he had worked alongside British, Welsh, Scotch-Irish, South African, and American scientists, learning from them and providing for them throughout his many research and science-management jobs. It is no surprise that upon returning to Canada he never wavered from the importance of working with the best scientists in allied countries for mutual interest. Indeed, one of Solandt's greatest values to the Canadian government was the confidence he had maintained with the British and established with Americans, and how this trust was transformed into access to information of mutual benefit.

Solandt had an insider's appreciation of British defence research, and his contacts with senior officials like Tizard, Cockcroft, and Wansbrough-Jones at DRPC and through the CACDS allowed for close coordination and sharing of information on defence research projects. Solandt did his best to maintain close cooperation with U.S. counterparts Bush, Compton, and Whitman, though the connection was not as strong as with his old wartime cronies in Britain. Still, all three nations worked hard to harmonize their defence research efforts and support each other's projects through tripartite agreements. These allowed Solandt to be an active observer of research critical to the

West, including in the dangerous field of biological warfare. However, Solandt's most prominent role among allies was in the field of atomic weapons. Since his days at Hiroshima and Nagasaki, Solandt was among a handful of Canadians actively investigating the nature of and defence against atomic weapons. As chairman of DRB, he struggled hard for ten years to make his non-nuclear nation relevant in the field of nuclear affairs. It was a frustrating struggle of success and failures against the shifting backdrop of growing nuclear arsenals and greater destructive yields, forcing the ever-pragmatic Solandt to slowly adjust his view on the nature of war in the thermonuclear age.

14 Atomic Realist: Omond Solandt and Nuclear Weapons, 1945–1955

Solandt was Canada's senior government expert on atomic weapons as the world entered the atomic age. The U.S. atomic monopoly ended in 1949, when the USSR detonated its first atomic device. In 1952, Britain had become the third atomic power. Over the next three years, both the United States and the USSR detonated hydrogen weapons, including thermonuclear bombs, with destructive yields in the ten to fifty– megaton- (ten to fifty millions of tons of TNT) range. President Dwight D. Eisenhower's "New Look" strategy emphasized an increased reliance on nuclear weapon to counter Soviet strength in conventional arms without massive rearmament. Nuclear weapons had been part of NATO's collective defence policy since the alliance's inception, and by 1953 the United States had deployed tactical nuclear weapons to bolster the alliance's defences, though it would not be until 1957 that the alliance fully embraced the "Massive Retaliation" deterrence strategy embedded in the "New Look."[1]

The atomic age dominated Solandt's career in government science. He had walked the atomic battlefields of Hiroshima and Nagasaki in November 1945, convinced that the awesome power of these weapons had to be treated rationally. While chairman of the DRB, he publicly stated his work in Japan had provided him a "realistic perspective concerning the importance of the atomic bomb."[2] Throughout this period, Solandt provided the Canadian government and Armed Forces with advice and strategies on how best to prepare for war in the nuclear world. As he stated in 1958, "Although Canada did not embark on an atomic weapons programme, and indeed actively supported proposals for atomic disarmament, nonetheless the Canadian Services had to be trained and equipped to fight in atomic wars should the need ever arise. Helping in this task was the central role of the Defence Research Board."[3]

There was one catch. He was to do this without the Canadian Armed Forces having nuclear weapons. It was a paradoxical and difficult task, but Solandt did it with tireless detective work. He and the DRB contributed to Canada's strategy regarding nuclear weapons in two spheres: preparation and information. Solandt firmly believed in the value of civil defence and supported it throughout his chairmanship. But his greatest role in nuclear affairs was information gathering. For a decade, Solandt harvested as much data on nuclear weapons and energy as possible so that the Canadian government and Armed Forces, to the best of their ability, could make informed decisions on the actual nature of these weapons. He focused the DRB's efforts where he could on nuclear research, including maintaining an often strained relationship with the Atomic Energy Board and Chalk River nuclear reactor.

But the majority of Solandt's nuclear efforts rested on gaining access to information on atomic weapons from Britain and the United States, through both official and personal channels. Solandt was Canada's scientific representative on the tripartite Combined Policy Committee on nuclear affairs, but when access to U.S. atomic secrets proved increasingly difficult, he pursued a stronger relationship with the British. As a member of Sir William Penney's entourage at Monte Bello, Solandt participated in the detonation of Britain's first atomic device in 1952.

These efforts provided Canada with unique insights into atomic affairs, but they never satisfied Solandt. Since his days with Charles Best, Solandt was convinced that meticulous and exhaustive research of any problem was fundamental to accurate and thus valuable analysis. By 1954, as Canada's allies and enemies entered the thermonuclear age, Solandt believed Canada could not rely on what limited information it could cajole out of Chalk River or Britain or the United States. Canada required its own nuclear weapons if it was to fully prepare for war in the nuclear age.

Civil Defence, 1946–1955

The DRB's policy discussion on civil defence began in 1946. Coordination between DND, the Ministry of Health and Welfare, and other agencies was critical. Minister of national defence Brooke Claxton, however, did not want to generate public interest in this arena. Fear and mystery regarding atomic weapons was still pervasive.

Solandt was a firm believer in the effectiveness of civil defence. His experience in London during the Blitz, in working on defences against the V1 and V2 rocket, and his casualty research in the atomic battleground of Japan had convinced him of its value against both HE and atomic bombs. In December 1947, he told the board that pre-emptive work on design and construction concerns, such as advice to hospitals and industry, was critical, as was locating a suitable architect or engineer to coordinate these efforts. Foulkes wanted a comprehensive plan for a civil defence organization before anything else, and that should wait until "a study of the new weapons and other considerations that exist at the present" had been done. Solandt had not suggested a single person would do the whole planning, as Foulkes implied, but that someone with these qualifications was needed at once to try to incorporate in new buildings "structural features that would mitigate the blast effects of atomic explosives."[4]

Foulkes shared Claxton's concerns and worried about alarmist statements in the media about there being no real defence against atomic, biological, and chemical weapons. He also feared a growing defeatist attitude in Canada. The public needed a person of authority to inform them that "steps are being taken to protect this country in the case of another war." Solandt took the job. His public addresses on atomic weapons and modern war were endorsed by the board as a valuable part of the DRB's work, and often included discussion of civil defence.[5]

By 1948 Solandt and Doctor Cameron, president of the Nova Scotia Technical College,[6] had begun work on research concerns of civil defence.[7] Solandt submitted their proposal on civil defence to the CDC on 15 April 1948. He noted, "Civil defence was interpreted to include all those defensive measures that should be taken by or on behalf of the civil population to ensure that, under attack, the will to resist was maintained and the economic and social organization of the community functioned effectively. The measures fell into two parts – protective and remedial. Preparation for both these involved a number of Federal government agencies and close liaison with Provincial authorities."[8] The Cabinet Defence Committee agreed to create a civil defence organization. By December, Claxton had selected Major-General F.F. Worthington as the Ministry of National Defence's civil defence coordinator. Mister W.J. MacCallus, the DRB director of research personnel, was lent to General Worthington to organize civil defence work. The DRB handled administrative arrangements for Worthington and provided experts to Worthington's organization.[9] Solandt again pressed the board on the importance of civil defence. Even a modest research program would yield large dividends in case of war and also keep Canada abreast of allied work in the field. It was the DRB policy to assist Worthington in solving research problems as they arose. Standardization of fire fighting equipment was one of these problems and the Board agreed to work on it.[10] Solandt and others also attended the U.K. civil defence trial Exercise Britannia, 23–8 May 1949, to assess Britain's research on civil defence, though their reports on the event have not been found.[11]

Civil defence research remained a DRB priority throughout the Korean War and rearmament campaign of the early 1950s. For Solandt, it became a frustrating topic. From 1949 to 1950, the DRB had created a Civil Defence section to aid the civil defence coordinator. Solandt backed the research, though he felt it had not progressed in step with the actual requirements of civil defence. A Civil Defence School was to be established, and the DRB's new section would work alongside it.[12] For 1950–1, Solandt again was frustrated with the lack of progress with medical and fire-fighting fields. But a development program for radiation-detecting instruments was now initiated, including a chemical method of measuring radiation intensity.[13] By 1953, the ratio of effort between active (air defence) and passive (civil defence) defence was out of balance, and Solandt made this clear to the COSC. Close to $2 billion had been spent on active defence and only $6 million on passive. He pushed for civil defence to be made a priority. Services should also be made responsible for bomb disposal, he argued, and reports on civil defence should reflect this. However, little appears to have come from these efforts while Solandt was chairman, though they laid the groundwork for the renewed interest in civil defence during the Diefenbaker administration.[14]

On 12 March 1954, Foulkes outlined the strategic situation for Major-General Worthington and the COSC. Reassessing the Soviet threat required reassessment of policies being developed by the chiefs of staff, which were to prevent destruction or serious disruption of war potential of North America, and to prevent the USSR from overrunning Western Europe without causing serious damage to Western industrial potential. Worthington believed that in light of new nuclear weapons, civil defence concepts were broadening. If a civil defence organization was in the target area, it would be rendered useless if hit by a thermonuclear weapon. As such, civil defence required resources outside of a target area. But the biggest weakness now was apathy from provincial and municipal authorities. U.S. policies now involved survival and mass population movements. Canada had not kept up.[15]

Solandt reiterated his concerns of 1952. The present scale of active to passive defence was based on civil defence requirements for 1946–7 with no revision. The "ratio of emphasis seemed to be out of phase with the need."[16] A year later, he challenged chief of the general staff Lieutenant-General Guy Simonds's own civil defence assessment, which did not emphasize its importance in war. If the theory that a major threat was from thermonuclear weapons was accepted, Solandt said, civil defence now had all of the original problems as well as new ones. A central civil defence organization needed to maintain firm control and direction in an emergency. Radiation detection equipment was required throughout the country, as well as effective communication and intelligence systems to determine the nature and extent of any thermonuclear explosion. A command system for population movement was also essential. Since hundreds and thousands of people were involved, civil defence "should possibly rate as second to active air defence in the Canadian defence program." Solandt well knew that lack of civil defence measures at Nagasaki accounted for a very substantial number of casualties and this would be only more so with thermonuclear weapons. He told Simonds he considered "the armed forces as an investment in the prevention of war, but civil defence as an investment in survival in the event of war."[17] But since the DRB was responsible only for research on civil defence, the best Solandt could do was make it clear that what Simonds proposed was poor in thought and execution. This likely did not sit well with the general.

The DRB and Chalk River, 1946–1955

However, civil defence was a subsidiary concern of Solandt's main focus on atomic affairs: how best to prepare the Canadian government and military for the realities of atomic warfare. He provided expertise, based on his experience in Japan, to the COSC in 1946,[18] led an ad hoc committee with NRC members to create a report on "probable effect of the atomic bomb on typical Canadian targets" (an unclassified version was used in his addresses on the weapons effects), and was also DND's member on the Cabinet Advisory Panel on Atomic Energy Questions.[19]

But as atomic weapons grew in number and sophistication, Solandt knew he could not rest on his laurels. He needed access to any and all atomic research, at home and among allies. Since 1943, Canada had been intimately involved in atomic affairs and was a junior partner in the development of the atomic weapon through the U.S.-led Manhattan Project.[20] As part of that contribution, Canada had developed an atomic nuclear reactor outside Ottawa near Chalk River, under the original directorship of Solandt's friend Sir John Cockcroft. Canada's nuclear energy program, however, was under the auspices of C.J. Mackenzie, through his presidency of a Crown corporation, the Atomic Energy Control Board (AECB).[21]

Solandt had contacts at Chalk River, including Cockcroft and Doctor W. Lewis, head of research, but "Jack" Mackenzie was his gateway to Canadian atomic research. In 1948 Solandt secured an agreement with Mackenzie for the DRB to have a military liaison committee working with the AECB, "to keep the Armed Services in touch with the developments at Chalk River. The military liaison committee would be modeled after the committee performing a similar function in the United States."[22]

These early attempts did not produce great dividends. The issue of coordination was raised again a year later. In early May 1949, A.D.P. Heeney, secretary to the Cabinet, raised concerns with the COSC about the relationships between DND, AECB, and Chalk River on atomic affairs. The Advisory Panel on Atomic Energy, the body for discussing military, political, and economic policy maters of atomic energy, had no representatives from the COSC. On technical matters involving Chalk River and DND, the DRB should be the coordinating body. Solandt disagreed. All military research matters concerning atomic energy that required liaison with Chalk River could be done by the DRB and no coordination was needed above this level. The COSC agreed. The DRB would coordinate with NRC and AECB on atomic energy matters.[23]

At the following DRB meeting in March, Solandt announced this change to the board.[24] Liaison work, such as meetings with Chalk River personnel, was soon initiated.[25] On 26 March, a board meeting was held on technical aspects of atomic energy, with all interested parties in attendance. Solandt told the board that "in connection with the proposals put forward at this meeting that Dr Lansgstroth [a senior scientists and future superintendent of the Suffield Experimental Station] had recently visited Chalk River to arrange for a proposed field trial programme in cooperation with the Experimental Station, Suffield." Seconding of officers to Chalk River was also pursued.[26]

The relationship with AECB was never easy or smooth. At a March 1949 COSC meeting, General McNaughton, Canada's representative at the United Nations Atomic Energy Commission (UNAEC), had asked the COSC for their opinion on the atomic bomb. In particular, was it "a unique weapon of mass destruction"? No one had a clear and definite answer, so a special committee was formed to assess this topic. Foulkes blamed their inability to answer the question on Canada's deficit of U.S. intelligence and information on atomic weapons. Solandt blamed AECB.[27] Strict regulations had prohibited AECB from giving information to DND and this needed to change. Solandt and a working group attempted to solve this problem, and modifications in regulations were initiated.[28]

But by September 1949, Solandt was tired of talk on coordination. He reported to the board that progress regarding Chalk River would be accelerated by putting a small DRB staff there. "Even though the work at Chalk River is often thought of as being entirely peaceful in character," Solandt said, "it really makes a substantial contribution to Canada's military strength and is, indirectly, an important factor in the trading of defence information with other friendly countries." He concluded that ABC[29] warfare research would continue. "There should be no reduction in priority on research in these fields, but in general the work is going well."[30] The efforts of this staff remain unknown.[31]

Relations between the DRB and Chalk River remained strained. In 1950, the board agreed to Mackenzie's suggestion that he brief the board periodically on atomic energy and Solandt should arrange a special meeting of the board at Chalk River in September or October.[32] Solandt was not pleased with this arrangement at year's end, because Mackenzie rarely made these briefs, and progress on working with Chalk River, and relating their work on defence, had been "disappointingly slow." So Solandt instead placed a DRB liaison officer at Chalk River to accelerate information exchange.[33] Mackenzie's recalcitrance towards DRB might have been more fallout from his confrontation with Solandt in 1947. The relationship between the two organizations never improved. So while Solandt fought with Mackenzie to gain access to Canada's only stock of atomic materials, he turned to his own organization's scientific acumen to gain valuable knowledge on atomic weapons and use it with allies.

The DRB and Atomic Warfare

In September 1948, Solandt moved DRB's atomic policy forward. The first step was education. "The Chairman reported that he felt that the time has now come when the Services should start to train atomic warfare officers and to learn the techniques of atomic warfare just as they had for chemical warfare. Because of the nature of atomic warfare, the Services will need expert assistance in this work from the Defence Research Board."[34] As noted in chapter 12, DRB created the first postwar courses on atomic, biological, and chemical warfare for the Canadian Armed Forces. The Suffield Experimental Station (SES) would also conduct limited trials on radiological and radioactive weapons to gauge this facet of atomic weapons.[35]

The DRB also had great success in monitoring the effects of atomic explosions. This work required cooperation from the multiple parties the DRB harnessed for defence research. After the Soviet atomic detonation of 1949, Solandt initiated a collection system without allied prompting. As he told an audience in 1958,

> We had not made any advance preparation for airborne sampling but had all the facilities to mount a very competent programme extremely quickly. The chemical warfare lab in Ottawa had the equipment and skills to design suitable filters for continuous airborne sampling ... the RCAF were ready and willing to fly on a moment's notice and Canadian scientists were at the forefront of mass spectrography ... what Canada totally lacked was the knowledge to translate the results into useful information. DRB very quickly had a series of flights underway from the West Indies to the Arctic ... [Canada passed them to the Americans and the British and] the AEC and US intelligence discovered that our results were far better than theirs. They subsequently came to depend heavily on our filters ... [T]here is no question that we received an ample return of information from them.[36]

This filter analysis continued during the Korean War as the United States and USSR tested both ground-based and airborne nuclear weapons. The RCAF flew flights over the Northwest Territories where atomic particles were most often found. The DRB designed the filters to be placed on airplane wings to collect atmospheric particles of uranium and plutonium. Radio chemical analysis of the filters was done at Chalk River. Particle analysis of the filters and their particles was done at McMaster University in Hamilton, Ontario, by Doctor Harry Thode, who had constructed his own mass spectrometer and was supported by DRB grants. Thode was convinced that the samples taken and analysed in 1953–4 indicated the Soviets were likely detonating hydrogen weapons.

Later, it was concluded that Thode and McMullen had in fact discovered the first Soviet hydrogen bomb test of 12 August 1953. The DRB received all their reports and Solandt used these data to "trade secrets" with the Pentagon. They also provided Solandt with an advanced view of the changing strategic situation in Soviet nuclear capabilities on par with U.K. research on this area.[37]

In 1951, Solandt reported "that following the recent atomic explosions in Nevada," the Physics Division of NRC (after the superintendent of DRCL and others suggested it) "monitored the local snowfalls for radioactivity." The results were significant and the DRB soon provided $1,000–$2,000 to any universities interested in the research, though it is unclear which universities, besides McMaster, participated.[38] Similarly, by 1951 field dispersal and testing of radioactive materials was carried out at SES to check the theoretical dosages that should be obtained, and to investigate methods of decontamination. All of these "chips" allowed the DRB to negotiate for information in Washington, though the specifics of what Solandt and the DRB gained through this process is largely unknown.[39]

But while the DRB used the information garnered from Chalk River or through its own clever use of scientific assets, the best and most relevant data on atomic weapons came from the nations who had them. Between 1946 and 1954, Solandt secured as much atomic data as he could from both U.S. and British scientists and organizations.

Solandt and the Combined Policy Committee

In 1946, the United States passed the McMahon Act, which made it illegal to share or exchange atomic research and information with foreign powers. Britain and Canada, who had contributed to the Manhattan Project, were now locked out of U.S. atomic weapons research. The Combined Policy Committee (CPC), the wartime body created by the Quebec Agreements of 1943 to coordinate Anglo/American/Canadian atomic weapons research, was suspended. This led to greater cooperation between Britain and Canada (see below), but by 1949, in the wake of the Soviet detonation, U.S. authorities slowly warmed to the idea of working once again with both their wartime allies.[40]

On 13 September 1949, Heeney, undersecretary of state for external affairs, reviewed the status of the Combined Policy Committee (CPC) to the COSC. The first post-war CPC meeting was to be on 20 September 1949, and the United States had set up a subcommittee to consider the strategic and military aspects of atomic energy. C.D. Howe and Mackenzie were to go, and Howe requested that Solandt attend for the Chiefs of Staff Committee.[41] At the 3 October 1949 Special Meeting of the COSC, Solandt reported on the CPC meeting and the post-war history of US/British/Canadian relations on atomic matters. The United States now had a forum for dealing with the United Kingdom and Canada within the parameters of the McMahon Act. In January 1948, a modus vivendi had been developed to continue cooperation and collaboration with the United States and Canada within existing legislation. Exchange of information on raw materials worked well, but the information machinery was cumbersome and the modus vivendi never fully implemented. In late 1948, a group of U.S. authorities, including leading scientists, had held a series of discussions. They wanted to ensure atomic energy development in the United States was not drifting outside the scientific advancement in this field because of the McMahon Act's restrictions. The present modus vivendi would expire at the end of the year, and a tripartite agreement with the United Kingdom and Canada was being assembled.[42]

Solandt then summarized the CPC meeting. U.S. manufacturing facilities could not utilize all the raw materials, so expansion of manufacturing facilities in Britain and Canada made sense and was approved by U.K. and Canadian representatives. The United Kingdom, which had a strategic need for atomic weapons, maintained an active and aggressive atomic research policy. Production would be undertaken utilizing the available supply of raw materials until 1953, and then recycling might begin. There was broad discussion of improved information exchange, though the United States was still enthusiastic about compartmentalization of atomic energy research for security purposes. Canada agreed with the others that Washington would be the centre for atomic energy information exchanges.[43] U.S. dispersal strategy for storage establishments was also discussed.[44] With regard to the feasibility of U.K. bombs being stored in Canada, Solandt said, "The Canadian Government would not be prepared to be host to an entirely British project and would be unlikely to accept anything of this kind that could not be accomplished as a Canadian enterprise."[45]

Air Vice-Marshal Frank Miller reported on military aspects of the meeting. The Americans did not want U.S. bombs stored in the United Kingdom, nor did they want anyone having bombs manufactured outside of North America. They did want closer relations with the United

Kingdom and Canada on launch sites, carriers, and methods of delivery, and problems of target selection, but had not indicated the need for Canadian bases for launching atomic bombs in an emergency.[46]

Despite further meetings in Washington that November, the CPC largely ended as a body of discussion in 1949, with tripartite conferences taking their place. Solandt recalled actually being at a CPC meeting when the arrest of Soviet agent and atomic spy Klaus Fuchs was announced in February 1950. After that, lips were sealed in Washington.[47] With CPC relations at a seeming dead end, direct relations between the DRB and its equivalents in the United States and United Kingdom became more critical. U.S. Atomic Energy Commission general manager Carroll L. Wilson provided Solandt with information directly, but never too much of substantial use.[48] On 3 April 1952, Solandt introduced a major policy question regarding defensive aspects of atomic war to the COSC. "Canada was not receiving any information from the US on the results of the recent atomic bomb tests. It was understood that the US military chiefs were not opposing the release of such information to Canada but that existing US legislation almost entirely precluded exchange of atomic data … Until a freer exchange of atomic information became possible," Solandt said, "Canadian scientists could not do as thorough a job as desired for the armed forces."[49] This was the first step towards Solandt's realization that Canada needed its own atomic weapons.

As the Korean War triggered Canada's rearmament and raised fears of war in Europe, Solandt told the board that the DRB's focus was now on atomic weapons. More facilities and testing in this area would be supported.[50] "Even though Canada is not involved in the production of the atomic bomb there are a number of pressing military needs related to atomic defence which involve research and development. During the past year DRCL has been working on a number of radio-chemical problems, SES has commenced a program of investigations on the field dispersal of radioactive materials, the Physics Department of the University of British Columbia has contained its work on radiation counters which will function in extreme cold, and the Chemistry Department at the University of Saskatchewan has commenced an investigation of chemical detectors for radioactivity."[51]

Vice-Chairman Davies wrote the 1952 annual chairman's report. While discussing it with the board, Solandt made a few observations. There needed to be greater emphasis on weapons of the future which are the common interest of Canada and its allies. These were still BW and CW munitions, guided missiles, and defensive measures against atomic weapons. The DRB's top priority this next year would be to help the services increase their knowledge of, and prepare against, atomic attack.[52]

Solandt would back this new focus by three activities: he would urge the COSC to place greater emphasis on defensive aspects of atomic war; the DRB scientists would provide more aid to the services on atomic matters; and, through the chiefs of staff, Solandt would obtain government approval for a plan "which would enable Canada to send substantial groups (50 or so) of scientific and Service personnel to help on future atomic trials. This would be done on a tripartite basis and would enable a cadre of personnel having a first hand knowledge of atomic explosions to be formed through the Services and DRB." The board gave unanimous support to this proposal.[53] A year later, Solandt reported success in these fields. Discussions with Sir William Penney, Britain's chief atomic weapons scientist, had translated into working proposals with the British. Research on defending against atomic attacks on Canada or the Armed Forces in the field, respirators, protective equipment, radiation detection equipment for services and civil defence had all continued.[54]

Solandt and British Atomic Weapons Research

Perpetually frustrated with the American restrictions on atomic matters, Solandt turned to his British connections, who included Cockcroft, who had returned to England to work on Britain's atomic energy program, and Sir William Penney, future director of Atomic Weapons Research Establishment. Solandt also participated as an observer on the Royal Air Force's 1948 Exercise Pandora, which included planning for the use of atomic weapons.[55]

Britain also looked to Canada as a chief source for plutonium for their atomic weapons and energy research. In 1950 Cockroft increased his request for Canadian plutonium as part of a program on purification and extractions of plutonium metals. This increased to five kilograms halfway through the year. According to the official history of Britain's atomic weapons program, Cockroft's request was "put in the context of greater Canadian interest in the military aspects of atomic energy and a desire for a rapprochement with Britain in this field. This interest could be encouraged not only by the request for plutonium but also by the loan of Canadian scientists to the British weapons research organization and the provision of an atomic weapons test range for trials subsequent to Monte Bello. The Canadians could only provide half the British need, but they promised to do so as long as this did not interfere with their plutonium agreements with the United States."[56] Solandt himself would be among the handful of Canadians to witness Britain's emergence as an atomic power.

Solandt personally participated in Operation Hurricane, Britain's first atomic weapons trials, in Australia, 3 October 1952. As no observers were allowed, Solandt attended as a physician selected by Penne to treat team members for radiation. Canada's role on tripartite research and Solandt's actual experience in atomic weapons for the British government soothed any concerns about "colonial" participation in London, though not in Canberra. Indeed, according to the Australian history, "The Canadians would have access to classified information and the Australians would not. [Australian prime minister Robert Menzies] was not told either that the work to be undertaken by Australian scientists was likely to be of a very low security classification and that they would not be close enough to the actual trial to be able to draw any scientific conclusions from it."[57] The same was not said of Solandt, who, according to the Australian history, had been involved in the British atomic test program from the beginning and was far from being a "passive observer."[58]

At Monte Bello, Solandt witnessed a twenty-four-kiloton atomic explosion. The bomb was similar in construction to the "Fat Man" bomb that was dropped on Nagasaki and fuelled by plutonium Canada likely provided. We do not know Solandt's thoughts upon seeing an explosion similar to the one that had created the devastation he had witnessed in Japan. It was no doubt a fascinating reversal of cause and effect: witnessing, in a moment, the destructive power that had created the casualties and devastation he had seen seven years prior. Even amidst this powerful symmetry, Solandt retained the atomic pragmatism, saying for years afterward that the atomic bomb was a quantitatively not qualitatively superior weapon.[59]

Solandt reported on the British atomic bomb tests to the CDC on 14 November 1952, though the content of this report is unknown.[60] It is likely, however, that he was able to provide a full and effective report on Britain's first atomic device. Later in life, Solandt recalled that his relationship to atomic tests in the Cold War was almost always personal rather than official, though it still provided him access to information and contacts of value in atomic warfare.[61]

After Monte Bello, Solandt maintained links with British atomic science and research. He hired Doctor Alec Longhair, who had worked on the British liaison staff on strength with the British Embassy in Washington, to be the DRB's special weapons advisor to the Army. According to Solandt, Longhair knew all the "characters" and had British top secret atomic clearance. As soon as Longhair was hired, the DRB gained effective "backstairs gossip" on the movement of personnel and information exchanged in Washington. None of these ideas or investigations, however, made it to the level of policy and as such none of the particulars were written down.[62]

The Hydrogen Bomb and Changing Perspectives, 1952–1954

Solandt's increasing efforts to provide the Canadian Armed Forces with first-hand research coincided with a shift in U,S,, British, and thus NATO policy regarding nuclear weapons. In 1952, the United States had elected Dwight D. Eisenhower as president. Eisenhower's "New Look" strategy emphasized the importance of nuclear weapons to Western defence against Soviet strength in conventional weapons. The British Global Strategy Paper (GSP) of 1952 outlined the need for nuclear weapons in NATO to reduce its general weakness in conventional forces in relation to the Red Army. The GSP in turn influenced the generation of NATO Military Committee paper MC14/1, the heart of NATO's nuclear strategy in 1954.[63] In 1955, with the Soviets now defended by the Warsaw Pact alliance, Canada supported NATO's decision to arm and equip their forces with nuclear weapons.[64]

Solandt's increasing desire for Canada to accept a role as a nation with nuclear capabilities evolved within this strategic context. Solandt had pursued personal connections in Washington on atomic weapons, but by February 1953 he informed the chiefs of staff that it was now essential to ask the U.S. government to permit Canadian participation on U.S. bomb trials so the services could gain experience with the problems of atomic defence. Foulkes felt the McMahon Act prohibited this now, but the new Eisenhower administration would likely lift these restrictions. Not keen to wait, Solandt received COSC approval to pursue Canadian participation in future U.K. bomb trials.[65] Solandt then pressed the board to approve the creation of a Committee on Defensive Aspects of Atomic War, which he chaired.[66] On 23 June 1953, he told the COSC he had met with Cockcroft, who asked Canada to duplicate some of Britain's atomic facilities. In considering a response, Solandt suggested a review of Canada's atomic energy program to decide future requirements in scientific and military fields. He believed there was likely no problem doing as Cockcroft suggested, if the program benefited Canada. Foulkes disagreed and wanted more discussion before any changes were made.[67]

One year after the first U.K. bomb test, Solandt again raised the issue of Canadian participation on U.K. atomic trials. Reports on it had been received and requirements were to be sent to the United Kingdom. Solandt warned that while Canadian service participation at U.K. trials would be a great value to Canada, Canada would need to assist or support the British effort to gain access to information. In short, Canada had to have something of value to trade for nuclear secrets. Foulkes agreed to talk to his equivalent in England, though nothing more was officially discussed that year.[68]

By 1953, Solandt wanted atomic policy changed. While work with atomic allies was critical, the "explosion of atomic weapons in Canada where a greater number of troops can gain experience more economically" should be pursued. This was the first time the use of atomic weapons in Canada was considered as policy by the DRB. This would be supplemented by initiating programs for the "production of atomic power for industry which could also lead to the provision of atomic weapons for the Canadian Services should the need arise." A preliminary study suggested the urgent need for early planning of such a program.[69] "In the atomic side, numerous projects dealing with the effect of radiation will be instituted and increased emphasis will be placed on the detection of fission products from foreign nuclear explosions."[70]

On 12 August 1953, the USSR exploded its first boosted-fission weapon. The high-yield explosion was reported as resulting from a hydrogen weapon. In October, Solandt focused the board meeting on what effect the detonations had on U.S. strategy. Current thinking suggested Russia would shortly have a significant stockpile of nuclear weapons and would in a position to launch an atomic attack against North America. "Canada should now put more emphasis into atomic weapons, particularly into atomic defence." Without their own atomic weapons to study and assess, however, "indoctrination of the Services and civilians in atomic matters was a difficult problem." Still, the DRB would do all it could.[71] The board agreed. Solandt said few Canadians had participated on U.S. atomic bomb tests, but that number could be increased.

He also noted the United Kingdom might ask Canada for space for field trials as a "joint indoctrination programme for their own and Canadian Forces." British authorities thought the current Australian site was too far away, and a spot in Churchill, Manitoba, might fit the need.[72] These suggestions were predicated on private discussions Solandt had with Penney on Canada as a site for nuclear power development and weapons testing. On the British side, Penney made two unofficial proposals for joint Canadian/U.K. atomic efforts. The British wanted more uranium-enriching capacity, but Penney feared the vulnerability of having more facilities in the United Kingdom and sought Canadian aid.[73] This was an atomic component of the general U.K. defence research policy push for industrial and defence research dispersal throughout the Commonwealth that had been long discussed at the DRPC and through the CACDS.[74] Solandt and Penney researched this issue quietly so as not to upset their respective governments before generating a proposal that might get anyone's "shirt in a knot."[75]

Penney also wondered if Canada might build a suitable test site for atomic weapons. Solandt took him to see Churchill, Manitoba, as well as Churchill Falls, Labrador, as possible sites. Although the appeals never came to anything, Solandt used his dialogue with Penney to gain new information on atomic energy and weapons for Canada's own defence and energy needs. In the end, Australia remained the more palatable option to both governments. No senior policy discussions were needed, and no secrets were being shared. Solandt recalled, however, that the COSC was annoyed at his initiatives. If anything more than a discussion had taken place, however, Solandt knew it would have become a national concern. Instead, the issue blew over quite comfortably. Solandt never felt concern about showing Penney anything or examining the possibility of a British test site in Canada. To have denied or ignored Penney's request would have been "fatuitous." Such conduct was no way to build allied partnership in defence research.[76]

By October 1953, however, nuclear weapons were of grave concern to the board. Both the United States and the USSR were creating thermonuclear arsenals and furthering research in advanced missiles, rockets, and high-speed bombers. "The subject of atomic power development as a joint programme with the production of fissionable materials was discussed at considerable length and the general opinion was that this phase of atomic planning was a natural evolution of early planning." This program would cover defence and training of Canadian Armed

Forces and civilians, and indoctrination of troops in actual combat with atomic weapons by participating in explosions in Canada. The evolution of industrial power in Canada would also assist military programs.[77]

A Year of Change

Between 1953 and 1954, Solandt's pursuit of greater Canadian participation in atomic bomb trials evolved to its logical conclusion. For Canada to take its place among its allies at the dawning of the thermonuclear age, it required its own nuclear weapons. Reaching this conclusion took the better part of a year when both the COSC and the DRB grappled with these technological changes.[78]

Then, on 1 March 1954, the largest U.S. thermonuclear weapon test, Bravo Shot, was detonated at Bikini Atoll as part of Operation Castle, with a yield of fifteen megatons. Solandt's view of atomic policy grew more serious, though he remained as pragmatic and cool as ever. Civil defence became critical, but the greatest shift in priorities in Solandt's mind was the requirement for the Canadian Armed Forces to be armed with nuclear weapons.

By June 1954, Solandt gave the board his frank opinion of defence research in the thermonuclear age. It was time for a change. "This aroused considerable comment, particularly from the Chiefs of Staff and after a lengthy discussion, the majority of those present at the meeting, were of the opinion that – the Defence Research Board should concentrate its future efforts towards meeting the threat of an all out atomic war and that the surest way to maintain peace was to ensure that this continent had a retaliatory potential which would act as a deterrent to a possible enemy."[79] For Solandt, research on atomic war would now be the DRB's top priority, even if limited wars, like Korea, broke out. Minor wars could be contained, Solandt surmised, so long as Canada and her allies maintained a technological advantage and lead. But the major threat was strategic use of atomic weapons against North American targets. This threat would take the form of Soviet high-flying and fast bombers. "This means that a start should now be made to the planning of the weapon system after the CF105." The DRB and the RCAF would lead this initiative on future systems and included weapons and their ground environment.[80]

The board agreed with Solandt's proposals.[81] The DRB would educate the public on the reality of these threats and defence, including civil defence. Operational studies of planning industrial communities, service camps, supply depots, and the value of dispersion of valuable targets was to be started.[82] Simonds, however, cautioned the DRB about its policies on air and civil defence and wanted planning on these issues reserved for later. Still, Mister Quarles, U.S. assistant secretary of defense (research and development), also attended the meeting. Quarles agreed with the DRB's arguments on minor wars and technological leverage, especially the value of atomic retaliation as established in the Eisenhower administration's "New Look" strategy.[83]

The board needed time to consider these developments and asked the COSC to give their own view on the future or war so that a special meeting could address commitment and priority concerns for the DRB.[84] Solandt wrote the COSC report for the 30 July 1954 COSC meeting. Foulkes said that this guidance was to be defensive. He warned against "panicking" Canada's strategic outlook on the nature of these weapons. Atomic and thermonuclear weapons must "continue to be considered merely as a type of weapon. Regardless of enemy capabilities in the thermonuclear field, we must still plan to carry out our assigned task." He wanted the policy revised so it was clearly stated that the Canadian services would carry out their duty where atomic or thermonuclear weapons are used. Simonds reiterated his desire for the DRB's role on these matters to be through advice and support, though it fell to Solandt to redraft the paper.[85]

By October 1954, Solandt was tired of scrambling for atomic crumbs or revising papers on atomic weapons Canada did not posses. His voice hardened on atomic affairs. Canada had gained as much as possible through its domestic atomic energy work and allied connections, but it was no longer enough. To make an effective contribution to national defence, Canada had to be armed with atomic weapons. He told the board he "strongly believes that atomic weapons should be obtained for the Canadian services. It was expected that within the next few years air-to-air and surface-to-air weapons with atomic warheads would be available in the U.S. It was ridiculous that Canada should be expected to play a large role in the defence of North America and not be equipped for that purpose with the same atomic weapons as the U.S. Canada was not even planning for the use of nuclear weapons, and, for example, had made no provision to use an atomic air-to-air rocket with the CF 105."[86]

He was certain that atomic anti-submarine weapons would soon be made in the United States, and Canada should reconsider its anti-submarine stance in the Navy, and pressed on these matters. But Solandt also knew the political concerns of these weapons, and the limit of his influence. "The Board could do no more than make a strong recommendation to the Chiefs of Staff that they inform the Canadian government of recent developments in atomic weapons and request that steps be taken to obtain some types of U.S. weapons for use by the Canadian Forces, or sufficient information about them to manufacture in Canada."[87]

The board agreed unanimously, believing the public could be properly informed to reduce any protest on procurement of nuclear weapons for defensive or offensive operations. Solandt believed procurement of American nuclear weapons would not be easy, though a good argument could be made if Canada used some geographic leverage. He suggested "US forces would have to obtain Canadian permission to use atomic weapons over Canadian territory. The growing participation of Canada in UK trials would be helpful in overcoming American reluctance to part with classified atomic information."[88]

A committee of the senior services in Canada, Britain and the US was scheduled to discuss the growing import of nuclear weapons in tripartite as well as NATO countries. This meeting would include scientists, military planners, then chiefs of staff to discuss problems and then find solutions them. Solandt felt that "all our thinking and planning concerning the course of a global war would be affected by the results of these meetings. He believed that the U.S. could not hold back information from its Allies in the face of knowledge of these weapons in the USSR, and particularly from Canada because of the common interests in continental air defence. DRB had already submitted calculations on fall-out effects to the U.S. for comment and had obtained qualitative but informative replies. Members were assured that excellent co-operation existed between DRB and Chalk River."[89]

While these meetings were being prepared, in November Solandt briefed the COSC, the civil defence coordinator,[90] and then the CDC regarding the deadly nature of surface-burst thermonuclear weapons. For the CDC, he described how "'fall out' was created and indicated the importance of weather conditions, particularly the direction and force of the wind in this regard. He dealt particularly with the effect of the gamma rays given off by fission particles. While defence problems in general had become much more difficult as a result of the development of these high-powered thermonuclear weapons, the civil defence problem was now more clearly defined, and it was possible to envisage and plan measures which would be of real assistance to the public if an attack with a megaton weapon occurred."[91] The CDC acknowledged that only a small number of these weapons were needed for devastating attacks on North America, "and that, technically speaking, air defence was now an area problem rather than a simple point defence question."[92] However, Solandt never raised the issue of Canada possessing its own nuclear weapons. Apparently, it was still too hot a topic. It remains unknown how Solandt proceeded after this date at the DRB, since the minutes of the DRB meetings after 1954 have been lost. He never raised the issue at the CDC himself.

Conclusion

By 1955, Solandt had made the decision to retire as chairman of the DRB. Thus, he left government service immediately before Canada's Armed Forces were armed with nuclear weapons. He also avoided the shifting and confusing nuclear policies of the Conservative government that came to power in 1956 under prime minister John Diefenbaker.[93] Whether or not Solandt could have helped maintain a clear and definitive nuclear policy for Canada's Armed Forces under the Conservatives is difficult to assess. What is clear from this chapter is that between 1945 and 1956, Solandt generated a steady and constant influence on Canada's atomic affairs.

This guidance was most successful in the realm of atomic information. As an expert on atomic bomb damage, Solandt provided the Canadian government and Armed Forces a valuable eye on atomic matters. Solandt used every means at his disposal to coalesce research on the nature of atomic warfare. He harnessed domestic resources from Chalk River and the DRB to the best of his abilities and took a personal role in gathering as much data as possible from U.S. and British allies. By 1953, however, such sources of data were not enough.

Canada was fortunate to have Solandt's steady hand on nuclear affairs for so long. This was no "atomic playboy," but an atomic realist. For most of the world, fear came in the wake of Hiroshima and Nagasaki – fear that, while justified, also clouded judgment and obfuscated facts, when facts could be found. Solandt knew all too well that fear, like most emotions, was an enemy of analysis. Unlike every other chief of staff, he had seen the nature of atomic warfare on human beings first hand. He had walked the shattered cities of Hiroshima and Nagasaki as a doctor and scientist. As chairman of the DRB, he was given the strange but understandable mandate of preparing the Canadian Armed Forces for atomic warfare without access to atomic weapons. Solandt brought much to this paradoxical task: his unique insight into atomic weapons, his contacts in the highest echelons of science, and his own detective-like zest for finding information of value on atomic affairs. But above these attributes, he brought his cool, rational intellect to bear on weapons many considered irrational, providing Canada a respected and pragmatic voice during the turbulent birth of the nuclear age.

15 Conclusion

Solandt had told himself he would dedicate five years to making the DRB work and work well.[1] The Korean War had upset his timetable, but in ten years as chairman he had accomplished his goal. By 1956, the DRB was a well-respected addition to Canada's Armed Forces, a pillar of government science, and a respected science organization among the nation's senior allies. It was an achievement that could not have happened without Omond Solandt.

But 1955 was a year for change. Chiefs of staff had come and gone, while Solandt and Charles Foulkes remained a constant presence. Foulkes had made the Chief of Staff Committee his home, but Solandt was less content being a "wise old man" in defence research at only forty-seven. New defence research challenges were emerging. The missile age was dawning, computers were increasing in importance, and space was seen as a new environment for warfare and not just science fiction. Meanwhile, nuclear weapons were becoming more efficient, deadly, and numerous. The DRB needed a fresh perspective to handle these affairs, one that did not originate with Omond Solandt. Personally, he had suffered a terrible blow when his elder brother, Don, who had so often led him to his own fields of interest, succumbed to terminal illness and died in 1955.[2] New challenges and memories would do him and his family good.

Solandt felt justified in leaving defence research behind. Since 1940, his life had been consumed with the once foreign world of military affairs. By 1956 he had served both Britain and Canada well in the realm of defence. He told lifelong friend Laurie Chute that he had also grown tired working on engines of destruction and wanted more positive and constructive objectives.[3] There was no end of opportunity for him. Job offers had been plentiful as he became an international figure through his work at the DRB. The most appealing offer came in February 1952 from R.H. Fitzgerald, chancellor of the University of Pittsburgh. Fitzgerald wanted Solandt to become vice-chancellor of the university's medical affairs. Solandt found the facilities impressive, the salary generous, and the challenge intriguing. But with war raging in Korea, and a war in Europe looming, Solandt declined. "In spite of all these obvious attraction I feel that I cannot leave my present job when there is such an urgent need to build up the armed strength of the free world."[4] Such service before self, at a time of crisis, would have made Reverend Donald M. Solandt proud.

By 1955, the Korean War was over, Soviet dictator Joseph Stalin was dead, and a relative stability accrued to international relations between the United States and the Soviet Union. It was a good time to exit the DRB. An attractive offer arrived from Donald Gordon, president of Canadian National (CN) and one of Canada's industrial titans. Gordon was looking for a successor to CN's legendary vice-president of research and development, Star Fairweather. He had originally approached Sir Henry Tizard, but Tizard said Gordon was wasting his time looking for a Brit when Solandt was right there in Canada.[5] Gordon approached Solandt with a formal offer in September 1955. The salary was $22,500 a year, almost double his initial DRB salary, and the job also provided the new and exciting challenge of bringing the national rail system into the modern era. It was a terrific opportunity.[6]

However, Solandt had no interest in leaving government service – just military affairs. He enjoyed the challenges of working on behalf of Canadians, providing for the nation's interests, and wanted to continue to do so. Before he responded, he discussed Gordon's offer and possible future options in government with the minister of national defence, Ralph Campney. Unlike with Claxton, who retired in 1954, Solandt had a workable but not warm relationship with Campney. Solandt suggested his potential at becoming president of NRC or deputy minister of industry, and made it clear he was not asking for a job but was interested in future vacancies. He did not want to leave Ottawa if he could find the right position.[7]

Campney's response left Solandt speechless: if Solandt left the DRB to find any other government position, Campney would fight him every step of the way.[8] Why? The same reason Solandt had been denied the position of deputy science advisor of the Army Council during the war: the government found him irreplaceable at his current post.

Keeping his powder dry, Solandt asked if he could meet with prime minister Louis St Laurent. Campney agreed, and the prime minister gave Solandt the exact same answer. St Laurent believed Solandt would make a fine president of NRC, but he could think of no one better suited to running the complex, multipurpose, and unique Defence Research Board than Omond Solandt. Any future government position Solandt desired required an Order-in-Council and thus the prime minister's approval. St Laurent would not give it – it was the DRB or nothing. Thus, for Solandt, he did not leave government service in 1956; he was "driven out" by myopic thinking.[9]

It was an ironic twist. Solandt had been selected as the only man suited to run the DRB. In the process of building the organization and making it effective, he stamped the DRB not only with his vision but his personality and capabilities. In doing so, he had created an organization that everyone believed required an Omond Solandt to make effective. This was perhaps his greatest success and failure as the DRB chairman. The palace for his most stunning intellectual achievements in government science had become a metaphorical prison.

The End of the Solandt Era at the DRB

Hartley Zimmerman succeeded Solandt at the DRB in 1956, and while the eleven-year Zimmerman period is not under review here, the end of the Solandt era requires comment. Zimmerman was a mining engineer who had been a member of the Board for the Department of Defence Production since 1952. He had worked in the munitions industry during the war but was not a scientist. In 1954, he replaced E. Ll. Davies as Solandt's vice-chairman when Davies became the DRB liaison officer in 1954.[10] Solandt never thought it was his place to pick his successor, but if he had known how "terrible" Zimmerman would be in the post, he would have fought for a different person. But Zimmerman became chairman in 1956. As soon as Solandt left, communication with the DRB ended. While Solandt maintained personal contact with some members, Zimmerman was not interested in the "Old Man's" opinion or advice.[11]

From what Solandt heard, Zimmerman came under the influence of Canadian chemist and NRC president, Doctor Ned Steacie. According to Arthur Fordyce, Steacie despised Solandt and viewed him as a "charlatan" and not a scientist at all.[12] Solandt heard how Steacie's forceful personality soon dominated all the scientific questions of the board.

He became Zimmerman's science advisor, which was not his job. As such, the board now had a chairman who could not trust his own judgment on matters of science policy. Many DRB scientists were either not impressed or actively depressed with Zimmerman's chairmanship. The work was still important, and terrific developments grew from this period, including the Alouette satellite program that the DRB spearheaded at CARDE, but much of this good work was built on the firm foundations of the Solandt era.[13]

Unlike Zimmerman, Solandt maintained a deep interest in the technical and scientific endeavours of almost every corner of the DRB.[14] He had fought hard for the board against government and military heavyweights and had been the nation's expert on the scientific and medical nature of nuclear warfare. Any successor had big shoes to fill. Zimmerman did not come close, especially for the working staff, whom Solandt had inspired with his wide knowledge and personal interest in DRB research. One anonymous DRB employee was so distraught about Zimmerman's leadership that he sent prime minister John Diefenbaker a letter dated 5 March 1960: "Although we are a body of expert scientists, we have as our leader and mouthpiece a man who is not a scientist and who is unable to give Science a proper influence on military judgement ... [Zimmerman] is a mining engineer, his doctoral degree is only an honorary one, and his acquaintance with defence research before 1956 was due only to his being Mr CD Howe's personal representative on the DRB ... We admire him as a man but we do not respect him as a scientist and we know that he does not speak up for our scientific conclusions in the Chiefs of Staff Committee."[15] Such an indictment could come only in the wake of Solandt's leadership. He had set the standard for what was expected as the DRB chairman, and in comparison Zimmerman was found wanting. However, DRB employee George Lindsey felt Zimmerman was a better and tougher administrator, one who had no qualms about firing people who did not do good work. For Lindsey, who gained prominence in the DRB and its successor organizations, Solandt was too kind in this regard. Still, Zimmerman lacked Solandt's scientific knowledge and forceful intellect, especially on the COSC or CDC.[16]

Solandt left the DRB with a bang. In March 1956, he addressed the press for ninety minutes in a "superlative swan song" regarding war and technology in the future. Nuclear weapons needed to be stored on Canadian soil for effective continental defence. ICBMs would be built, and the nation that builds effective anti-ballistic missile defence systems first would be in a position of strategic superiority while the monopoly lasted. Such systems would require scraping the entire continental radar defence system as we know it. The CF-105 Arrow "would be Canada's last jet fighter. Velvet Glove would be the first and last guided missile developed in Canada, with the Canadian armed services using the US sparrow." With a hint of ego, he warned that in modern and future warfare, scientific advances were a better deterrent than extra squads of fighters or divisions of militia, and Canada needed to invest in this deterrent.[17]

His tone for the swan song was quintessentially Solandt: clear, calm, but with gravitas. While he attempted to introduce Zimmerman, the new chairman was "all but ignored" as all the reporters asked Solandt questions.[18]

The Post-Solandt Era of the DRB

The DRB was not the same after Solandt left. Structurally, the organization changed in substance and style from the 1960s until being dismantled in 1974. The DRB managed to survive most of the changes demanded by the Glassco Royal Commission initiated in 1960, which attempted to reform and coordinate the civil service to make it more efficient and called for massive reforms in Canada's military and industrial policy. According to one source, the DRB was largely unaffected by the Glassco investigation, but subsequent budget cuts and changes in Canada's political leadership sealed the death knell for its unique place in government.[19]

In 1971, the Liberal government of Pierre Elliot Trudeau established a Management Review Group to investigate procurement cost overruns and duplication of effort in the Department of National Defence (DND), which resulted in the structural merging of the DND, Canadian Forces Headquarters, and the Defence Research Board. Exactly when this was to happen, and the structure it would take, came as a surprise. No timetable was given, but the new Trudeau administration found the DRB too independent and unresponsive. In 1974, an Order-in-Council was passed that effectively disbanded it as it had existed.[20] It was relieved of all executive, administrative, and directive responsibility for the defence research program in Canada and any direct responsibility to the establishments. The authority and funds for providing grants to universities was cancelled, as was the board's efforts with industry. Responsibilities for research were subdivided among deputy ministers instead of the DRB superintendents. It would now be headed by a chief of research and development under the assistant deputy minister (materials). No longer would the head of defence research in Canada be a chief of staff. The DRB's unique hiring system, outside the Civil Service Commission, was ended.[21] There was no clear and direct route between the head of the organization and its research. According to Gordon Watson, a veteran DRB employee since its inception, no scientists backed this plan, and the final Chairman, Doctor L.J. L'Heureux, was allegedly silenced from mentioning the government's decision to kill the board.[22]

Solandt was livid, and his normally pleasant demeanour went cold with anger. In a 1975 article in *Science Forum*, he called the reorganization of the DRB "an act of mayhem committed in the name of administrative tidiness."[23] He lambasted the government for not thinking through the effects this would have on Canada's role in science and industry, and militarily. Their small-minded, bureaucratic "tidiness" had struck a decisive blow against Canada's standing in scientific affairs.[24] The Trudeau government dismantled what he had spent ten years building. One can only imagine Solandt's cold anger at any chairman who did not stand up for the DRB. Courage to fight for the DRB was something Solandt never lacked.

Solandt after the DRB

Solandt stayed with CN as vice-president for R&D until 1963. He then held a series of prominent positions in industry before turning his hand to an active consultancy on a range of issues. In the early 1970s he headed a government commission on power transmission and its effect on the environment in Ontario, and a decade later was an expert consultant on the royal commission investigating the 1982 *Ocean Ranger* disaster.[25] He maintained a role as a nuclear weapons expert as the Canadian member of the 1958 Western Delegation to the Conference of Experts to Study Methods of Detecting Violations of Possible Agreements on the Suspension of Nuclear Tests in Geneva. That same year he was also elected the founding president of the Canadian Operational Research Society, an organization he helped create. He held the post for two years.[26] Solandt was also active in air defence research, though not in Canada: he was the only Canadian member of the American MITRE Board of Trustees, from 1959 to 1969, a position that gave him no end of headaches when the FBI challenged his need for a security clearance.[27]

He was also considered for the position of scientific advisor to NATO's supreme commander in 1961. Solandt turned down the high-profile and generous offer, feeling they needed a younger scientist. While he never knew it, there is some evidence that his candidacy was challenged by his old wartime colleague Solly Zuckerman. The reasons why, however, remain unclear.[28] Solandt was vice-president of R&D for DeHavilland Aircraft of Canada and chairman of DCF Systems Ltd, a DeHavilland subsidiary, until 1965. He left De Havilland for similar work at Hawker Siddley of Canada and also became chancellor of the University of Toronto in 1965. He briefly returned to government science as the inaugural chairman of the new Science Council of Canada in 1966, where his tenure was seen as the height of the organization's success.[29]

Canada again celebrated him by making him a Companion of the Order of Canada in 1970, and over the next decade until his retirement Solandt worked as a consultant on numerous international, national, and provincial projects. His reputation as a forceful manager on committees, first learned at Lulworth, remained intact. According to Doctor Walter P. Falcon, director of the Institute for Institutional Studies at Stanford, Solandt was a first-rate addition to any job. "Put quite simply, I would say that in twenty plus years that I have served on numerous boards, Omond Solandt was the best board member I ever knew."[30]

The Legacy of Omond Solandt

But as Solandt reflected on the variety of jobs his talents had brought him, the DRB took centre stage. He enjoyed the responsibility of leadership and the power to execute it in the nation's defence. The journey to this apex of his career was as unique as the man who lived it. Solandt's intellect and interests provided him opportunities. His absolute confidence in his abilities turned these opportunities into a stunning career in government science. The catalyst for this rise was the Second World War.

In six years, Solandt went from being a mammalian physiologist at Cambridge to an expert on the effects of atomic weapons. He started the Cold War as director-general of an organization that had never existed in peacetime and yet by 1956 he was a national figure on the future of war and technology. He had come a long way from the laboratory of Charles Best, though any that knew him back then would not have been surprised. Excelling was just part of his nature.

The legacy of Omond Solandt in government science from 1939 to 1956 stands within two fields: research, and leadership of government science organizations. As a researcher, Solandt was a pioneer in the application of scientific principles to military problems. His intellect was made for such challenges, even if his training was unconventional by the standards of operational research pioneers. His efforts to make the South West London Blood Depot systemically efficient bore the markings of a future operational research (OR) scientist. His research on physiological dimensions of tank problems at Lulworth and the tank section of the British Army Operational Research Group (AORG) cemented his reputation in OR, where he championed the use of battle data from operations as the bedrock of the team's research. The capstone of his research career, however, was providing casualty analysis for the British government on the victims of the atomic bombs at Hiroshima and Nagasaki. This research contributed to one of the earliest assessments of the reality of atomic warfare. All of these efforts rested on his extensive medical, technical, and scientific training he had soaked up at Central Tech, the University of Toronto, and Cambridge. Added to this was Solandt's own fierce intellect and capacity for "diagnosing" problems and finding solutions, no matter the field of inquiry.

But Solandt's greatest legacy was made in the unique role of leading and managing government science institutions. He did this with no training in the art of management. Instead, Solandt took observations, experience, and his own sense of order and efficiency and created his own leadership and management approach. At both Sutton and Lulworth, Solandt developed expertise in leading research organizations under war conditions, predicated on his harrowing experience as director of the polio ward during the 1937 epidemic in Ontario. His personal approach to leadership emphasized a demand for excellence, support for quality research, and a delicate balance between providing direction for projects without interfering in the research. His quiet but forceful personality made more friends than enemies in scientific, government and military circles, backed by a reputation of being a reliable scientist. Solandt referred to his leadership style as that of a maestro conducting a symphony: provide the music and guide the instruments of your orchestra, but let the players play.

As deputy and superintendent of AORG, Solandt's orchestra of science grew into diverse fields of army research. While some preferred his predecessor Basil Schonland, just as many appreciated how Solandt got better wages for scientists and took an active interest in the diverse fields AORG, from soil analysis and instrument mechanics to continental defence against German V1 and V2 rockets. From the Military Personnel Research Committee of the Medical Research Council to the British War Office, Solandt became a seasoned "committee man" in government science. He had been introduced to the difficulties of translating scientific data into government policy at Lulworth, meeting resistance to his advice by senior military and government officials. But at AORG he made a reputation for being tough on committees and persuasive in government circles, using powerful contacts, such as Sir John Cockcroft and Sir Robert Watson-Watt, to his advantage. While he never achieved the level of success he wanted in the War Office, the experience at AORG, both good and bad, formed in Solandt's mind a way forward for army science. Those experiences and thoughts became his blueprint for a peacetime defence research organization. His blueprint called for direct access to the senior ministers, a presence on senior military committees, dealing personally with industry and academia, and having enough independence to allow science under his purview the freedom it needed to bear the best fruit.

Solandt's ideas largely paralleled Canada's own vision of post-war defence research generated by Colonel Wally Goforth and his team. Goforth had produced Canada's initial vision of the DRB, and his far-seeing intellect was appreciated and tempered by Solandt's pragmatic realism and civil service savvy. Goforth's role as the original Canadian DRB visionary remains uncontested, though it was Solandt, who held many of the same views long before he got the call from Ottawa, who provided the spine and presence to turn the idea of the DRB into a reality.

Solandt's legacy as an effective and unique leader in government science was enshrined in the DRB. He was an active and critical chief of staff in an arena where he was not initially wanted, and he established the DRB's resources as pillars of support and insight within the senior services. He established valuable networks of government influence with industry and academia, which paid dividends in the field of electronics research and the development of the Mid-Canada Line's McGill Fence radar system. He provided guidance and support to the DRB's nine establishments. Through his leadership, Canada launched into the missile age with Velvet Glove, though never achieved its goal of being an active weapon system. DRB did maintain a reputation for excellence in the realm of chemical and biological warfare field trials, domestically and internationally. His own instincts for avoiding bad research projects helped the DRB sidestep the scandal associated with Doctor Ewen Cameron's "de-patterning" research for the CIA. Solandt also provided Canada unique access to the defence research world in Britain, where Solandt had initially made his name. He also established new links to the U.S. defence research world. And, for a decade, he was Canada's leading authority on nuclear weapons. His cool and fiercely rational intellect acted as a salve against the rising fears of the nuclear age. This rationale led Solandt to the conclusion that Canada required its own nuclear weapons if it was to truly be prepared of any nuclear confrontation. For years, he had done his best to accomplish the paradoxical: prepare the Canadian Armed Forces for the reality of nuclear weapons without having access to nuclear weapons. Such an experiment seemed designed for failure. Solandt cultivated every possible resource to bear on this issue, from the DRB liaison with the nuclear reactor team at Chalk River, to personal involvement during the British detonation of an atomic device at Monte Bello in 1952. All provided valuable insight, but by the time of his resignation he was convinced that Canada required its own nuclear weapons. Unfortunately, here Solandt was ahead of his time. But the ground he broke to prepare Canada for the realities of war in the nuclear age was fertile for future defence research strategies, should someone of high calibre take the reigns of the DRB.

Like C.J. Mackenzie before him, Solandt became a state-builder for the Canadian government in defence research. Unlike Mackenzie, Solandt accomplished this task within the context of peace, not war. Against a backdrop of small defence budgets, anti-militarist policies, demobilization, and the rise of the international and technical threats of the USSR, Solandt led the DRB through a period of growth from a handful of staff to a 1,000-strong research body, stretched across the nation to meet the research needs of the Canadian Armed Forces.[31] From the DRB, Solandt also ruled over the increase of the Canadian government's influence into industry and academia through grants, fellowships, and the DRB's committee system. The influence of the state via the DRB in both industry and academia is not yet well appreciated and remains to be established.[32]

At the time of his death in 1993, Solandt had no professional regrets and a cavalcade of positive memories of the work he accomplished and those he worked with and for. At the symposium held in his honour in 1994, Robert D. Havener, former director of the International Development Research Centre, who worked with Solandt on agricultural research for arid countries, remarked on the common traits everyone found when working with Omond. "Those of you who knew him personally will not be surprised by the litany of superlative characteristics. Words such as curiosity, photographic memory, reliability, sense of proportion, humour, kindness, and wise counsel were often mentioned. So were such words as energy, zest for life, courage, relentless logic, and the ability to express himself clearly and forcefully. But to me his greatest qualities were his strong sense of right and wrong, his uncommonly good sense, and his concern for the welfare of the world and all the people who inhabit it."[33] All of these attributes came to fruition during the Second World War and were made manifest during the Cold War. Solandt pioneered the art of science in government long before his old colleague C.P. Snow had written about the "two cultures."[34] In doing so, he contributed to the development and growth of applied science in Britain and Canada. Before Solandt died on 12 May 1993, he provided a foreword to the first book on his legacy, co-written by David Grenville. Even on the eve of his passing, Solandt found the energy to criticise current government policy on science and technology, and offer a solution. "The government's idea that you can buy technology and use it effectively is completely unreal. You can import technology and manufacturing skills to kickstart a new initiative, but it won't take off and keep going until the knowledge and skills become indigenous and there is continuity." This required scientists, engineers and other experts in important professions. He then added a poignant comment: "Problems of the real world are almost all interdisciplinary. It is rare that an important problem is encountered that fits exactly into one of the artificial disciplines into which universities are subdivided. Therefore, the idea of interdisciplinary work must spread through the whole scientific community and the biggest change is needed in universities. The development of mechanisms and policies to encourage this must be given priority."[35] Solandt's career was a testament to the value of an "interdisciplinary" approach. He brought an engineer's efficiency to managing research, a doctor's diagnostic skill to conducting technical research, and a capacity and facility to work with experts in almost any field of inquiry towards meeting an objective. The ability to see the connections among these fields with a high degree of precision allowed Solandt to become one of the best state scientists Canada has produced.

While he could not have known it, Solandt aptly described the perspective needed to research, analyse, and establish his place in the history of government science in both Britain and Canada. To paraphrase his view of being the DRB chairman, any biographer of Solandt attempting to situate his role in government science had to be prepared to study and appreciate medical, scientific, technical, and military history in Britain and Canada in the first half of the twentieth century.

This brief account is, hopefully, not the final word on Omond Solandt and his many legacies. Since this is the first such scholarly account, this work could not be a comprehensive biography of Solandt's entire life and experiences. Almost every sector of his life could be entertained as a serious work of historical scholarship, across a variety of historical perspectives. Much work remains to be done regarding Solandt's life and accomplishments. The view of allies of DRB's efforts, especially with regards to BW, CW, and atomic research has yet to be fully explored.

With a focus on his development as both a researcher and leader in state-sponsored science, this initial account and analysis of Omond Solandt should provide future scholarship with both a narrative spine and historical context from which to launch, whatever their final trajectory. DRB's work on Arctic research still needs a champion. A comprehensive view of allied views of Solandt and the DRB, especially from the American defence research elite, requires attention and detail. Any robust effort on Canada's cold war efforts on CW, BW, and atomic research would be a welcome addition to what is provided here. But the first priority was to establish the life and legacy of Omond Solandt in government science. Both are worthy of record.

Appendix 1:
Amendment to the National Defence Act of 1927 regarding Defence Research, 7 February 1947*

(1) The Governor-in-Council shall establish a Defence Research Board composed of such persons, not exceeding twelve in number, as may be appointed thereto by him, which shall carry out such duties in connection with research relating to the defence of Canada and the development of or improvement to Service equipment and material as the Minister may assign to it, and shall advise the Minister on all matters relating to scientific, technical and other research and development which affect national defence.

(2) The Governor-in-Council may appoint one of the members of the Defence Research Board to be Director General of Defence Research. The Director General of Defence Research shall be the Chairman and chief executive officer of the Defence Research Board. The Director General of Defence Research and the other members of the Board shall hold office during pleasure and shall be paid such salaries, remunerations and expenses as may be fixed from tine to time by the Governor-in-Council.

(3) Subject to the approval of the Governor-in-Council, the Defence Research Board may,

    (a) enter into contracts in the name of His Majesty and establish scholarships in connection with and make grants-in-aid for research and investigations for national defence;

    (b) establish and support a pension fund or make other pensions or superannuation arrangements for the benefit of all or any of the

---

*Quoted from Goodspeed, *A History of the Defence Research Board*, 66

permanent or temporary officers or employers of the Defence Research Board

(4) The Governor-in-Council may by regulation:-

(a) notwithstanding anything contained in the *Civil Service Act* and section five of this Act prescribe the manner of selection, remuneration and terms of appointment and service of the officers and employees engaged in the world of the Defence Research Board.

(b) co-ordinate the work of the Defence Research Board with the Honorary Advisory Council for Scientific and Industrial research and other organizations and corporations engaged in scientific research and investigation;

(c) make provision general for carrying out the purposes of this section

(5) All expenses of the Defence Research Board shall be paid out of the moneys appropriated by Parliament for the purpose received by the Board through the conduct of it separations, bequests, donations or otherwise and shall be paid by the Minister of Finance on the requisition of the Minister. The Minister may request the Minister of Finance to allocate any portion of the moneys appropriated by Parliament for the purpose of the Board for scholarships or grants-in-aid of research and investigations and thereupon the Minister of Finance shall hold such portion of said moneys in trust and may at any time on the requisition of the Minister disburse such moneys for scholarship or grants-in-aid of research and investigation.

Appendix 2:
The National Defence Act of 7 June 1950, Part III, Bill 133*

53. (1) There shall be a Defence Research Board which shall carry out such duties in connection with research relating to the defence of Canada and the development of or improvements in material as the Minister may assign to it and shall advise the Minister on all matters relating to scientific, technical, and other research and development that in its opinion may affect national defence.

(2) The Defence Research Board shall consist of a Chairman and a Vice-Chairman, appointed by the Governor-in-Council, the person who from time to time hold the offices of Chief of the Naval Staff, Chief of the General Staff, Chief of the Air Staff, President of the Honorary Advisory Council for Scientific and Industrial Research and Deputy Minister of National Defence, and such additional members representative of universities, industry and other research interests as the Governor-in-Council appoints

(3) The Chairman and Vice-Chairman shall hold office during pleasure and shall be paid such salaries as the Governor-in-Council determines.

(4) The members of the Defence Research Board, other than the Chairman, Vice-Chairman, or the *ex officio* members, shall hold office for a period not exceeding three years but shall be eligible for re-appointment, and shall be paid such remunerations, if any, as the Governor-in-Council determines.

---

*Quoted from Goodspeed, *A History of the Defence Research Board*, 67–8.

(5) Each member shall be paid his travelling and other expenses incurred in connection with the work of the Defence Research Board

(6) The Chairman shall be the chief executive officer of the Defence Research Board and, under the direction of the Minister and in accordance with policies approved by the Board, shall oversee and direct the officers, clerks, and employees of the Board, have general control of the business of the Board, have supervision over the work directed to be carried out by the Board, be charged with the organization, administration and operation of the defence establishment of the Board and perform such other duties as the Minister may assign him

(7) The Vice-Chairman shall perform such duties as may be assigned to him under the by-laws made by the Defence Research Board

(8) The Chairman shall have a status equivalent to that of a Chief of Staff of a Service of the Canadian Forces

54. The Defence Research Board may, with the approval of the Minister

(a) notwithstanding the Civil Service Act or any other section of the Act or any other statute or law, appoint and employ the professional, scientific, technical, clerical and other employees required to carry out efficiently the duties of the Board, prescribe their duties and subject to the approval of the Governor-in0Council prescribes their terms of appointment and service and fix their remuneration;

(b) make by-laws or rules for the regulation of its proceedings and for the performance of its functions;

(c) enter into contracts in the name of His Majesty for research and investigation with respect only to matters relating to defence; and

(d) make grants-in-aid of research and investigations with respect only to matters relating to defence and establish scholarships for the education or training of persons to qualify them to engage in such research and investigations

55. (1) All expenses of the Defence Research Board shall be paid out of moneys appropriated by Parliament for the purpose or received by the Board through the conduct of its operations, bequests, donations or otherwise and shall be paid by the Minister of Finance on the requisition of the Minister

(2) the Minister may request the Minister of Finance to allocate any portion of moneys appropriated by Parliament of the purposes

of the Defence Research Board for scholarships or grants-in-aid of research and investigations and thereupon the Minister of Finance shall hold that portion of the moneys in trust and may at any time on the requisition of the Minister disabuse that portion of the moneys for scholarships or grants-in-aid of research and investigations

(3) Any moneys allocated by the Minister of Finance under this section that, in the opinion of the Minister, are not required for the purposes for which they were allocated shall cease to be held in trust.

Notes

1 Introduction

1 Michael Bliss, *William Osler: A Life in Medicine* (Toronto: University of Toronto Press, 1999), xii.
2 This fact was confirmed by the NAC staff in a personal communication with the author. After an exhaustive search by both the author and NAC staff, the minutes of the DRB meetings after 1954 have not been found in the DRB's main fonds, or those of its successor organization. Email from Ghislain Malette to Jason S. Ridler, 8 April 2008.
3 For interview details, see Bibliography.
4 David Grenville interview by Jason Ridler, 15 December 2007.
5 David Grenville, "Interviews with Solandt and Colleagues: Background Notes – 27 May 2006." Paper provided by the author.
6 David Grenville, George Lindsey, and Cecil Law, eds., *Perspectives of Science: The Legacy of Omond Solandt* (Kingston: Queen's Quarterly, 1994).
7 Andrew Kindsvatter, *American Soldiers: Ground Combat in the World Wars, Korea, and Vietnam* (Kansas: University of Kansas Press, 2003), xv.
8 Solandt, in his early eighties, recalled Paul Hellyer was minister of defence when he left the DRB. This is incorrect. Hellyer was assistant to minister of defence Ralph Campney at the end of the Solandt era.
9 Michael Shermer, *The Borderlands of Science: Where Sense Meets Nonsense* (Oxford: Oxford University Press, 2001), 264, italics in the original.
10 For further discussion of this theory of genius, see Robert Weisberg, *Creativity: Genius and Other Myths* (New York: W.H. Greeman, 1986).
11 Omond McKillop Solandt interview by David Grenville, 10 May 1986, part 6.

12 Laurie Chute interview by David Grenville, 12 September 1985, part 1.

13 Solandt interview, 10 May 1986, part 6.

14 Ibid.

15 G. Pascal Zachary, *Endless Frontier: Vannevar Bush, Engineer of the American Century* (Boston: MIT Press, 1999).

16 Grenville interview with G.S. Hattersley-Smith, 21 November 1985.

17 Sigurd Olsen website, "Diary of Dennis Collican 1957 Reindeer Lake Diary," http://www4.uwm.edu/letsci/research/sigurd_olson/voyageur/1957_reindeer%20lake_diary.htm.

18 National Archives, United Kingdom (hereafter referred to as NAUK), FD 1/6392, Carmichael to Mellanby, 5 April 1941.

19 Wilfrid Laurier Archive, Ronnie Shepard Papers and Collection (hereafter referred to as WLA RSC), Johnson to Schonland, 14 May 1945.

20 WLA RSC, transcription of interview with Lord Swann by Terry Copp, 1989.

21 Solandt interview, 24 February 1986, part 1.

22 A.M. Fordyce interview by David Grenville, 20 April 1986, part 1.

23 Henry M. Best, *Margaret and Charley: The Personal Story of Dr Charles Best, Co-Discoverer of Insulin* (Toronto: Dundurn Group, 2003), 265n558.

24 W.S. Feldberg interview by David Grenville, 18 November 1985.

25 Maggie Mollison interview by David Grenville, 16 October 1985, part 1; and Patrick Mollison interview by David Grenville, 19 November 1985.

26 H.A. Sargeaunt interview by David Grenville, 4 December 1985, part 2.

27 George Lindsey interview by David Grenville, 23 August 1985, part 2.

28 National Archive of Canada (hereafter referred to as NAC), MG 32, B5, vol. 221, file BC 1946–1954, reel 4, Claxton Memoirs, 977-1985.

29 Maggie Mollison interview, part 1.

30 Solandt interview, 9 September 1985, part 4.

31 Ibid., part 5.

32 E.A. Treadwell interview by David Grenville, 8 December 1985, part 1.

33 Cecil Law interview by Jason S. Ridler, 27 May 2007.

34 Solandt interview, 1 March 1986.

35 Chute interview, part 1. Directorate of History and Heritage Canada (hereafter referred to as DHH), Raymont Fonds 733/1223, box 61, series 3, file 1307, Minutes of the Chiefs of Staff Committee (Regular and Special) 1952–1953, Minutes for Special Meeting of the COSC, 16 April 1953.

36 Hydro One – Hurontario Switching Station Environmental Report, Appendix A, "Backgrounder: Report of the Solandt Commission," http://www.hydroonenetworks.com/en/community/projects/transmission/hurontario_station/FINAL_ESR/Appendix_A.pdf; and Solandt CV, Personal Papers of David Grenville.

37 Maggie Mollison interview, part 1.
38 Solandt interview, 9 September 1985, part 4.
39 Ibid.
40 Solandt to Michael Bliss, n.d., provided by Bliss.

2 Idyllic Childhood

1 David Grenville, "Omond Solandt: A Biographical Sketch," in *Perspectives in Science: The Legacy of Omond Solandt*, ed. Cecil Law, George Lindsey, and David Grenville (Kingston: Queen's Quarterly, 1994), 1–9; "Donald Solandt" and "Omond McKillop Solandt," *The Canadian Who's Who* (Toronto: Trans Canada, 1948), 4:879–80. Solandt interview, 10 September 1985, part 5.
2 This is true even of historical scholarship where Solandt's name appears. One example will suffice. In the Guy Hartcup and T.E. Allbione biography on physicist John Cockcroft, a friend and colleague of Solandt's from his Cambridge days, the authors refer to the November 1943 British mission to the United States to discuss mobile radar sets for the "forthcoming campaign in North-West Europe, how duplication of Allied radar equipment could be avoided, and to decide on radio and radar requirements for the Far Eastern Theatre." The mission was headed by Sir Robert Watson-Watt and included "Osmund [sic] Solandt, a Canadian scientist attached to the British Army Operational Research Group." Guy Hartcup and T.E. Allbione, *Cockcroft and the Atom* (Bristol: Adam Hilger, 1984), 118.
3 Queen's University Archive (hereafter referred to as QUA), Queen's Student Register, vol. 5, call #1161, file 2885, "Donald McKillop Solandt." Oddly, other biographical references to Donald M. Solandt, including the *Canadian Who's Who* entry for 1935, do not include his date of birth.
4 "Archibald McKillop (1824–1905): The Blind Bard of Megantic," *Township Heritage WebMagazine*, http://townshipsheritage.com/article/archibald-mckillop-1824-1905-blind-bard-megantic.
5 Solandt interview, 6 May 1986, part 2.
6 QUA, Queen's Student Register, vol. 5, call #1161, file 2885, "Donald McKillop Solandt."
7 Solandt interview, 10 May 1986, part 5. For more on the HMS *Lily*, see Port Arthur Lighthouse: Shipwrecks, http://www.pointamourlighthouse.ca/home/10.
8 QUA, Queen's Student Register, vol. 5, call #1161, file 2885, "Donald McKillop Solandt."
9 A.B. Baird, "Donald McKillop Solandt, M.A., D.D., Business Manager," *The Chronicle of a Century, 1829–1929: The Record of One Hundred Years of*

Progress in the Publishing Concerns of the Methodist, Presbyterian and Congregational Churches in Canada, ed. Lorne Pierce (Toronto: Ryerson, 1929), 168; "Solandt, Rev. Donald McKillop, B.A., B.D., M.A., D.D." *Who's Who in Canada, 1934–1935* (Toronto: International, 1936), 1672.

10 Grenville, "Omond Solandt," 1–9.

11 "Solandt, Rev. Donald McKillop," 1672.

12 Solandt interviews, 8 May 1986, part 4; 28 November 1986; 29 November 1986, part 2.

13 "Solandt, Rev. Donald McKillop," 1672.

14 D.M. Solandt, "Manitoba," in *The Canadian Patriotic Fund: A Record of Its Activities from 1914 to 1918,* ed. Philip H. Morris, 107–22 (1920).

15 Ibid., 112.

16 Solandt interview, 10 September 1985, part 6.

17 In light of his success with the fund, in 1917 Solandt was also appointed temporary pension commissioner for Manitoba as well. He also took on the duties of organizer of the Interchurch Forward Movement in Manitoba. Maintaining the work ethic that had driven him from the farm, through university, and church duties, it is a small wonder that his sons respected though did not really know their busy father and gravitated toward their mother. Baird, "Donald McKillop Solandt," 168; "Solandt, Rev. Donald McKillop," 1672.

18 Solandt interview, 10 September 1985, part 6.

19 George L. Parker, *The Beginning of the Book in Trade in Canada* (Toronto: University of Toronto Press, 1985).

20 Charles Gordon, *Postscript to Adventure: The Autobiography of Ralph Connors* (New York: Farrar Rinehart, 1938), 361–2; Baird, "Donald McKillop Solandt," 168; "Solandt, Rev. Donald McKillop," 1672.

21 Gordon, *Postscript to Adventure,* 359–60.

22 Ibid., 362.

23 Solandt interview, 10 September 1985, part 6.

24 QUA, Adam Shortt Correspondences, call #2047, box 5, files 29–30.

25 Ibid.

26 Presbyterian Publications was one imprint of the newly named Ryerson Press. The Ryerson Press was created out of the imprints of the William Briggs and "Methodist Book and Publishing House" in 1919, and by 1920 also included Presbyterian Publications. Ryerson Press had achieved notable successes in the late nineteenth century under Edgerton Ryerson and S.W. Fallis, but in 1920 they began their accession in Canadian publication under the leadership of Reverend Lorne Pierce. Pierce, like Solandt, was a Queen's graduate and former student of John Watson, though they did

not attend at the same time. Sandra Campbell, "Nationalism, Morality, and Gender: Lorne Pierce and the Canadian Literary Cannon, 1920–1960," *Papers of the Bibliographical Society of Canada* 32, no. 2 (Fall 1994): 138; Lorne Pierce, *The House of Ryerson 1829–1954* (Toronto: Ryerson and United Church Publishing House, 1954), 41–2.

27 Solandt interviews, 9 September 1985, part 4; 10 September 1985, part 6.

28 Solandt makes this clear himself in the interviews with Grenville. Ibid.

29 Solandt interview, 10 September 1985, part 5.

30 Barbara Griffin interview by David Grenville, 26 February 1986.

31 Solandt interviews, 8 September 1985, part 1; 10 September 1985, part 6.

32 Solandt interview, 10 September 1985.

33 Solandt interview, 8 May 1986, part 3.

34 Solandt interview, 10 September 1985, part 5.

35 Ibid.

36 One anecdote was a family legend. That summer, while the family enjoyed a picnic on the boat after visiting some wealthy friends in the region, Edith dropped her valuable sunburst broach in the water. They could see it at the bottom. Donald's first attempt to climb down the anchor's chain failed, as did both sons' attempts. Finally, Donald hooked two buckets of sand to his feet and let them drag him down to recover the broach. Ibid., part 6.

37 Ibid.

38 Solandt interview, 9 September 1985, part 4.

39 Solandt interview, 28 November 1986; 29 November 1986, part 2.

40 Solandt interview, 10 September 1985, part 6.

41 Jason S. Ridler, "The Road to Ryerson: Reverend Donald McKillop Solandt's Journey to Canadian Publishing, 1889–1936," unpublished 2007.

42 Solandt interview, 10 September 1985, part 6.

43 Ibid.

44 University of Toronto Archives, Omond McKillop Solandt Fonds (hereafter referred to as UT OMS), B93/0041/0033, file DRB Addresses, 49–52, Address to Central Technical School Convocation, Toronto, Ontario, 17 November 1950.

45 Solandt interview, 10 September 1985, part 6.

46 Chute interview, part 1.

47 Solandt interview, 10 September 1985, part 6. Don's interest in literary affairs led him to write a popular health book for grade and high school students not long after he finished his master's degree under Charles Best. *Highways to Health* was published in 1934 by, not surprisingly, Ryerson Press. The book showed Don had a flair for literary work and boiling down the essentials of health science to a popular audience.

48 Ibid.; Chute interview, part 1.
49 Solandt interview, 1 March 1986, part 6.
50 Solandt interview, 10 September 1985, part 6.
51 Ibid., 1985, part 6; UT OMS, B93-0041/007, file Canadian Aeronautics, untitled biographical sketch, 18 September 1987.
52 Solandt interview, 10 September 1985, part 6.
53 Solandt was tested on a ½ Marconi Cabinet Ship Station and the Standard Marconi Emergency Aerial. UT OMS, B88/0016/001, file 1, Certificate of Proficiency in Radiotelegraphy, 23 April 1927.
54 Solandt interview, 10 September 1985, part 6.
55 Ibid.
56 UT OMS, B88/0016/001, file 1, Royal Life Saving Society Certificates, 19 August n.d.
57 It also stated, "A Candidate may take either the third or fifth method of rescue. The other three are compulsory. The drowning subject in the rescue methods must be carried at least ten yards." Every candidate had to swim one hundred yards in the breast and fifty yards on the back, the latter without use of hands. They also had to dive to a depth of five feet or more from the surface of the water and bring up a weighted object of a minimum of two or a maximum of five pounds. The Royal Life Saving Society, *Illustrated Handbook for the Rescue, Release from, and Resuscitation of the Drowning, Teaching, Teaching Swimming, &c., &c., &c.*, 14th ed. (London: Royal Life Saving Society, 1922), 96.

3 Protégé

1 Chute interview, parts 1 and 2.
2 UT OMS, B93-0041/007, file Canadian Aeronautics, untitled biographical sketch, 18 September 1987.
3 Ibid.
4 Pearson, the future Canadian diplomat, Nobel Prize winner, and prime minister, cherished coaching. "I loved doing this. It took up only two of three hours a day, and I had some talent in getting the most out of my players; working them hard but keeping the fun in playing, without all the pressures that now have become common in competitive sport." UT OMS, B93-0041/007, file Canadian Aeronautics, untitled biographical sketch, 18 September 1987. Lester B. Pearson, *Mike: The Memoirs of the Rt Hon. Lester B. Pearson* (Toronto: University of Toronto Press, 1972), 1:50.
5 See Jack Granatstein, *The Ottawa Men: The Civil Service Mandarins, 1935–1957* (Toronto: Oxford University Press, 1982).

6 "Omond McKillop Solandt," *Canadian Who's Who*.

7 Jonathan Vance, *High Flight: Aviation and the Canadian Imagination* (Toronto: Penguin Canada, 2002), 124–5.

8 Ontario Public Archives (hereafter referred to as OPA), RD 1-116, b268616, file Ontario Air Service; A.P. Heathcrote, "Wings over the Wilderness: Ontario Provincial Air Service Aircraft of Yesterday and Today," *Topic* (Nov./Dec. 1965): 4.

9 UT OMS, B93-0041/007, file Canadian Aeronautics, "Aerospace Connection in the Career of Omond Solandt," 2 October 1987.

10 Solandt interview, 8 May 1986, part 3.

11 Both Solandt brothers paid for all extracurricular activities themselves, the family providing support for school, and room and board at home. While times were tight, there is no evidence that Reverend Solandt was out of work at what was then United Church Publishing during the worst of the Depression, so Omond would not think of taxing the family's resources for his own activities.

12 UT OMS, B93-0041/007, file Canadian Aeronautics, untitled biographical sketch, 18 September 1987.

13 Solandt interview, 8 September 1985, part 1; UT OMS, B93-0041/007, file Canadian Aeronautics, untitled biographical sketch, 18 September 1987, n.p.

14 Indeed, Solandt would reflect throughout his career that the one advantage his British-trained colleagues had, especially those he worked with during the war, was a more thorough appreciation of mathematics rooted in solid training in public school. UT OMS, B93-0041/007, file Canadian Aeronautics, untitled biographical sketch.

15 Ibid.

16 Charles Best and Norman Taylor, *The Human Body and Its Functions* (New York: Henry Holt, 1932).

17 UT OMS, B93-0041/007, file Canadian Aeronautics, untitled biographical sketch.

18 Interview notes from Michael Bliss interview of Solandt, 24 September 1980. Transcribed copy provided by Doctor Bliss.

19 Solandt interview, 9 September 1985, part 4.

20 UT OMS, B93-0041/007, file Canadian Aeronautics, untitled biographical sketch.

21 By then, Hill was a Nobel laureate and war veteran who had pioneered the use of scientific methods toward anti-aircraft fire during the Zeppelin raids of 1916–18, a precursor to the operational research Solandt and others would invent during the war, *Margaret and Charley*, 98–9; Maurice Kirby,

Operational Research in War and Peace: The British Experience from the 1930s to the 1970s (London: Imperial War College Press), 34–5.

22 Both men gravitated toward the more congenial Best. Dale suggested he should get an advanced degree. Best agreed and finished his medical degree in Toronto, then, on a Rockefeller fellowship, headed to Britain to work with Dale at the National Institute for Medical Research, London. Best's work with Dale and Hill concerned the treatment of hypertension, and Dale saw to it that all of Best's work would be counted toward an advanced degree. Best finished his doctorate in 1928 at the University of London, his examination done by Dale, Hill, and Professor John Leathes. Best, *Margaret and Charley*, 114–16, 118–19, 121, 123, 125.

23 James A. Marcum, "The Development of Heparin in Toronto," *Journal of the History of Medicine* 52 (July 1997): 334.

24 According to Solandt, Best's early lectures were conscientious and systematic, but not very exciting. UT OMS, B93/0041/007, file Canadian Aeronautics, untitled biographical sketch. Laurie Chute recounted stories of Best administering different substances to fellow students and himself on their Friday afternoon secessions. See Chute interview, part 1.

25 Paul Hellyer interview by Jason S. Ridler, 26 April 2007.

26 Solandt interview, 8 September 1985, part 1.

27 Ibid.

28 Chute interview, part 1.

29 Grenville, "Omond Solandt," 1–9.

30 Chute interview, part 1.

31 Solandt interview by Michael Bliss, 24 September 1980. Transcribed copy provided by author.

32 Solandt interview, 8 September 1985, parts 1 and 2.

33 As mentioned, Solandt also singled out Lawrence Irving's influence as critical to his abilities in OR.

34 Michael Bliss recounts Best's first explosive workday with Fred Banting. Best had done a horrible job cleaning the lab equipment, and Banting angrily dressed down the young Best: any measurement made with the dirty equipment would be worthless. He demanded Best clean everything again. According to Banting, Best reacted to anger with anger, but soon cooled and did as he was told, leaving everything perfect for the next day's work, his mistake never to be repeated. Best's chief responsibility after insulin was discovered was perfecting the method for insulin extraction, a difficult job requiring painstaking measurement. Best toiled through the insulin extraction process first, initially with little positive result. J.B. Collip, another member of the team whose work contributed to insulin's discovery,

made a more purified extract before Best, but refused to share his method. Banting responded with anger again and came close to confronting Collip with physical violence. When Collip's extraction method began to fail, it was Best and his team who "rediscovered" it and provided it to Banting. From such hard lessons was Best's appreciation of detail forged and passed on to the Solandt brothers. Banting's formidable anger, however, was absent from Best's professional methodology. Michael Bliss, *Banting: A Biography* (Toronto: McClelland and Stewart, 1984), 63–5, 82–3, 88, 93, 98.

35 For Solandt, that was the nature of variables in research. OR work in the field often challenged original assumptions about the importance of any variable. One had to observe, measure, and experiment carefully and avoid the intuitive solution, which is usually wrong. Look at the data cold first, since common sense often ignores critical data. Scepticism was their mandate. Solandt interview, 8 September 1985, part 2; Grenville, "Omond McKillop Solandt," 5; Omond Solandt, "Observation, Experiment, and Measurement in Operational Research," *Journal of Operational Research* 3, no. 1 (February 1955): 1–14.

36 Feldberg interview.

37 Sargeant interview, part 2.

38 O.M. Solandt, "The Duration of the Recovery Period, and the Respiratory Quotient of the Excess Metabolism, Following Strenuous Muscular Exercise," MA thesis, University of Toronto, 1932.

39 UT OMS, B93-0041/007, file Canadian Aeronautics, untitled biographical sketch.

40 Solandt, "Duration of the Recovery Period."

41 Ibid., 4–9.

42 Ibid., 76.

43 Jacalyn Duffin, *History of Medicine* (Toronto: University of Toronto Press, 2004), 41.

44 Cecil Law, "Dr Omond McKillop Solandt: Operational Research Pioneer," *INFOR* 31, no. 2 (May 1993).

45 Solandt, "Duration of the Recovery Period," 47; UT OMS, B93-0041/007, file Canadian Aeronautics, untitled biographical sketch.

46 Solandt, "Duration of the Recovery Period," 47; UT OMS, B93-0041/007, file Canadian Aeronautics, untitled biographical sketch.

47 "Omond McKillop Solandt."

48 UT OMS, B93-0041/007, file Canadian Aeronautics, untitled biographical sketch.

49 O.M. Solandt, "The Interpretation of Respiratory Quotients," *University of Toronto Medical Journal* 19, no. 6 (1932): 214–18.

50 O.M. Solandt and Jesse H. Ridout, "The Duration of the Recovery Period Following Strenuous Muscular Exercise," *Proceedings of the Royal Society,* B, 113 (1933): 327–44.

51 O.M. Solandt and G.C. Ferguson, "The Effects of Strenuous Exercise of Short Duration upon the Sugar Content of the Blood," *Transactions of the Royal Society of Canada,* section 5 (1932): 173–9. This is in large part the publication of the majority of Solandt's MA thesis, with a literature review of relevant articles from Best and Ridout.

52 O.M. Solandt, D.Y. Solandt, and R.W. Gerard, "Methemoglobin and Methylene Blue as Cyanide Antagonists," *Proceedings of the Society for Experimental Biology and Medicine* 31 (1934): 539–41; O.M. Solandt, "Recent Advances in the Treatment of Cyanide Poisoning," *Canadian Public Health Journal* (December 1934): 592–8.

53 Solandt, "Recent Advances in the Treatment of Cyanide Poisoning."

54 UT OMS, B93-0041/007, file Canadian Aeronautics, untitled biographical sketch.

55 L. Irving, D.Y. Solandt, and O.M. Solandt, "Nerve Impulses from Branchial Pressure Receptors in the Dogfish." *Journal of Physiology* 84, no. 2 (1935): 187–90.

56 L. Irving, O.M. Solandt, D.Y. Solandt, and K.C. Fisher, "The Respiratory Metabolism of the Seal and Its Adjustment to Diving," *Journal of Cellular and Comparative Physiology* 7, no. 1 (October 1935): 137; Lawrence Irving, O.M. Solandt, D.Y. Solandt, and K.C. Fisher, "Respiratory Characteristics of the Blood of the Seal," *Journal of Cellular and Comparative Physiology* 6, no. 3 (August 1935): 393–403.

57 Irving et al., "Respiratory Characteristics of the Blood of the Seal," 401.

58 "Donald Young Solandt"; M.R. Bennet, "Obituary: Sir Bernard Katz (1911–2003)," *Journal of Neurocytology* 32 (2003): 423.

59 A.V. Hill gave his first paper there in at the ninth congress in Groningen, 1913, nine years before he won his Nobel award in physiology with O. Meyerhoff. He would become a critical member of the congresses' later committees as well as Don Solandt's mentor in England. At the eleventh congress in Edinburgh, 1923, J.J.R. MacLeod delivered a lecture on insulin. According to one account, "The presence of Banting, the actual discover of insulin, at the Congress made the story revealed by Macleod even more vivid in the minds of his hearers." Banting also gave a short paper on insulin. For a condensed yet thorough review of history of these congresses, see K.J. Franklin, "A Short History of the International Congresses of Physiologists," *Annals of Science* 3, no. 3 (15 July 1938): 241–335.

60 Franklin, "Short History of the International Congresses of
   Physiologists," 297.
61 Solandt interview, 8 September 1985, part 3.
62 Franklin, "Short History of the International Congresses of
   Physiologists," 315.
63 Ibid.
64 Solandt interview, 8 September 1985, part 3.
65 Ibid.
66 Best, *Margaret and Charley*, 146n361. This is excerpt from an interview
   Henry Best did with Solandt on 20 February 1991.
67 I.P. Pavlov, "Address by Professor I.P. Pavlov," *Proceedings of the XVth
   International Physiological Congress, Leningrad–Moscow August 9th to 16th
   1935*. Sechenov Journal of Physiology of the USSR (Moscow: State
   Biological and Medical Press, 1938), 11–12.
68 Ibid., 12.
69 See Solandt's speeches from the early Cold War, UT OMS, B93-0041/033,
   DRB Addresses, 1946–1948.

4 Commencement

1 UT OMS, B88-0016/001, file 2, Medical Council of Canada Membership, 2
   July 1936.
2 QUA, Queen's Historical Collection D, Alumni Files 3599, "Solandt,
   Donald M; Young, Edith," death recorded on alumni status card.
3 The story of their romance needs mention. Solandt had met Elizabeth
   McPhedran at Cambridge, where she was visiting friends in the spring of
   1939. She was the daughter of Harris McPhedran, professor of medicine
   at St Michael's hospital in Toronto, and former professor of Solandt and
   Chute. Solandt was taken with her charms and high energy. She stayed in
   Britain after all of Solandt's friends had left after exams, but she returned
   to Canada in 1940. Laurie Chute always felt this had been an incredibly
   lonely time for Solandt. Not long after, Solandt wrote McPhedran, asking
   her to return. Chute believed he might have proposed to her via this letter.
   In late 1940, she returned and the two were wed on 25 January 1941. See
   Solandt CV, and Chute interview, part 1.
4 Solandt interview, 3 March 1986, part 5.
5 Solandt interview, 8 September 1985, part 3.
6 UT OMS, B93/0041/0033, file DRB Addresses, 54–5, Address to University
   of Toronto Convocation, 16 June 1954.
7 "Omond McKillop Solandt."

8 R.A. Kerwick, "Sir Alan Nigel Drury 3 November 1889–2 August 1980," *Biographical Memoirs of the Fellows of the Royal Society* 27 (1981): 173–98. On Drury and Solandt's collaborative efforts, see A. Drury, C. Luttwak-Mann, and O.M. Solandt, "The Inactivation of Adenosine by Blood with Especial Reference to Cat's Blood," *Quarterly Journal of Experimental Physiology* 27 (1938): 215–36.

9 UT OMS, B88-0016/001, file 2, Certificates, Toronto General Hospital Staff Certificate, 30 June 1938.

10 Solandt interview, 8 September 1985, part 1.

11 Ibid.

12 Duffin, *History of Medicine*, 160–1.

13 Christopher Rutty, "The Middle-Class Plague: Epidemic Polio and the Canadian State, 1936–1937," *Canadian Medical Bulletin* 13 (1996): 280.

14 Ibid.

15 H.H. Hyland, W.J. Gardner, F.C. Heal, W.A. Ollie, and O.M. Solandt, "Acute Anterior Poliomyelitis (A Review of Sixty-Six Adult Cases Which Occurred in the 1937 Ontario Epidemic," *Canadian Medical Association Journal* 39, no. 1 (1938): 1–12.

16 Grenville interview by Jason S. Ridler, 26 April 2006.

17 Solandt interview, 8 September 1985, part 1.

18 Ibid.

19 Rutty, "Middle-Class Plague," 296.

20 Ibid.

21 Solandt interview, 9 September 1985, part 4. On the relationships of Banting, Hall, Graham, and Franks, see Bliss, *Banting*, 79.

22 Hyland et al., "Acute Anterior Poliomyelitis."

23 Ibid., 15.

24 Ibid., 11–12.

25 Ibid.

26 Ibid., 19.

27 On Taylor and his influence in the twentieth century, see Robert Kanigal, *The One Best Way: Frederick Winslow Taylor and the Enigma of Efficiency* (London: Penguin, 1997).

28 Hyland et al., "Acute Anterior Poliomyelitis," 24.

29 Ibid., 43–5.

30 O.M. Solandt, "Massive Collapse of Lung," *American Journal of Medical Science* 3, no. 7 (September 1937): 428–43. See also Ronnie Shephard, "The Influence of Solandt on the Development of Operational Research in Britain," in Law, Lindsey, and Grenville, *Perspectives in Science*, 30–47; and Terry Copp, ed., *Montgomery's Scientists: Operation Research in Northwest*

Europe, the Work of No. 2 Operational Research Section with 21st Army Group, June 1944–July 1945 (Waterloo, ON: Laurier Centre for Military Strategic and Disarmament Studies, 2000), 11–12.

31 Solandt, "Observation, Experiment, and Measurement in Operational Research," 3.

32 UT OMS, B88-0016/001, file 2, Ontario College of Physicians and Surgeons Certification, 28 May 1938.

33 Such degrees were titled "Neither Honourary Nor Earned."

34 Drury, Lutwak-Mann, and Solandt, "Inactivation of Adenosine by Blood."

35 W.S. Feldberg and O.M. Solandt, "The Stimulating Effect of Glucose and Pyruvate on the Rabbit's Guts," *Proceedings of the Physiological Society* (4 May 1940): 97–100.

36 UT OMS, B93/0041/0033, file DRB Addresses, 54–5, Address to University of Toronto Convocation, 16 June 1954.

37 Solandt interview, 8 September 1985, parts 3, 4; Feldberg interview.

38 Feldberg interview.

39 Solandt interview, 8 September 1985, part 3.

40 Ibid., part 1.

41 Feldberg interview.

5 First Taste of War

1 Solandt attended sessions of the MRC Blood Transfusion Sub Committee, the first one being fourth on 24 May 1940. All the directors had contributed to reports on clinical trials regarding the relative value of fresh and stored blood for transfusion purposes. The committee was in agreement about the good work done to answer the questions on this topic. The meeting also discussed wound shock and the use of morphine, a topic Solandt would deal with in MRC subcommittee on AFVs. It was at this meeting that Doctor J.O. Oliver had relinquished his directorship at Sutton and that his place would be taken by Solandt. NAUK, FD 1/5892, "Medical Research Council Blood Transfusion Research Sub Committee Minutes, 24 May 1940."

2 There were four depots, all of which had their own independent director: Doctors Janet Vaughan (Slough), H.F. Brewer (Luton), M. Maizels (Maidstone) and, successively, Doctors J.O. Olvier, Omond Solandt, and J.F. Loutit (Sutton).

3 F.H.K. Green and Gordon Covell, *Official Medical History of the War of 1939–1945: Medical Research* (London: Her Majesty's Stationery Office, 1953), 110–37.

4 C.L. Dunn, *Official Medical History of the War of 1939–1945: The Emergency Medical Services.* Vol. 1, *England and Wales* (London: His Majesty's Stationery Office, 1952), 335.

5 Ibid.

6 John Keegan, *The Second World War* (London: Viking, 1989), 88–103.

7 Maggie Mollison interview, 16 October 1985.

8 Patrick Mollison interview, 16 October 1985, part 3.

9 Ibid.

10 Ibid.

11 Maggie Mollison interview, 16 October 1985, part 3.

12 Solandt interview, 9 September 1985, part 4.

13 Patrick Mollison interview, 16 October 1985, part 3.

14 Cecil Law Private Papers, Solandt Symposium Archives, Omond Solandt to Edith Solandt, 18 August 1940.

15 Ibid.

16 Ibid.

17 O.M. Solandt, "The Work of a London Supply Depot," *Canadian Medical Association Journal* 44 (1941): 189–91.

18 "That's the disadvantage for the commander fighting a battle on home ground. People can see what he is up to. His mistake cannot be covered up by security silences and carefully worded *communiqués* and the subsequent thoughtful whitewash of his Public Relations Officer." Frederick Pile, *Ack-Ack: Britain's Defence against Air Attack during the Second World War* (London: George G. Harrap, 1949), 48.

19 Solandt, "Work of a London Supply Depot," 1.

20 Ibid., 2.

21 Ibid.

22 A simple gravity method, using a standard MRC pint bottle, was used to great effect. Doctors rarely failed to get a full bottle of blood in more than 2 cases in 100, even though no attempt has been made to select donors with good veins. After bleeding, donors rested for fifteen minutes, then were given tea or coffee. Donors were bled once every three to six months. Again, Solandt noticed patients' reactions. Donors were "agreeably surprised" at the ease of the procedure. The response was always better from those who had been bled before "than among those who do not know what they are in for." The "human" factor is always observed in cases of human beings being treated. About one in fifty donors fainted, and there were rare cases of "vaso-vagal" attacks. Recovery of these cases is always quick, and the donor "is usually keen to come again in spite of advice to the contrary." The blood was then taken to lab where the bleeding set was

removed and a sterile screw cap put on. Blood was refrigerated at 4 to 6 degrees Centigrade until required. Outside the depot a cubicle for sterile capping was used, and a refrigerator truck was used to store blood and transport it back to the depot. Ibid., 32.

23 Ibid.

24 Ibid.

25 Ibid., 4.

26 Ibid.

27 The transfusion set mentioned used a gas-mantle filter. Three thousand sets were distributed. Equipment was also sent to hospitals in case the depot was out of action. The depot's equipment maintenance was a critical function. They sharpened over one thousand needles a week. Ibid., 5–6.

28 Blood transfusion had been used as early at the Franco-Prussian War, but the results of haemorrhaging made it a dangerous affair. In the First World War, it stopped haemorrhage and shock. Viennese physiologist Karl Landsteiner discovered the different blood groupings in 1900, the ABO system of four groups: A, B, AB, and O, differentiated by the presence of certain proteins (antigens) and blood plasma (antibodies).

> There are two types of antigens A and B; and two kinds of antibodies anti-A and anti-B. Type A blood has A antigens in red cells and anti-B antibodies in plasma. If type A blood is given to a person with type B blood, the anti-A antibodies in the recipient's plasma react with the donor's A antigens in the red blood cells. The donor red cells, "attacked" by the recipient's antibodies, then agglutinate and reduce or even block blood flowing through the capillaries, causing a severe reaction and often death. The agglutinated donor red cells are soon broken down and haemoglobin is released (the process is known as haemolysis) into the plasma of the recipient's blood. The crucial factor producing agglutination and subsequent haemolysis is the type of antigen in the donor's red cells. In the above example, blood types A and B are incompatible because of the way their red cell proteins clump with each other's antibodies. Blood type AB contains no antibodies, so persons of this group can receive any type. Group 0 persons have no antigens and thus can donate blood to any other group.

Richard W. Kapp, "Charles H. Best, the Canadian Red Cross Society, and Canada's
First Blood Donation Programme," *Canadian Bulletin of Medical History* 12 (1995): 30–1.

29 Best led and organized the establishment of the Canadian Red Cross Society's Blood for the Wounded campaign. Between 1940 and 1945, the

Red Cross collected 2,338,533 voluntary donations from a population of 11,507, 000 (United States 13,323,242 pints; Great Britain 3 million). Ibid. 27.
30 Solandt, "Work of a London Supply Depot," 5–6.
31 Green and Covell, *Official Medical History of the War of 1939–1945*, 362.
32 NAUK, FD 1/5892, Research at the Sutton Blood Supply Depot, n.d.
33 Sodium alginate was prepared from seaweed and available under the trade name Manucol. It has a wide range of molecular weights, including the weights of plasma protein. "This suggested that sodium alginate might be a suitable substance for use in the treatment of shock when blood or plasma was not available for transfusion. Experiments soon showed sodium alginate is toxic when injected intravenously and is quite unsuitable for this purpose." Tests done on rabbits, mice, and cats and in blood experiments showed the toxicity when injected was much greater than expected, given its chemical composition. O.M. Solandt, "Some Observations upon Sodium Alginate," *Quarterly Journal of Experimental Physiology and Cognate Medical Sciences* 31, no. 1 (1941): 25–31.
34 Green and Covell, *Official Medical History of the War of 1939–1945*, 88.
35 Ibid., 89.
36 R. Maynon-White and O.M. Solandt, "A Case of Limb Compression Ending Fatally in Uraemia," *British Medical Journal* 1 (22 March 1941); reprinted in *Crush Injuries and Kidney Function: Reprint of Three Articles from the "British Medical Journal," March 22, 1941, with Reports of Cases Collected by the M.R.C. and Editorials* (Tavistock Square: Office of the British Medical Association, 1941), 25–8.
37 Ibid.
38 It is unclear from this article whether Solandt treated her at the depot or the nearest hospital to the depot, St Thomas.
39 Maynon-White and Solandt, "Case of Limb Compression," 28.
40 E.G.L. Bywaters and D. Beall, "Crush Injuries with Impairment of Renal Function," *British Medical Journal* 1 (1941): 427–32.
41 Solandt, "Work of a London Supply Depot," 7.
42 UT OMS, B93/0041/0033, file DRB Addresses, 1954–55, Address to University of Toronto Convocation, 16 June 1954.
43 Solandt, "Work of a London Supply Depot," 8.
44 UT OMS, B93-0041/033, file DRB Addresses, 1946–48, Address to the Defence Medical Association Meeting by OMS, 22 November 1946.
45 UT OMS, B93/0041/0033, file DRB Addresses, 1954–55, Address to University of Toronto Convocation, 16 June 1954.
46 Green and Covell, *Official Medical History of the War of 1939–1945*, 110–37.

6 Tank Doctor

1 Solandt CV.

2 M.M. Postan, D. Hay, and J.D. Scott, *Design and Development of Weapons: Studies in Government and Industrial Organization* (London: Her Majesty's Stationery Office and Longmans, Green, 1964).

3 NAUK, FD 1/6392, Medical Research Council Extracts from Meeting on 22 November 1940. The documentary evidence seems to refute some of the cause and effect in the official history of the MRC's war effort, which states that in October 1940, the MRC approached the Directorate of the Royal Tank Corps through Brigadier C.G. Ling, the representative of the Army Council on the MRC's Body Protection Committee (BPC), with the offer of constructing a laboratory to do research on the fighting efficiency of AFVs and their crews. See Medical Research Council, *Medical Research in War: Report of the Medical Research Council for the Years 1939–1945* (London: His Majesty's Stationery Office, 1947), 140–1.

4 NAUK, FD 1/6392, Notes on Memoranda from Meeting at the War Office 4 December 1940. In attendance were Carmichael, Ling, Brigadier C.G. Bouchier (ADTS), and Lieutenant-Colonel J.A.M. Bond (AFV 1).

5 NAUK, FD 1/6384, Body Protection Committee Minutes, 9 December 1940.

6 Other military members included Captain H.B. Wright, RAMC, Captain R.D. Harkness, RAMC. Lieutenant P. Garrard, RAMC. Lieutenant B.M. Wright, RAMC, worked there for a short period. See Medical Research Council, *Medical Research in War*, 141.

7 Solandt interview, 8 September 1985.

8 The following are all the Solandt reports from the Lulworth period dealing with gas fumes. Mayon-White and Solandt, "Summary of Gassing Trials in Cruiser Tanks, Mark IV CRUSADER," series 1868, no. 88, 1941; Solandt, "Field Gassing Trials in Cruiser Tanks Mark VI and Mark VIA," series 1855, no. 75, August 1941, n.a., linked to Solandt, "Report on Gassing Trial: Infantry Tank Mark III (Medium Machine Gun BESA 7.62 mm)," series 1840, no. 60; Mayon-White and Solandt, "CHURCHILL Armed with 3 in 20 cwt Gun: Report on Gun Fume Trials," series 2042, no. 42/27, 1942; Solandt, "Carbon Monoxide Contamination of Air in Churchill III Tanks from Fumes of 6pr QF and Co-ax BESA Guns," series 2023, no. 42/8, March 1942; Mayon-White and Solandt, "Humber Armoured Car Mark IV Gun Fumes," series 2043, no. 42/28, March 1942; Solandt, "Self-Propelled 25pr Field Gun: Gun Fume Trials on 25 March and 20 April 1942, at Larkhill," series 2044, no. 42/29, April 1942; Mayon-White and Solandt, "CO Contamination of Air in AFVs by

Fumes of 6pr QF Gun," series 2045, no. 42/30, April 1942; Solandt, "Danger from Exhaust Leakage in an Enclosed TETRARCH LT VII," April 1942," series 2046, no. 42/31.

All documents listed from Ronnie Sheppard, *A Bibliography of (Mainly) Army Operational Research, 1939–1948*, Department of Management Sciences Working Paper OR/WP/21 (Shrivenham: Royal Military College of Science).

9 UT OMS, B93 0041/031, file Notes on US Tri Partite Operations; Omond Solandt, "Notes on CO Poisoning," n.d.

10 The most valuable being Esther M. Killick, "The Effects of Repeated Exposure to Carbon Monoxide Poisoning" (1933); Killick, "The Acclimatization of the Human Subject to Atmospheres Containing Low Concentration of Carbon Monoxide" (1936); Killick, "The Effects of Repeated Exposure to Carbon Monoxide Poisoning," *Transaction of the Institution of Mining Engineers*, 84, part 4 (1933): 268–78; and Killick, "The Acclimatization of the Human Subject to Atmospheres Containing Low Concentration of Carbon Monoxide," *Journal of Physiology* 87, no. 1 (1936): 41–55. Collected in UT OMS, B93 0041/031, and stamped "Medical Research Council: AFV School Lulworth Camp Dorset."

11 Commonly used propellants such as cordite and nitro cellulose produced 800–900 cc of gas per grams of explosives under normal conditions, and 30–40 per cent of this is CO. UT OMS, B93 0041/031, file Notes on US Tri Partite Operations, Notes on CO Poisoning.

12 UT OMS, B93 0041/031, file Notes on US Tri Partite Operations, Lecture on CO Poisoning in Tanks.

13 NAUK, FD 1/6392, Report from Dr O.M. Solandt, 5 March 1941.

14 Ibid.

15 Ibid.

16 Ibid.

17 NAUK, FD 1/6392, Solandt to Carmichael, regarding Report from Dr O.M. Solandt, 5 March 1941.

18 NAUK, FD 1/6392, Carmichael to Mellanby, 5 April 1941.

19 Ibid.

20 Ibid.

21 Ibid.

22 NAUK, FD 1/6392, Mellanby to Carmichael, 10 April 1941.

23 After the war, Solandt wrote that one had to know how the military did their business before any suggestion of merit from an outside actor could be discussed. Solandt, "Observation, Experiment, and Measurement in Operational Research."

24 On 23 December he reported to Mellanby that "Mr Ramsbothom from
   Rolls Royce is now in control of all tank design. I think it would greatly
   assist our work if you could arrange for me to see him." Ramsbothom was
   believed to be directly under Lord Beaverbrook at the Ministry of Supply.
   Mellanby contacted H.J. Gough at the Ministry of Supply about Solandt's
   request and stipulated, "You can be assured that Solandt is a first class
   scientific worker, and I am quite certain that anything he has got to say to
   Ramsbothom would be well worth hearing." Mellanby informed Solandt
   that Ramsbothom's particular job is "to design new tanks and I told Gough
   that it was more important therefore that you should meet him and Gough
   agreed." NAUK, FD 1/6392, Mellanby to Solandt, 31 December 1941.
25 Lovett-Evans was also a friend of Charlie Best and mentor to Laurie
   Chute. See Chute interview.
26 NAUK, FD 1/6392, Report, 27 May 1941, "Gassing Trial–Infantry Tank
   Mark III; Medium Machine Gun-Besa 7.92 mm." Headaches and general
   nausea were felt by the crew after the trial, though had the trial continued
   for fifteen minutes the results would have been serious, especially if the
   conditions were the more physical strains of battle. There had been no
   research yet on if taking breaks where fresh air could be taken would effect
   results, but Solandt's lab was working on this concern.
27 Ibid.
28 NAUK, FD 1/6392, 11 June 1941, Report MRC Physiological Laboratory,
   AFV School, Lulworth, "The Effect of Repeated Short Exposures to Carbon
   Monoxide."
29 Ibid.
30 Ibid.
31 The lab had pre-selection of personnel in their mandate.
32 NAUK, FD 1/6392, "Some Notes on Armoured Fighting Vehicles," 12 June
   1941.
33 This included vision, posture and instrument design, food, and pre-selection
   of personnel. Ibid.
34 NAUK, FD 1/6391, Report on Meeting with DTD, 29 May 1941. It discussed
   the value of increasing AFV ability to fight before daybreak and at night,
   comparative analysis with German telescopes, and the impossibility of
   escape from the Cruiser Mark VI when the turret is reversed. He requested
   equipment tests on luminous paints on dials. Solandt would leave most of
   the visual research to the resident expert, Doctor Poachin.
35 NAUK, FD 1/6391, Carmichael to Mellanby, 7 June 1941.
36 The original meeting consisted of Sir Edward Mellanby (chairman), Doctor
   T. Bedford, Colonel J.A.M. Bond (Directorate of Armoured Fighting

Vehicles), Professor C. Lovett Evans (Chemical Defence Station: Porton), Colonel Rycroft (Directorate of Tank Design), Colonel A.H. Sandford (RAMC), Doctor K.J.W. Craik (Lulworth), Doctor O.M. Solandt (director of Lulworth), and Doctor E.A. Carmichael (secretary). Mister A.H.R. Smith, Canadian liaison officer, Canadian Tank Corps, also attended.

37 NAUK, FD 1/7057, AFV Sub Committee Minutes, 7 July 1941.

38 NAUK, FD 1/6393, Carmichael to Mellanby, 5 February 1941.

39 NAUK, FD 1/7057, AFV Sub Committee Minutes, 18 August 1942.

40 NAUK, FD 1/7057, AFV Sub Committee Minutes, 29 September 1941.

41 NAUK, FD 1/7057, AFV Sub Committee Minutes, 24 December 1941.

42 NAUK, FD 1/7057, AFV Sub Committee Minutes, 9 February 1942.

43 Ibid.

44 NAUK, FD 1/7057, AFV Sub Committee Minutes, 4 May 1942.

45 NAUK, FD 1/7057, AFV Sub Committee Minutes, 6 July 1942.

46 "Operational Research in War and Peace," *Nature* 160, no. 4072 (15 November 1947): 661.

47 NAUK, FD 1/6394, Carmichael to Mellanby, 28 January 1941.

48 NAUK FD 1/6398, Lieutenant-Colonel H.L. Marriott, RAMC, "Military Personnel Research Committee of the Medical Research Council, Report on Visit to Middle East Forces, 14 July 1941." Marriott was instructed by Brigadier Ling to stay with the injured because he was "anxious not to bother the tank authorities at the front." A Lieutenant-General McArthur was contacted about Marriott's arrival, since he was going on the auspices of the War Office to Eastern Command, not the RAMC. "This would circumvent delays and difficulties out East and would give Marriott a much better standing with the operational people out there," argued Carmichael.

49 7th Armoured Division combatants interviewed were Major G.C. Webb, DAQMG, Major G.P.B. Roberts, Major J.G. Stephens, Captain P.R.C. Hobart, Captain C. de Lisle, Lieutenant H.G. Heathrall, Second Lieutenant P.K. Earl, Colonel Q. Wallace, MC, ADMS, and Major E.F.S. Morrison, DADMS. Marriott also interviewed Captain T.V. Somerville, DSO, MC, RMO, 3rd Hussars, who had Western Desert experience and was awarded his DSO for attending wounded under fire.

50 A Major Stephens was caught in an A9 tank with a broken track on the Abyssinian border and fought with it closed for three-quarters of an hour, during which were fired 750 rounds of Vickers and 33 rounds of the two-pounder with no ill effects. NAUK, FD 1/6398, Lieutenant-Colonel H.L. Marriott, RAMC, "Military Personnel Research Committee of the Medical Research Council, Report on Visit to Middle East Forces, 14 July 1941."

51 Marriott complained that getting direct evidence of battle casualties regarding the dead was also difficult. There was no diagnosis of soldiers killed in action that had not reached a field ambulance. They were simply listed as "Killed by enemy action." See NAUK, FD 1/6398, Lieutenant-Colonel H.L. Marriott, RAMC, "Military Personnel Research Committee of the Medical Research Council, Report on Visit to Middle East Forces, 14 July 1941."

52 NAUK, FD 1/6398, E.A. Carmichael, "The Need for Scientific Observers in Battle Areas," 28 August 1941, draft.

53 NAUK, 1/6392, Solandt to Mellanby, 8 September 1941. Solandt then suggested that if no one in England was suitable or available, they could use Captain Thomas Armstrong, who had studied with Drury at Cambridge, who was then in the Middle East and was with the tank corps for the last eighteen months.

54 NAUK, FD 1/7057, AFV Sub Committee Minutes, 29 September 1941.

55 NAUK, FD 1/6392, Solandt to Mellanby, 3 October 1941.

56 NAUK, FD 1/6424, Bedford to Major Sieff, War Office, 14 January 1942.

57 NAUK, FD 1/7057, 18 August 1941. Chute's leaving would also facilitate a manpower shortage at Lulworth, already overworked, and lead to minor conflict between Carmichael and Solandt about Chute's replacement and the need for a physicist at the lab. See NAUK, FD 1/6393, Carmichael to Thomson, 20 January 1942.

58 It is unclear if Solandt had anything to do with Spooner's selection.

59 Shepard, "The Influence of Solandt on the Development of Operational Research in Britain," 30–44.

60 Chute met resistance from the senior commanders for merely being "another bloody tourist" in their theatre of operations. He also contended with Red Cross teams who would not drive him to the front, since they felt any medical research on casualties could be used for weapons design and thus broke the Geneva Convention and medical ethics. Chute thought both claims were idiotic, since he was more concerned with saving future tank crews and winning the war against the fascists than anything else. Chute reported on burns because of enemy action, the use of anti-gas equipment, casualties in AFV crews and factors effecting efficiency, temperature, vision devices, fatigues in AFV crews, assorted concerns of AFV crews, use of tinted-red anti-gas shields, the effects of HE shells on personnel in AFVs, and anti-aircraft casualties. Chute interview.

61 See NAUK, WO 222/73, Major L. Chute, "Reports on Burns Due to Enemy Action," 11 May 1942.

62 NAUK, FD 1/7057, AFV Sub Committee Minutes, 6 July 1942. Chute interview.

63 University of East Anglia Archives (hereafter referred to as UEA), Solly
  Zuckerman fonds, SZ/OEMU/21/3/41, Omond Solandt, "Comments on
  'Reports on Casualties in AFVs (B.P.C. 42/115 by Major A.L. Chute),"
  6 October 1942.
64 Ibid.
65 NAUK, FD 1/7057, AFV Sub Committee Minutes, 1 October 1942.
66 NAUK, FD 1/6393, Mellanby to Parkinson, 26 August 1942.
67 NAUK, FD 1/6393, Mellanby to assistant director of tank design, 26
  August 1942.
68 NAUK, FD 1/6393, Mellanby to Oliver Lucas, 23 September 1942.
69 Ibid.
70 NAUK, FD 1/7057, AFV Sub Committee Minutes, 12 November 1941.

7 Managing Science

1 See Solandt interview, 9 September 1985, part 4.
2 Solandt interview, 8 September 1985, part 1.
3 P.M.S. Blackett, "Scientists at the Operational Level" (1941), reprinted
  in "Operational Research," *Advancement of Science* 17 (1948), collected
  in *Studies of War: Nuclear and Conventional* (Edinburgh: Oliver & Boyd,
  1962), 169.
4 UT OMS, 93/0044/033, file Addresses 1946–1952, Operational Research.
5 The original "circus" included such scientific luminaries as physiologists
  D.K. Hill, son of A.V. Hill, and A.F. Huxley, physicists A. Proter and F.R.N.
  Nabarro, astrophysicist H.E. Butler, general physicist I. Evans, and surveyor
  G.W. Raybould. Latecomers would include physiologist L.E. Bayliss, and
  mathematicians A.J. Skinner and M. Keast. Many of these would be critical
  members of Solandt's AORG team. Brian Austin, *Schonland: Soldier Scientist*
  (London: London Institute of Physics, 2003), 212.
6 Ibid., 215.
7 Schonland's rise in operational research is well covered by Brian Austin's
  biography. See ibid., 224–6.
8 Ibid., 251.
9 NAUK, WO 291/22, "Reconstitution of Operational Research Group,
  Ministry of Supply," 26 January 1943.
10 Solandt had worked with Ellis's OR unit in July 1942 on a report based on
  his analysis of Chute's casualty report. NAUK, WO 291/73, Operational
  Research Group Report No. 78, "Analysis of AFV Casualties – 'I': Tanks
  Mark III," December 1942.
11 Solandt interview, 8 September 1985, part 1.

12 Copp, *Montgomery's Scientists*, 11. Solandt was also instrumental in establishing the Tank Armament Research Establishment at Porton in October 1943, which worked with ORS4 throughout the war. See Kirby, *Operational Research in War and Peace*, 119.

13 UT OMS, B93-0041/31, file US Army Tripartite Operations, O.M. Solandt, "Lecture on Tank Operational Research."

14 Ibid.

15 Lieutenant-Colonel Barlow of the Directorate of Artillery investigated the internal risk of fire due to poor armour on ammunition, and deduced that those who died in tanks with severe burns had in fact been killed first by bullet wounds and not the burns themselves. After these investigations, research on petrol-proof clothing stopped. Ibid.

16 This anecdote has been recalled by Solandt and Sargeaunt. See Solandt interview, 8 September 1985, part 1; and H.A. Sargeaunt interview, 4 December 1985.

17 Solandt interview, 8 September 1985, part 2.

18 Many reports originated from his Lulworth research on the efficiency of tank gunnery in a series of operational conditions, including firing while on the move as well as at night, the nature of AFV casualties, turret design, vision, and crew gear, The following reports by Solandt at AORS 4 were identified by Ronnie Shepard. Solandt, "Fire on the Move from Tanks," series 96, no. 96, October 1943; D.M. Solandt (*sic*), "Tank Turret Ring Diameter," series 389, no. 1009, April 1943; O.M. Solandt, "All Round Vision Device," series 491, no. 11, March 1943; Solandt, "Firing Tests for Tank Gunners," series 506, no. 26, March 1943; Solandt, "Firing Gear for Tank Guns (Progress Report)," series 550, no. 70, May 1943; Solandt, R.J. Whitney, and J.G. Wallace, "The Problems of Magnification or Tank Sighting Telescope," series 554, no. 74, May 1943; Solandt, "Analysis If Results Obtained in Battle with 6pr 7cwt Guns," series 670, no. 190, October 1943; O.M. Solandt and T. Lewis, "Observations on Tank Casualties," series 742, no. 262, March 1943.

19 Along with Sargeaunt, Solandt's initial staff in 1943 at ORS 4 included E. Benn, J.M. Mitchison, G. Waugh, P.E. Williams, E. Harrison, G.V. Price, J.G. Wallace, and R.J. Whitney. See Shepard, 43.

20 Solandt complained to Patrick Mollison while at Sutton that he felt badly about his limited knowledge of the arts. Patrick Mollison interview.

21 Sargeaunt interview.

22 Ibid.

23 When the problems were raised again six months later, Sandys argued it was a pity that all the reports had been burned when, to his astonishment, six appeared to survive. Solandt's short, clear, forceful memoranda

unfortunately did not bear fruit, as the Sherman tanks had come to be seen as the panacea to British tank concerns, but Sargeaunt felt it was one of the best reports from the era. Sargeaunt interview.

24 In January 1944, AORS4 released a memorandum on the progress of British tank design that was highly critical, offering suggestions for rectifying many failures, not only of the tanks but the system that provided them. Schonland viewed the report as helpful and constructive, but Doctor H.J. Gough, director of scientific research and development, Ministry of Supply, felt it was not AORG's purview to criticize his HQ's organization, and should stick to technical matters. Solandt participated in these discussions and was in part responsible for hunting down and destroying the memorandum. There is no indication that Solandt disagreed with the memorandum and, indeed, Schonland initially endorsed it. Solandt likely did not relish the destruction of critical thinking from his old OR section. His opinion of Gough throughout the war would be largely negative as well. It is also unclear if this is the report Sargeaunt referred to, though many copies of the report were not destroyed and located in the National Archives. NAUK, WO 185/229, Gough to secretary of the Directorate of Armoured Fighting Vehicles, 11 January 1944; Solandt to Paris, 7 February 1944.

25 Neville Mott, *A Life in Science* (London: Taylor and Francis, 1986), 63.

26 G.D. Kaye, an actuary who worked at AORS1 and later at the DRB, often worked with Mott on calculations for the lethality of weapons, and they often referred themselves in this fashion. G.D. Kaye interview by David Grenville, 18 September 1985.

27 Mott's calculations on fragmentation eventually became known as "Mott's Law."

28 Solandt interview, 8 September 1985, part 2.

29 Kirby, *Operational Research in War and Peace*, 119–1920.

30 Ibid.

31 Ibid. He also cited biologist Doctor Reginald Whitney's work at Larkhill on improving anti-tank gun accuracy as first-rate. "The secret of Dr Whitney's success was his knowledge of how to design and experiment to eliminate uncontrollable variables. This had been an integral part of his normal advance training in biology. The gunners were keenly aware of the fact that the experience of the gunner had a very large effect on the accuracy of fire and so had made sure that all firings were done by the most highly experienced anti-tank gunners that they could find. They had assumed that this would totally eliminate the effect of learning. Reg as a newcomer observed that even the most experienced anti-tank gunner was visibly taken aback by the huge bang, kick, flash and blast that happened when

the new gun was fired. Even the most experienced gunner was apprehensive on his first few rounds and did not aim as well as he should. All of the gunners had followed in the same sequence, each starting with the targets at the shorts range and working out to longer ranges. In Reg's experiment he properly randomized the opening, and subsequent ranges of the effect of learning was eliminated as far as possible." UT OMS, 0041/31, file International Symposium on Military Operational Research, "Notes for an After Dinner Address by O.M. Solandt," 4 September 1986.

32 At the Defence Research Board, Doctor Cecil Law was always impressed with this approach. Solandt would always be interested and ask terrific questions, but he left it to the establishments to get the job done, and done well. Solandt had no tolerance for poor performances or sloppy research. Law interview.

33 Sargeaunt interview.

34 The control meetings were important but infrequent affairs. Few of their minutes have been found. They were attended by representatives of the senior scientific organizations within the War Office and the Ministry of Supply. Among them was Cockcroft and Solandt's Cambridge friend Lieutenant-Colonel Owen Wansbrough Jones, representing the Directorate of Special Weapons and Vehicles, War Office. See NAUK, WO 233/22, Minutes of the Control Meeting, Army Operational Research Group, 9 September 1943.

35 The third meeting on 20 May 1943, where he announced he would work with Lieutenant-Colonel Nigel Balchin at DBR on laying and firing of the seventeen-pounder. That day they also discussed the revised plan for sending OR sections abroad, likely music to Solandt's ears. NAUK, WO 233/22, Minutes of the Third Control Group Meeting, Army Operational Research Group, 20 May 1940.

36 Solandt interview, 9 September 1985, part 4.

37 Frank Nabarro, section head for AORS 7, was making the same salary as a graduate student in peacetime, roughly 600 pounds, despite the importance of his work and the size of his staff, often with military members receiving much higher salaries. Solandt argued this point with physicist and novelist C.P. Snow, then responsible for scientific manpower at the Ministry of Labour, who had no idea this was happening. Wages improved, but never to Solandt's satisfaction. He blamed in part the "bastard" nature of AORG having two parents, as much as Schonland, for the state of affairs. Solandt interview, 9 September 1985, part 4.

38 NAUK, WO 291/595, Army Operational Research Group memorandum no. 261, "Report on a Visit to the U.S.A. and Canada by O.M. Solandt."

39 Austin, *Schonland*, 267–8.

40 Solandt mentioned his deputies only in September 1944, and there is no indication he had a deputy until then. WLA RSC, Solandt to Schonland, 25 September 1944; Imperial War Museum Archives (hereafter IWM), Basil Schonland Papers, 86/63/01, "Some Recollections of My Time with 21st Army Group," draft, n.d. OR scientist Doctor Ivor Evans believed Bayliss to have been deputy to Ratcliffe at Petersham, then to both Schonland and Solandt at AORG. IWM, Ivor Evans Papers, box P426, file A2/A2, Evans to Mrs J.B. Alton, 28 February 1978.

41 WLA RSC, transcription of interview with Swann by Copp. On Solandt's "weaker hand," see IWM, Basil Schonland Papers, 86/63/01, Lewis to Schonland, 26 April 1946.

42 Treadwell interview.

43 Sargeaunt interview.

44 Treadwell interview; UT OMS, B93-0041/31, file UK AORG Correspondences and Reports, "Operational Research," written for the final AORG Coversaziones, 26–7 July 1945.

45 Solandt, "Observation, Experiment, and Measurement in Operational Research."

46 See Solly Zuckerman, *From Apes to Warlords* (London: Harper and Row, 1978); J.G. Crowther and R. Whiddington, *Science at War* (London: His Majesty's Stationery Office, 1947), 99–113.

47 Kaye interview.

48 Treadwell interview, part 2.

49 Solandt interview, 8 September 1985, part 2.

50 UT OMS, B93-0041/0031, Address to University of Toronto Convocation, 1954.

51 Treadwell interview.

52 WLA RSC, Solandt to Schonland, 21 December 1944.

53 Churchill was both prime minister and minister of defence.

54 Treadwell interview, part 2; Sargeaunt interview, part 2; Solandt interview, 14 March 1990.

55 Solandt interview by Grenville, 13 November 1985.

56 WLA RSC, Solandt to Schonland, 12 August 1944.

57 "Not only was this condition met but I was also made a full member of the Chiefs of Staff Committee so had no possible excuse for failure." UT OMS, 0041/31, file International Symposium on Military Operational Research, "Notes for an After Dinner Address by O.M. Solandt," 4 September 1986.

58 Austin, *Schonland*, 260–70.

59 UT OMS, 0041/31, file International Symposium on Military Operational Research, "Notes for an After Dinner Address by O.M. Solandt," 4 September 1986.

60 WLA RSC, Solandt to Schonland, 13 September 1944.

61 UT OMS, 0041/31, file International Symposium on Military Operational Research, "Notes for an After Dinner Address by O.M. Solandt," 4 September 1986.

62 IWM, Basil Schonland Papers, 86/63/01, "Some Recollections of My Time with 21st Army Group," draft, n.d. For Solandt, Johnson was affable but a useless OR team leader. His team succeeded despite, not because of, his leadership. According to Solandt, Schonland agreed with him, but he could not remove Johnson unless he found him a position of equal significance. Solandt interview, 8 September 1985, part 1.

63 Solandt interview, 8 September 1985, part 1.

64 Solandt said that while the Indian OR section was set up, they lost three men to it who had only recently been recouped. WLA RSC, Solandt to Schonland, 12 August 1944.

65 WLA RSC, Schonland to Solandt, 13 August 1944.

66 UT OMS, B93-0041/0031, file UK AORG Solandt and OR in 21st Army Group, Sargeaunt to Solandt, 26 July 1944, 25 September 1944.

67 WLA RSC, Solandt to Schonland, 6 September 1944.

68 He wrote to Johnson about the need to get Sargeaunt back to Lulworth so that the tank section could be put back on the right lines. WLA RSC, Solandt to Johnson, 16 October 1944.

69 This exchange is well described in Austin, *Schonland*, 289–90.

70 Austin also does not note that Schonland also apologized for having the AORG scientists waiting so long. WLA RSC, Schonland to Solandt, 7 October 1944.

71 WLA RSC, Solandt to Schonland, 25 September 1944.

72 Ibid.

73 Ibid.

74 Ibid.

75 WLA RSC, Solandt to Johnson, 16 October 1944

76 WLA RSC, Solandt to Schonland, 12 October 1944.

77 UT OMS, B93-0041/0031, file 11, Solandt to Schonland, 4 March 1945.

78 UT OMS, B93-0041/0031, file 11, Solandt to Schonland, 26 March 1945.

79 WLA RSC, Johnson to Schonland, 14 May 1945.

80 Solandt often sought his old boss's advice, including on whether or not he should accept Canada's initiative to run a peacetime defence research organization, and also attempted to secure Schonland work in Canada,

should he be interested. UT OMS, B93-0041/0031, file 11, Solandt to Schonland, 5 September 1945.

81 WLA RSC, Solandt to Schonland, 15 August 1945.

82 Austin, *Schonland*, 300.

83 WLA RSC, Schonland to Solandt, 19 August 1945.

84 Solandt interview, 8 September 1985, part 1.

85 Austin, *Schonland*, 290.

86 Ibid. Sargeaunt replaced Johnson and led No. 2 ORS to great effect. Ibid.; Solandt interview, 8 September 1985, part 1.

87 WLA RSC, memorandum, "The Beginning of the End" and "SA/AC Department," 31 January 1945.

88 WLA RSC, Solandt to Schonland, 21 December 1944. In early February, Tony Sargeaunt congratulated Solandt on his "promotion to DSA. Let's hope you will be given the appropriate rank to lend more power to your elbow. You most certainly deserve it." Indeed Sargeaunt was hopeful that now someone "with a clue" would be in the War Office and hoped Solandt would soon become scientific advisor for the Army Council. WLA RSC, Sargeaunt to Solandt, 2 February 1945.

89 UT OMS, B93-0041/0031, file 11, Solandt to Schonland 4 March 1945.

90 IWM, Basil Schonland Papers, 86/63/01, D.K. Hill to Schonland, 7 February; Spencer Humbey to Schonland, 23 July 1945.

91 WLA RSC, Solandt to Schonland, 6 June 1945.

92 UT OMS, B93-0041/31, file U.S. Army Tripartite Operations, "Army Operational Research in South East Asia," n.d.

93 Ibid.

94 The itinerary of the trip included visits to Q ALFSEA, Barrackpore; the OR sections at 14th Army HQ in Meiktilla, Burma (9 May); SEAC HQ, Kandy (15 May); GHQ I in New Delhi (May 16–17); and Dehradun (May 19). He also attended a conference held by Brigadier Welch on future work of ORS and user trial establishments (UTE) as well as post-war plans (20 May). He witnessed trials by UTE on their method of work and experiment, then visited the Ordnance Laboratories at Awnpore (May 21–2), before returning to sections of GHQ in New Delhi before heading home (29 May). UT OMS, B93-0041/31, file U.S. Army Tripartite Operations, "Army Operational Research in South East Asia, Appendix B, Itinerary," n.d.

95 This included the SA not being a member of the Army Council, but being under the deputy chief of the Imperial General Staff, two different parent organizations for AORG, and the separation of the ORS teams in Europe from AORG.

96 UT OMS, B93-0041/31, file US Army Tripartite Operations, "Science in the Army in India," n.d.

97 Ibid.

98 UT OMS, B93-0041/31, file US Army Tripartite Operations, "Army Operational Research in South East Asia."

99 Ibid.

100 Ibid.

101 WLA RSC, Solandt to Schonland, 6 June 1945; Sargeaunt to Solandt, 2 February 1945.

102 UT OMS, B93-0041/0031, file 11, Solandt to Schonland, 4 September 1945.

103 Balchin was amazingly competent, Solandt felt, but a "menace" in his own way, putting across things to Ellis that they had not fully considered. UT OMS, B93-0041/0031, file 11, Solandt to Schonland, 4 September 1945.

104 UT OMS, B93-0041/31, file UK AORG Correspondence and Reports, Solandt to Sir Edward Appeleton, 9 October 1945.

105 UT OMS, B93-0041/31, file UK AORG Correspondence and Reports, Solandt to Lansborough Thomson, 9 October 1945.

106 These were annual get-togethers that included informal reports for AORG. Started by Schonland, the meetings were celebratory but also dealt with highs and lows. The "Conversazions" pamphlet included words from
the superintendent, jokes, and songs, as well as observations on the organization and its goals for the next year.

107 Solandt apologized to Schonland for not putting 21 Army Group on a higher plateau but felt Schonland had not been there long enough to make the same mark he had with AA command. UT OMS, B93-0041/0031, file 11, Solandt to Schonland, 4 September 1945.

108 Ibid.

109 UT OMS, B93-0041/31, file UK AORG Correspondence and Reports, "Operational Research" written for the final AORG Conversazions, 26–7 July 1945.

110 UT OMS, B93-0041/0031, file 11, Solandt to Schonland, 4 September 1945.

111 Mountbatten was considered a pro-science theatre commander, having utilized scientists, both good and bad, while operating the Combined Operations Headquarters. See Zuckerman, *From Apes to Warlords*, 150–71.

112 UT OMS, B93-0041/0031, file 11, Solandt to Schonland, 5 September 1945.

8 Atomic Battlefield

1 Solandt interview, 10 May 1986, part 6.
2 John Hersey, *Hiroshima* (London: Vintage, 1946).
3 Donald Avery, *The Science of War: Canadian Scientists and Allied Military Technology during the Second World War* (Toronto: University of Toronto Press, 1998), 176–202.
4 UT OMS, B1991-0015/011, file 5, Solandt, "Canadian Involvement with Nuclear Weapons – 1946–1956," draft, 21 February 1987, 1–3.
5 Brian Austin interview by Jason S. Ridler, 5 May 2006.
6 Solandt is referred to as "Osmund [*sic*] Solandt, a Canadian scientist attached to the British Army Operational Research Group." Hartcup and Allbione, *Cockcroft and the Atom*, 114–16.
7 The literature on the USSBS is vast, including more than 200 reports of the survey. For an introduction to their content and organization, see *The United States Strategic Bombing Survey Summary Report (European War)* (30 September 1945); David McIsaac, *Strategic Bombing in World War Two: The Story of the United States Strategic Bombing Survey* (New York: Garland, 1976). Other works of note on the USSBS are Paul Nitze, with Ann M. Smith and Steven L. Rearden, *From Hiroshima to Glasnost: At the Centre of Decision* (New York: Grove Weidenbfeld, 1989), 25–45. For a critical view of the USSBS work in the Pacific, see Gian P. Gentile, "Shaping the Past Battlefield 'For the Future': The United States Strategic Bombing Survey's Evaluation of the American Air War against Japan," *Journal of Military History* 64 (October 2000): 1085.
8 UT OMS, B1991/0015/011, file 6, Truman to D'Olier, 15 August 1945 (emphasis added).
9 McIsaac, *Strategic Bombing in World War Two*, 37, 64, 183.
10 Margaret Gowing, *Britain and Atomic Energy, 1939–1945* (London: MacMillam, 1965).
11 See table 2; and UT OMS, B1991-0155/011, file 4, Solandt, "The Atomic Bomb," September 1980, incomplete manuscript, 100–1.
12 Solandt interview, 8 May 1986, part 5.
13 UT OMS, B1991-0155/011, file 4, Solandt, "Atomic Bomb."
14 UT OMS, B1991-015/011, file 6, "Directive to Colonel Solandt," U.S. Assistant Chief of the Air Staff (Operations) Air Ministry, on Behalf of the Chiefs of Staff, n.d.
15 Solandt is likely referring to General Kenneth Nichols, an Army engineer who had served as deputy to General Leslie Groves throughout the war. See Richard Rhodes, *The Making of the Atomic Bomb* (New York: Touchstone, 1988) 426–7, 429, 490, 548, 610, 768.

16 UT OMS, B1991-0015/011, file 16, Solandt speech, "Jacob Bronowski in Japan – 1945," n.d. It was given shortly after Bronowski's death in 1975.
17 UT OMS, B1991-015/011, file 4, Solandt, "Atomic Bomb," 102.
18 UT OMS, UT OMS, B1991-0015/011, file 16, Solandt speech, "Jacob Bronowski in Japan – 1945."
19 Ibid. It is possible that the scientist with whom the mission spoke was Doctor Yoshio Nishina, the leader of the "Rikken" group of Japanese scientists tasked with developing atomic power during the war. See UT OMS, B1991-0155/011, file 6, "Intelligence Memorandum, 10 November 1945: Japanese Survey of Atomic Bombing of Hiroshima and Nagasaki."
20 UT OMS, B1991-0155/011, file 4, Solandt, "Atomic Bomb."
21 UT OMS, B1991-0015/011, file 16, Solandt speech, "Jacob Bronowski in Japan – 1945."
22 The British Mission to Japan consisted of two teams, one under Professor Thompson (RAF Field Team 1, Group 1) and the other under Solandt (RAF Field Team 1, Group 2).
23 UT OMS, B1191-0015/001, file 6, diary of Colonel O.M. Solandt on Chiefs of Staff Mission to Japan, October to December 1945.
24 Jacob Bronowski, *Science and Human Values* (London: Hutchinson of London, 1961), 13. Bronowski first presented these views on 26 February 1953 at the Massachusetts Institute of Technology.
25 UT OMS, B1991-0155/011, file 16, "Jacob Bronowski in Japan – 1945," 6.
26 UT OMS, B1991-0155/011, file 13, Solandt letter from Nagasaki, 6 November 1945.
27 UT OMS, B1991-0155/011, file 19, "Casualties Due to the Atomic Bomb at Hiroshima and Nagasaki," n.d., 4–5, likely written at the end of November.
28 UT OMS, B1991-0155/011, file 7, "Diary: Monday Nov 5 and Tuesday Nov 6 – Casualties" (1945), 7–8.
29 Ibid.
30 Ibid., 10.
31 Ibid.
32 UT OMS, B1991-0155/011, file 7, "Diary: Sunday Nov. 11" (1945), 24.
33 Ibid.
34 UT OMS, B1991-0155/011, file 19, handwritten notes for typist of report.
35 *The Effects of Atomic Bombs at Hiroshima and Nagasaki: Report of the British Mission to Japan* (London: His Majesty's Stationery Office, 1946), 78–88.
36 UT OMS, B1991-0155/011, "Casualties Due to the Atomic Bomb at Hiroshima and Nagasaki," chapter draft, n.d. Many versions of this chapter, which appears in the final report, maintain that one reason for the extensive

damage done at both Hiroshima and Nagasaki was Japan's poor civil defence system.

37 *Effects of Atomic Bombs at Hiroshima and Nagasaki*, 78–88.
38 UT OMS, B91-0015/011, file 19, "Report on Japanese Scientific and Technical Development," n.d., 1. It is likely this report was sent in early December 1945.
39 Jay Hancock, "Determined Victor: Canada's Role in the Prosecution of Class 'A' Japanese War Criminals," MA thesis, Royal Military College of Canada War Studies Programme, 2002, 98–101. Hancock mistakenly refers to Solandt as an officer of the Department of Reconstruction and Supply.

9 The Only Man for the Job

1 John Swettenham, *McNaughton*. Vol. 1, *1887–1939* (Toronto: Ryerson, 1968), 318–42.
2 For a critical evaluation of Mackenzie's leadership in government-led naval research, see David Zimmerman, *The Great Naval Battle of Ottawa* (Toronto: University of Toronto Press, 1989).
3 The best biography on Howe remains Robert Bothwell and William Kilbourn, *C.D. Howe: A Biography* (Toronto: McClelland and Stewart, 1979). By 1943 alone, the DMS had created twenty-six branches and twenty-eight Crown companies to meet Allied war needs, making the industrial front one of Canada's most important wartime contributions. M.A. Hennessy, "The Industrial Front: The Scale and Scope of Canadian Industrial Mobilization during the Second World War," in *Forging a Nation: Perspectives on the Canadian Military Experience*, ed. Bernd Horn (St Catharines, ON: Vanwell, 2002), 135.
4 The NRC's departments conducted research within their own fields such as mechanical engineering, chemistry, radio and radar work.
5 The army maintained Canadian Armament Research and Development (CARDE) at Valcartier, Quebec, and the Vehicle Design and Development Establishment in Ottawa (which dealt with the Weasel). The Chemical Warfare Laboratories (eventually assuming formal control from the NRC), the Canadian Signals and Radar Development Establishment and Inter-Service Research and Development Establishment (General Stores); the No. 1 Airborne Research and Development Centre at Shilo, Manitoba. The RCN initially used the NRC as their research and development body for their Naval Research Establishment (NRE) at Halifax, but by 1943 assumed more direct control of these functions. The RCAF had five major research centres: a Test and Development Establishment, Rockcliffe, near Ottawa,

a Winter Experimental Establishment at Edmonton, a Photographic Research Establishment and a Radio Wave Propagation Unit at Ottawa, and the Institute of Aviation Medicine in Toronto. The RCAF also became interested in the Flight Research Station at Arnprior, ON, and operated by the NRC. Goodspeed also mentions the joint Canadian/U.S. research project known as the War Disease Control Station at Grosse Ile., Quebec, and the Chemical Warfare Experimental Station in Suffield, "which had, for most of the war, been the joint responsibility of Canada and the United Kingdom." D.J. Goodspeed, *A History of the Defence Research Board of Canada* (Ottawa: Queen's Printer and Controller of Stationery, 1958), 7–9.

6 This included work on radar, chemical warfare, and explosives research (especially RDX), ballistics, naval science, and the creation of the atomic bomb, to name only a few. The full story of Canada's wartime defence research effort cannot be dealt with in detail here. The best official history on the subject is C.P. Stacey, *Arms, Men and Government: The War Policies of Canada, 1939–1945* (Ottawa: Queen's Printer, 1970). The best single-volume history on the organization of this effort is Avery, *The Science of War*.

7 Despite his desire to have NRC out of military research, NRC cooperated on a series of military ventures, most notably with the RCAF on aeronautical design. They also maintained their units' involvement in code-breaking and analysis. See Avery, *The Science of War*, 228–55; and Goodspeed, *History of the Defence Research Board*, 110.

8 This was just one of the many facts that came to light in wake of the defection of Soviet cipher clerk Igor Gouzenko in September 1945. Most spectacular among the Gouzenko revelations were the NRC scientists involved in atomic research that spied for the USSR. These included Alan Nunn May and the German-born, British-residing physicist Klaus Fuchs. Avery, *The Science of* War, 203–55.

9 A.M. Fordyce interview by David Grenville, 20 April 1986, part 2.

10 These included "Post War Research and Development in Canada" (14 August 1945) and "The Future of Canadian Army Research and Development" (17 September 1945). See Goodspeed, *History of the Defence Research Board*, 28–46.

11 Goforth, quoted in ibid., 34.

12 Ibid.

13 A.F. Fordyce made the uncharitable comment that D.J. Goodspeed, author of the official history of the DRB, "prostituted himself" to Foulkes by attributing so many of the DRB's concepts to the CGS. While the official DRB history is certainly positive about the role of Charles Foulkes toward supporting the Goforth group, it did not attribute many of the engaging and innovative aspects of the DRB to him. However, Goodspeed also did

not delve too deeply on Fordyce's or Mendel's role in DRB's creation, offering scant mention of their efforts. Fordyce interview, 20 April 1986, part 2; Goodspeed, *History of the Defence Research Board*.

14 Foulkes's wartime record as an operational commander has been debated by historians. His conduct and leadership were detested by many of the soldiers under his command. However, historians of his post-war work as a "political general" in Ottawa have been largely positive in their assessments of the Foulkes era. See Jack Granatstein, *The Generals* (Calgary: University of Calgary Press, 2005), 173–8; James Eayrs, *In Defence of Canada*. Vol. 3, *Peacemaking and Deterrence* (Toronto: University of Toronto Press, 1972), 61–6; Sean Maloney, "General Charles Foulkes: A Primer on How to Be CDS," in *Warrior Chiefs: Perspectives on Senior Canadian Military Leaders*, ed. Bernd Horn and Steve Harris, 219–36 (Toronto: Dundurn, 2001); Maloney, *Learning to Love the Bomb: Canada's Nuclear Weapons during the Cold War* (Dulles, VA: Potomac Books, 2007), 1–100.

15 Foulkes studied machine guns and qualified as an instructor of small arms in the 1920s, "where he won unusual praise from the Commanding Officer and was promoted captain in his regiment." He also took the first course offered on machine-gun carriers in 1931 at Kingston and Petewawa before taking on staff work. "Lt Gen. Foulkes, C.B.E., D.S.O.," *Canadian Army Journal* 49 (April 1945), 12–14.

16 See Maloney, "General Charles Foulkes," 219–24.

17 Goodspeed, *History of the Defence Research Board*, 19–21.

18 Ibid.

19 Ibid., 15.

20 This body was created with Cabinet approval on 3 October 1944 in response to proposals for a subcommittee to study post-war defence research issues by Air Vice-Marshall E.W. Stedman, director general of air research. Ibid.

21 UT OMS, B93-0041/033, file 4, Mackenzie to Solandt, 20 March 1956.

22 As seen in the previous chapter, Solandt had been approached in early September about being scientific advisor to the chief of the General Staff.

23 Arthur Fordyce noted that during the initial post-war discussion that led to the DRB, there was some fear that McNaughton might attempt to become director-general and run defence research in Canada, a fate Fordyce was glad never arrived. Howe was also an early consideration, because it required the chairman to be a member of Cabinet and closely tied to industry, but it appears Howe never wanted the job. Fordyce interview by David Grenville, 20 April 1986.

24 Quoted in Best, *Margaret and Charley*, 265n558.

25 Solandt interview, 9 September 1985, part 5.
26 Ibid., and 24 February 1986, part 1.
27 Goodspeed, *History of the Defence Research Board*, 46.
28 Fordyce interview, part 2.
29 Solandt interview, 9 September 1985, part 5.
30 Ibid.
31 NAC, RG 24, vol. 11996, file 1-0-34-2, vol. 1, DRB Minutes 1946, Minutes of the First Meeting of the DRB, 16 April 1947.
32 Solandt interview, 9 September 1985, part 5.
33 Fordyce interview, part 2.
34 Solandt interview, 24 February 1986, part 1.
35 Goodspeed, *History of the Defence Research Board*, 25.
36 The Air chief of staff critiqued the proposed scope of the new organization. A revised version of Solandt's proposal was submitted on 7 May. It emphasized the changing nature of research and the need for flexibility in structure versus a static organization, that Canada was not large enough to support separate research, development, and production in all the fields it required, so research and development should operate under a single authority. If there were overlapping interests, those with the majority interest should administer the establishment. Long-range research goals were to be emphasized as opposed to the war-time norm of immediate needs-based research. However, a firm hand was needed to direct this research toward valuable and pragmatic ends and not lapse solely into fundamental research.
37 Goodspeed, *History of the Defence Research Board*.
38 Ibid., 46.
39 Ibid.
40 Ibid., 61.
41 Ibid.
42 Director of the Department of Chemistry, Laval University, Quebec.
43 Vice-president of the Northern Electric Co. Ltd.
44 Goodspeed, *History of the Defence Research Board*, 67–8.
45 On 7 February 1947 Claxton introduced a draft bill as an amendment to the Department of National Defence Act of 1927. Reorganization within Claxton's department in 1947 led to the incorporation of all existing legislation on defence within a single National Defence. See appendix 2.
46 See appendix 1.
47 Goodspeed, *History of the Defence Research Board*, 67–8.
48 See Avery, *The Science of War*, 203–53.
49 Daniel Lee Klienman, *Politics on the Endless Frontier: Post-War Research Policy in the United States* (Durham, NC: Duke University Press, 1995),

74–100. See also Gregg Herkenn, *Cardinal Choices: Presidential Science Advisors from the Atomic Bomb to SDI* (Stanford: Stanford University Press, 2000), 17–34.

50 Ibid.

51 These included the Scientific Advisory Committee to the War Cabinet; the Ministry of Supply's Advisory Committee on Scientific Research and Technical Development; and the Central Register of Scientific Manpower under the Ministry of Labour. There was also the Directorate of Scientific and Industrial Research, the oldest government science body, which organized and directed scientific research, as well as various research councils within the services. See Tom Wilkie, *British Science and Politics since 1945* (Oxford: Blackwell, 1991), 50.

52 Philip Gummett, *Scientists in Whitehall* (Manchester: Manchester University Press, 1980), 218–20; Jon Agar and Brian Balmer, "British Scientists and the Cold War: The Defence Research Policy Committee and Information Networks," *Historical Studies of the Physical Sciences* 28, no. 2 (1998): 209–52.

53 Michael Dockrill, *British Defence since 1945* (New York: Basil Blackwell, 1988), 51.

54 Brian Balmer, *Britain and Biological Warfare, Expert Advice and Science Policy, 1935–65* (London: Palgrave, 2001), 29–53.

55 Quoted in Henry Tizard, "Science and the Services," *Royal United Service Institute Journal* 91, no. 563 (August 1946): 346.

56 According to G. Pascal Zachery, Bush's adversarial temperament and aggressive attitude created many enemies who were kept at bay only by President Roosevelt's support. For Ronald Clark, Tizard's early dismissal of the feasibility of atomic energy and weapons early in war made him a perennial outsider of atomic affairs during the Attlee government, and his long-standing rivalry with Lord Cherwell worsened matters with the return of a Churchill government in 1952. See Zachary, *Endless Frontier*, 279–310; Daniel Lee Klienman, *Politics on the Endless Frontier: Postwar Research Policy in the United States* (Durham, NC: Duke University Press, 1995), 75–100. Ronald W. Clark, *Tizard* (London: Methuen, 1965), 368–403.

57 These meetings were usually attended by Norman Robertson (secretary to the Cabinet), C.M. Drury (deputy minister), Foulkes, Mister M.W. Mackenzie (deputy minister of defence production), as well as a series of ex officio members from DEA, Department of Finances, and the Bank of Canada. DHH 733/1223, box 65, series III, file 1325, CDC Meetings, Minutes, Declarations, 1951. Minutes from PEADQ, 26 June 1951.

58 As policy historian James Eayrs noted, after the resignation of J.L. Ralston as minister of defence, Mackenzie King filled this Cabinet position with

men tasked largely with a single objective. For McNaughton, who served briefly as minister, it was to drum up support for volunteers to reduce the need for conscription. For Doug Abbot, the immediate post-war minister, it was efficient demobilization of Canada's military in peace. For Brooke Claxton, it would be to force reduction and budget cutting. Eayrs, *In Defence of Canada*, 3:16–28.

59 C.P. Stacey, *Canada in the Age of Conflict.* Vol. 2 (Toronto: MacMillan, 1981).

60 NAC, Mackenzie King Diary online, 14 November 1946, page 2. http:// www.bac-lac.gc.ca/eng/discover/politics-government/prime-ministers/ william-lyon-mackenzie-king/Pages/item.aspx?IdNumber=30214&.

61 Ibid.

62 Mackenzie King interposed himself during Sir Henry Tizard's famous mission to bring the cavity magnetron to the United States. Tizard had stopped by Ottawa first to share the news with C.J. Mackenzie, C.D. Howe, and the prime minister. Mackenzie King then, on his own, took great joy in telling U.S. President F.D. Roosevelt that Tizard was bringing a most critical piece of equipment to Washington. Being able to participate in such celebrated events, even in a small way, tickled Mackenzie King's fancy. Solandt would remember this later when he sought to get external validation for the DRB from British and U.S. science tsars (see chapter 13). Avery, *The Science of War*, 12, 52.

63 Reginald Whitaker and Gary Marcuse, *Cold War Canada: The Making of a National Insecurity State, 1945–1957* (Toronto: University of Toronto Press, 1974), 251.

64 NAC, MG 32, B5, vol. 221, file BC 1946–1954, reel 4, Claxton Memoirs, 980–1.

65 Ibid., 964.

66 At the time of his memoir, Claxton worried that cutting the DRB's efforts would cripple any effective chance the armed services had of fighting in future wars. "Are we to be the only major power depriving ourselves of access to the secret weapons of the future?" Ibid., 984–5.

67 Ibid., 965.

68 Solandt spoke to the Empire Club regarding "Exercise SWEETBRIAR" on Claxton's behalf, and occasionally reviewed Claxton's own speeches on technical matters. UT OMS, B93/0041/033, Addresses 1949–1952, Solandt, "Exercise: Sweetbriar" An Address to the Empire Club of Toronto, 30 March 1950.

69 He believed Canada was fortunate to have Solandt to work with and as head of the DRB, especially in his selection of members of the board. "It would be impossible to pick a better group than Omond Solandt had done." NAC, MG 32, B5, vol. 22, file BC 1946–1954, reel 4, Claxton Memoirs, 978.

70 Solandt interview, 24 February 1986, part 3.
71 This included several senior colonels at CARDE misusing the facilities for personal needs, including using CARDE resources for making camper vans. They were given the option by McNaughton of resignation or court martial. Solandt interview, 24 February 1986, part 1.
72 Ibid.
73 NAC, RG 24, vol. 11996, file 11996, Minutes of the 1st Meeting of the DRB, 17 April 1946.
74 Solandt interview, 24 February 1986, part 1.
75 Ibid.
76 At the second meeting of the DRB on 14 May 1947, Solandt explained to the civilian members his vision for defence research. "Dr Mackenzie told the Board that he considers the views expressed by the Chairman in the discussion a comprehensive and sound analysis of the general direction in which the Board should direct its efforts, and suggested the substance of the Chairman's remarks at this informal session be written down and circulated to all members of the board." NAC, RG 24, vol. 11996, file 11996, Minutes of the Second Meeting of the Defence Research Board, 14 May 1947.
77 They also kept each other abreast of scientific controversies. On 26 November 1948, Mackenzie wrote to Solandt about the stir created by the publication of Blackett's controversial book on atomic weapons and strategic bombing, which was highly critical of the Western allies. Mackenzie informed Solandt that the Russians were using it to lambast the Western powers during UN deliberations. McNaughton had provided Mackenzie with the official critique of the book from the British Foreign Office, and provided it for Solandt's edification. Solandt's response to the book was intriguing: "It is such a great pity that a person of such ability should have such a warped outlook on life." NAC, RG 24, vol. 4243, file Solandt/Mackenzie, Solandt to Mackenzie, 3 December 1948.
78 Solandt interview, 9 September 1985, part 5.

10 A Doctor among Soldiers

1 UT OMS, B93-0041/0033, file DRB Addresses 1952–1955, Address to the Canadian Club of Ottawa, 7 January 1953.
2 Solandt had an early hand in the formation of this organization. He was part of the initial informal discussion with the deputy ministers of trade and commerce as well as finance on the need for coordination of economic interest in defence across government departments. On 15 April 1948, Solandt prepared a proposal to the CDC for creating a "Defence Economics Committee"

that would have all the senior government big shots on board, including the minister of labour, trade and commerce as well as external affairs. Heeney was to be the chair. The CDC reserved judgment at that time, with Heeney's position being that it was too soon to form such a body. But the forming of NATO in 1949 brought the issue to the fore again, and the Privy Council Panel on Economic Aspects of Defence Questions was formed in large part to deal with NATO defence arrangements. NAC, RG 2, vol. 2748, file CDC Conclusions, vol. 8, Minutes of the Forty-Third Meeting of the CDC, 15 April 1948; DHH 72/1223, series III, file 2020, Panel on Economic Aspects of Defence Questions, 6 January 1950.

3 The origins of the Cold War continue to consume the minds of many historians. For critical appraisals of American foreign policy and the Cold War, see Walter LeFerber, *America, Russia, and the Cold War, 1945–1984* (New York: Knopf, 1997); and Gabriel Kolko, *The Politics of War: The World and United States Foreign Policy, 1943–1945* (New York: Random House, 1968); John Lewis Gaddis, *The US and the Origins of the Cold War, 1945–1947* (New York: Columbia University Press, 1972). The best comprehensive survey of current post–Cold War literature on the origins of the Cold War is John Lewis Gaddis, *We Now Know: Rethinking the Cold War* (Oxford: Oxford University Press, 1996).

4 Christopher Andrew and Vasili Mitrokhin, *The Sword and the Shield: The Mitrokhin Archive and the Secret History of the KGB* (London: Basic Books, 1999), 162–76.

5 The expansionist desires of Marshal Joseph Stalin's policies are discussed in Vojtech Mastny, *Russia's Road to the Cold War: Diplomacy, Warfare, and the Politics of Communism, 1941–1945* (Columbia University Press, 1979); Mastny, *The Cold War and Soviet Insecurity: The Stalin Years* (New York/ Oxford: Oxford University Press, 1996).

6 Bruce Kuniholm, *The Origins of the Cold War in the Near East: Great Power Conflict and Diplomacy in Iran, Turkey and Greece* (Princeton: Princeton University Press, 1980).

7 Karel Kaplan, *The Short March: The Communist Takeover in Czechoslovakia, 1945–1948* (London: C. Hurt, 1987).

8 Gaddis, *We Now Know*, 26–53.

9 Richard Rhodes, *Dark Sun: The Making of the Hydrogen Bomb* (New York: Touchstone, 1995), 27–48.

10 Paul Y. Hammond, "NSC-68: Prologue to Rearmament," in *Strategy, Politics, and Defense Budgets*, ed. Warner R. Schilling, Paul Y. Hammond, and Glenn H. Snyder, 267–378 (New York: Columbia University Press, 1962).

11 Stacey, *Canada in the Age of Conflict*. Vol. 2.

12 Ibid.

13 Eayrs, *In Defence of Canada*, 3:95.

14 Ibid.

15 Sean Maloney, "The Mobile Striking Force and Continental Defence, 1948–1955," *Canadian Military History* 2, no. 2 (1993): 75–89.

16 Wilfred G.D. Lund, "Vice-Admiral Grant: Father of the Post-War Royal Canadian Navy," in Horn and Harris, *Warrior Chiefs*, 193–7.

17 Eayrs, *In Defence of Canada*, 3:95.

18 See chapter 9.

19 Maloney, "General Charles Foulkes."

20 Lund, "Vice-Admiral Grant."

21 Jack Granatstein and David Bercuson, "Robert Leckie," *Dictionary of Canadian Military History* (Toronto: Oxford Universty Press, 1992), 113.

22 For Claxton's efforts in this regard, see Eayrs, *In Defence of Canada*, 3:1–90.

23 Foulkes served as chief of the General Staff, 1945–51, and then created the post of chairman, Chief of Staff Committee, to assure a degree of continuity on NATO defence planning when Lieutenant-General G.G. Simonds became CGS in 1951. Foulkes was CCOSC until 1960. Simonds was replaced by Lieutenant-General H.D. Graham, who served as CGS from 1955 to 1958. During Solandt's tenure there were similar changes in service chiefs for the Naval and Air Staff. For the navy: Vice Admiral (VAdm) C.G. Jones (1944–6); VAdm H.E. Reid (1946–7); VAdm H.T.W. Grant (1947–51); and VAdm E.R. Mainguy (1951–6). For the RCAF, it was Air Marshal (AM) Leackie (1944–7); AM W.A. Curtis (1947–53); and AM C.R. Slemon (1953–7). David Bercuson and J.L. Granatstein, *Dictionary of Canadian Military History* (New York: Oxford University Press, 1992), 242.

24 See Maloney, "General Charles Foulkes."

25 For instance, the DRB never moved fast enough on CW and BW research for Foulkes, and Solandt had to constantly reassure him that the work being done was the most efficient within DRB's resources and as part of alliance obligations. NAC, RG 24, vol. 11996, file 0-34-2, vol. 3, Minutes of the Eleventh Meeting of the DRB, Halifax, 12 June 1949.

26 NAC, MG 32, B5, vol. 221, file BC 1946–1954, reel 4, Claxton Memoirs, 982–83.

27 This idea was essentially a flying version of Simonds's own wartime army-innovation, the "Kangaroo" armoured personnel carrier. However bad the idea, Otto Maas, Simonds's scientific advisor, at least found the idea interesting. NAC, RG 24, vol. 4243, file Maas/Solandt, Maas to Solandt, 11 May 1951.

28 Solandt interview, 9 September 1985.

29 Solandt interviews, 27 and 29 November 1986, part 1.

30 Eayrs, *In Defence of Canada*, 3:102.

31 Air Vice Marshal Stedman and Group Captain Adams were suggested because of the qualifications, though Leckie failed to mention Stedman's role in defence research discussions before the war. Goodspeed, *History of the Defence Research Board*, 25.

32 It was thought inadvisable for a serving member of the armed services to be a member when the Chiefs of Staff were ex-officio members. Stedman was being considered as an aeronautical advisor to Solandt and raised the issue of advisability of a board member also being an employee. NAC, RG 24, vol. 11996, file Minutes of the First Meeting of the DRB, 16 April 1947.

33 According to Solandt, James "never gave an inch" on cooperating with DRB. Solandt interview, 24 February 1986, part 1.

34 Solandt interview, 9 September 1985, part 5.

35 Both parties accepted full credit for the research they did. So long as the RCAF didn't say "Those stinkers in DRB think this is theirs" or vice versa, both groups "smelled of roses." Solandt interview, 9 September 1985, part 5.

36 Solandt harshly criticized Crawford Gordon, president of A.V. Roe Canada, for not emphasizing RCAF contributions to jet design in an article Gordon's company wrote for *Jet Age* that included discussion of the CF-100 and CF-105. NAC, RG 24, vol. 4243, file Solandt/Gordon, Solandt to Gordon, 29 December 1955.

37 Eayrs, *In Defence of Canada*, 3:355; George Lindsey, "Canada-US Defence Relations in the Cold War," in *Fifty Years of Canada–United States Defence Cooperation: The Road from Obdensburg*, ed. Joseph Jockel and Joel Sokolsky (Lewiston: Edwin Mellen, 1992), 63.

38 C. Conliffe, "The Permanent Joint Board on Defence: Fifty Years after Roosevelt's Kingston Declaration," *Canadian Defence Quarterly* 18, no. 1 (Summer 1988), 58–9.

39 Eayrs, *In Defence of Canada*, 3:339.

40 On Kennan and the Marshall Plan, see Wilson Miscamble, "The Foreign Policy of the Truman Administration: A Post Cold War Appraisal," *Presidential Studies Quarterly* 24, no. 3 (Summer 1994): 479–94; George F. Kennan, *Memoirs, 1925–1950* (New York: Pantheon Books, 1967); David Mayers, "George F. Kennan and the Soviet Union, 1933–1938: Perceptions of a Young Diplomat," *International History Review* 5, no. 4 (November 1984), 525–49; Thomas H. Etzold and John Lewis Gaddis, eds., *Containment: Documents on American Policy and Strategy, 1955–1950* (New York: Columbia University Press, 1978), 50–64, 84–90; John Lewis Gaddis, *Strategies of Containment: A Critical Appraisal of Postwar American National Security Policy* (Oxford: Oxford University Press, 1982), 25–53; Nitze, with Smith and Rearden, *From Hiroshima to*

Glasnost, 82–101; Ernest May, ed., *American Cold War Strategy: Interpreting NSC 68* (Boston: Bedford Books, 1993); Michael Hogan, *The Marshall Plan: America, Britain, and the Reconstruction of Western Europe, 1947–1952* (Cambridge: Cambridge University Press, 1987).

41 Eayrs, *In Defence of Canada*, 3:356.

42 Ibid., 385.

43 Ibid.

44 DHH Raymont Fonds, 73/1223, box 59, file 130, Chiefs of Staff Committee Minutes 1947, Minutes of the Chiefs of Staff Committee (hereafter referred to as COSC), 4 February 1947.

45 DHH Raymont Fonds, 73/1223, box 59, file 130, Chiefs of Staff Committee Minutes 1947, Minutes of the COSC Meeting, 3 June 1947.

46 DHH Raymont Fonds, 73/1223, box 59, file 130, Chiefs of Staff Committee Minutes 1947, Minutes of the COSC Meeting, 4 September 1947.

47 DHH Raymont Fonds, 73/1223, box 59, file 130, Chiefs of Staff Committee Minutes 1947, 4 February 1947.

48 Ibid.

49 DHH Raymont fonds, 733/1223, box 61, series 3, file 1307, Minutes of the Chiefs of Staff Committee (Regular and Special) 1952–1953, Minutes of the COSC Meeting, 9 February 1953.

50 NAC, RG 2, vol. 2748, file CDC Conclusions, vol. 8, Minutes of the Forty-Fourth Meeting of the CDC, 2 June 1948.

51 They were "inclined to view that planes carrying atomic bombs would have to be part of a large bombing force (250 planes) using conventional bombs, as single planes would be intercepted. This technique would reduce the advantage of using atomic bombs because of the necessity of having to provide man and aircraft to constitute the conventional bombing force." NAC, RG 2, vol. 2748, file CDC Conclusions, vol. 1, Minutes of the Forty-Fourth Meeting of the CDC, 2 June 1948.

52 NAC, RG 2, vol. 2748, file CDC Conclusions, vol. 8, Minutes of the Special CDC Meeting with U.S. Secretary of Defence, 16 August 1948.

53 DHH Raymont Fonds, 73/1223, box 61, file 1307, Chiefs of Staff Committee Minutes 1949, Minutes of the COSC Meeting, 23 August 1949.

54 The value of understanding arctic warfare would also provide insights on how best to fight on Russian soil, though this was a secondary concern. Cecil Law claims that most of the research done at the laboratory at Churchill was designed with the intent of fighting in arctic conditions outside of North America. Law interview, 21 September 2004.

55 DHH Raymont Fonds, 73/1223, box 59, file 130, Chiefs of Staff Committee Minutes 1947, Minutes of the COSC Meeting, 7 January 1947.

56 DHH Raymont Fonds, 73/1223, box 59, file 130, Chiefs of Staff Committee Minutes 1946, Minutes of the COSC Meeting, 9 April 1946.

57 DHH Raymont Fonds, 73/1223, box 59, file 130, Chiefs of Staff Committee Minutes 1946, Minutes of the COSC Meeting, 1 October 1946.

58 DHH Raymont Fonds, 73/1223, box 59, file 130, Chiefs of Staff Committee Minutes 1947, Minutes of the COSC Meeting, 13 May 1947.

59 U.S. authorities had agreed to this and would install the radio equipment. Personnel would be taken from the Department of Trade. Cost would be non-recurring $257,000 for buildings and installation and recurring expenditure of $142,000. CDC agreed. NAC, RG 2, vol. 2748, file CDC Conclusions, vol. 8, Minutes of the Thirty-Ninth CDC Meeting, 21 November 1947.

60 DHH Raymont Fonds, 73/1223, box 59, file 130, Chiefs of Staff Committee Minutes 1947, Minutes of the COSC Meeting, 13 May 1947.

61 DHH Raymont Fonds, 73/1223, box 59, file 130, Chiefs of Staff Committee Minutes 1947, Minutes of the COSC Meeting, 23 December 1947.

62 DHH Raymont Fonds, 73/1223, box 59, file 130, Chiefs of Staff Committee Minutes 1947, Minutes of the COSC Meeting, 15 April 1947.

63 DHH Raymont Fonds, 73/1223, box 61, file 1307, Chiefs of Staff Committee Minutes 1949, Minutes of the COSC Meeting, 15 November 1949.

64 Eayrs, *In Defence of Canada*. Vol. 4, *Growing Up Allied* (Toronto: University of Toronto Press, 1980.

65 Sean Maloney, *War without Battles: Canada's NATO Brigade in Germany, 1951–1993* (Toronto: McGraw-Hill-Ryerson, 1997), 17–22; Desmond Morton, *A Military History of Canada*, 4th ed. (Toronto: McClelland and Stewart, 1999), 232–3.

66 See Maloney, *Learning to Love the Bomb*, 1–18; Eayrs, *In Defence of Canada*. Vol. 4, *Growing Up Allied*.

67 Ibid.

68 DHH Raymont Fonds, 73/1223, box 61, file 1307, Chiefs of Staff Committee Minutes 1949, Minutes of the COSC Meeting, 8 June 1949.

69 Maloney, *War without Battles*, xxxiii–5.

70 DHH Raymont Fonds, 73/1223, box 61, file 1307, Chiefs of Staff Committee Minutes 1949, Addendum to COSC Meeting, 2 February 1949.

71 DHH Raymont Fonds, 73/1223, box 59, file 130, Chiefs of Staff Committee Minutes 1947, Minutes of the COSC Meeting, 21 March 1947.

72 DHH Raymont Fonds, 73/1223, box 59, file 130, Chiefs of Staff Committee Minutes 1947, Minutes of the COSC Meeting, 28 April 1947.

73 DHH Raymont Fonds, 73/1223, box 61, file 1307, Chiefs of Staff Committee Minutes 1949, Minutes of the COSC Meeting, 25 September 1949.

74 Granatstein and Bercuson, *Dictionary of Canadian Military History*, 111.

75 The most comprehensive account of the origins of the Korean War remains Bruce Cummings, *The Origins of the Korean War*. Vol. 1, *Liberation and the Emergence of the Separate Regimes, 1945–1947* (Princeton: Princeton University Press, 1981); *The Origins of the Korean War*. Vol. 2, *Roaring of the Cataract, 1947–1950* (Princeton: Princeton University Press, 1990). The most accurate history of Stalin's role in the origins of the Korean War is reviewed in Gaddis, *We Now Know*, 54–84.

76 David Bercuson, *True Patriot: The Life of Brooke Claxton, 1898–1960* (Toronto: University of Toronto Press, 1993), 250.

77 Maloney, *War without Battles*, xxxiii–63.

78 Paul Y. Hammond, "NSC-68: Prologue to Rearmament," in *Strategy, Politics, and Defense Budgets*, ed. Warner R. Schilling, Paul Y. Hammond, and Glenn H. Snyder, 267–378 (New York: Columbia University Press, 1962).

79 Gaddis, *Strategies of Containment*, 151–2.

80 While Eisenhower was always privately cautious about executing any policy that might lead to a nuclear exchange, "there is no question that in [the Eisenhower administration's] effort to lower the costs of containment it was willing to take on, through its reliance on nuclear weapons, greater risks than its predecessors had been willing to [face]." Gaddis, *Strategies of Containment*, 151–2.

81 Maloney, *Learning to Love the Bomb*, 21–2.

82 Maloney, *War without Battles*, 69–73.

83 Rhodes, *Dark Sun*; Margaret Gowing and Lorna Arnold, *Independence and Deterrence: Britain and Atomic Energy, 1945–1952* (London: MacMillan, 1974).

84 David Holloway, *Stalin and the Bomb: The Soviet Union and Atomic Energy, 1939-1956* (New Haven, CT: Yale University Press, 1994).

85 H.F. Wood, *Strange Battleground: The Official History of the Canadian Army in Korea* (Ottawa: Queen's Printer, 1966), 34.

86 Ibid., 157.

87 On Sutherland's strategic vision in Canada, see Andrew Richter, *Avoiding Armageddon: Canadian Military Strategy and Nuclear Weapons, 1950–1963* (Vancouver: University of British Columbia Press, 2002); and James Lee and David Bellamy, "Dr R.J. Sutherland: A Retrospective," *Canadian Defence Quarterly* (Summer 1987): 41–6.

88 NAC, RG 24, vol. 11996, file 1-0-34-2, vol. 5, Annual Report of the Chairman, Defence Research Board, September 1951; NAC, RG 24, vol. 4208, file 270-180-105-1, vol. 2 2 Canadian Army OR Field Team in Korea, Guy Simonds to Omond Solandt, 29 January 1952.

89 NAC, RG 24, vol. 4208, file 270-180-105-1, vol. 2, 2 Canadian Army OR Field Team in Korea, Cecil Law for Superintendent of Defence Research Northern Laboratory to Operational Research Headquarters, 23 December 1953.

90 K. Pennycuick, who commanded the British OR effort in Korea, was critical of all the OR operations in Korea, including the 1-CAORT. Brent Byron Watson, however, has shown that 1-CAORT's casualty analysis in particular was critical to analysing the modern small arms in battle. K. Pennycuick, "Army Operational Research in the Far East, 1952–54," *Operational Research Quarterly (ORQ)* (1954): 121–9; Brent Byron Watson, *Far Eastern Tour: The Canadian Infantry in Korea, 1950–1953* (Montreal and Kingston: McGill-Queen's University Press, 2007), 121, 122, 133, 169.

91 NAC, RG 2, vol. 2748, file CDC Conclusions, vol. 8, Minutes of the Sixty-Fifth Meeting of the CDC, 19 July 1950.

92 Ibid.

93 NAC, RG 2, vol. 2748, file CDC Conclusions, vol. 6, Minutes of the Sixty-Ninth Meeting of the CDC, 28 December 1950.

94 Eayrs, *In Defence of Canada*, 4:102.

95 NAC, RG 2, vol. 2748, file CDC Conclusions, vol. 5, Minutes of the Eightieth Meeting of the CDC, 8 November 1951.

96 Ibid.

97 Ibid.

98 NAC, RG 2, vol. 2748, file CDC Conclusions, vol. 6, Minutes of the Eighty-Sixth Meeting of the CDC, 19 May 1952.

99 Ibid.

100 Ibid.

101 The Mid Canada Line consisted of ninety-eight radar stations, most unmanned and built along the 55th parallel, in 1954. The McGill Fence technology was developed at McGill University and was supported by DRB funds. It employed an automatic audible alarm system making it unnecessary for the station to be manned at all times. See chapter 11; Bercuson and Granatstein, *Dictionary of Canadian Military History*, 127.

102 Joseph Jockel, *No Boundaries Upstairs* (Vancouver: University of British Columbia Press, 1985), 62.

103 For the Mid Canada Line and McGill Fence, see ibid., 60–91. For SAGE, see Thomas Hughes, *Rescuing Prometheus: Four Monumental Projects That Changed the Modern World* (New York: Vintage Books, 1998), 15–68.

104 On 3 April 1952 the COSC noted that the minutes of the last meeting said they had reversed the leaders in an OR study of air defence. It would now be RCAF led and the DRB supported. DHH Raymont fonds, 733/1223, box 61,

series 3, file 8 (1307S), Minutes of the Chiefs of Staff Committee (Regular and Special) 1952–1953, Minutes of the COSC Meeting, 3 April 1952.

105 DHH Raymont fonds, 733/1223, box 61, series 3, file 8 (1307S), Minutes of the Chiefs of Staff Committee (Regular and Special) 1952–1953, Minutes of the COSC Meeting, 4 February 1952.

106 DHH Raymont fonds, 733/1223, box 61, series 3, file 8 (1307S) Minutes of the Chiefs of Staff Committee (Regular and Special) 1952–1953, Minutes of the COSC Meeting, 11 June 1952.

107 NAC, RG 2, vol. 2748, file CDC Conclusions, vol. 8, Minutes of the 91st CDC, 10 February 1953.

108 Ibid.

109 Ibid.

110 Ibid.

111 DHH Raymont fonds, 733/1223, box 61, series 3, file 1307, Minutes of the Chiefs of Staff Committee (Regular and Special) 1952–1953, Minutes of the COSC Meeting, 16 February 1953.

112 DHH Raymont fonds, 733/1223, box 61, series 3, file 1307, Minutes of the Chiefs of Staff Committee (Regular and Special) 1952–1953, Minutes of the COSC Meeting, 6 October 1953.

113 DHH Raymont fonds, 733/1223, box 61, series 3, file 1307, Minutes of the Chiefs of Staff Committee (Regular and Special) 1952–1953, Minutes of the COSC Meeting, 3 November 1954.

114 Joseph Jockel's stellar work on Canada and the United States regarding continental defence offers no insight regarding the role of Solandt on these affairs, and only limited insight on the influence of the DRB on the Mid Canada Line. See Jockel, *No Boundaries Upstairs.*

115 Goodspeed, *History of the Defence Research Board,* 105.

116 DHH 733/1223, box 65, series III, file 1325, CDC Meetings, Minutes, Declarations, 1951, Minutes of the Meeting of Panel on Economic Aspects of Defence Questions, 18–19 (April?) 1951.

117 Goodspeed, *History of the Defence Research Board,* 105.

118 DHH Raymont fonds, 733/1223, box 61, series 3, file 1307, Minutes of the Chiefs of Staff Committee (Regular and Special) 1952–1953, Minutes of the COSC Meeting, 6 October 1953.

119 DHH Raymont fonds, 733/1223, box 61, series 3, file 1308, Minutes of the Chiefs of Staff Committee, Minutes of the COSC Meeting, 11 February 1955.

120 Ibid.

121 Ibid.

122 DHH Raymont fonds, 733/1223, box 61, series 3, file 1308, Minutes of the Chiefs of Staff Committee, Minutes for 11 February 1955.

123 Solandt interview, 24 February 1986, part 1.
124 In his interviews, Solandt did not mention the specifics of this unofficial support.
125 What remains of the Solandt/Gordon letters centres on Crawford's assurances about production, fears of disinformation about the company, and Solandt's sharp rejoinder regarding the Avro's company history on the CF-100 and its failure to mention the RCAF's role in establishing accurate forecasts of its continental defence requirements. According to Solandt, "The CF-100 is a successful airplane because your Company did a good job of design and production and also because the RCAF did a good job of foreseeing their requirements seven or eight years in advance. This is true of all military aircraft and I feel that those who have the vision to write requirements that later prove to be inspired should share with the designers and manufacturers the credit for the success of final airplane. I feel that Art James had more to do with the basic concept of the CF-100 than any other one person and I am very sorry that he was not given credit for this in your story." Solandt suggested he correct this error. NAC, RG 24, vol. 4243, file Solandt/Gordon, Solandt to Gordon, 29 December 1955.
126 Solandt interview, 24 February 1986, part 1.
127 Ibid.; Jack Granatstein, *Canada 1957–1967: The Years of Uncertainty and Innovation* (Toronto: McClelland and Stewart, 1986), 107–12.
128 DHH Raymont fonds, 733/1223, box 61, series 3, file 1307, Minutes of the Chiefs of Staff Committee (Regular and Special) 1952–1953, Minutes for Special Meeting of the COSC, 16 April 1953.
129 Ibid.
130 DHH Raymont fonds, 733/1223, box 61, series 3, file 8 (1307S), Minutes of the Chiefs of Staff Committee (Regular and Special) 1952–1953, Minutes of the Meeting of the COSC, 21 January 1952.
131 See chapter 13.
132 However, he stated that his comments were general and not specific to the document under discussion. DHH Raymont fonds, 733/1223, box 61, series 3, file 8 (1307S), Minutes of the Chiefs of Staff Committee (Regular and Special) 1952–1953, Minutes of the Meeting of the COSC, 21 January 1952.

11 A Decade of Leadership

1 Law interview; Archie Pennie interview by David Grenville, 21 March 1986, part 1.

2 Pennie interview, part 1.

3 Alex Roland, *The Military-Industrial Complex* (American Historical Association and the Society for the History of Technology Booklet, 2001).

4 Jockel, *No Boundary Upstairs*, 60–91.

5 The best account of this episode is Annie Collins, *In The Sleep Room: The Story of CIA Brainwashing Experiments in Canada* (Toronto: Key Porter Books, 1988).

6 Pennie interview, part 1.

7 Ibid.

8 UT OMS, B93-0041/0033, Addresses 1946–1948, Address to the Industrial and Scientific Research Conference, CMA, 27 May 1948.

9 Harkness would serve two terms as the DRB's member for industry, returning in 1953. NAC, RG 24, vol. 10339, file DRB 1953, untitled news release, 11 May 1953.

10 Diamond became a board member in 1954. NAC, RG 24, vol. 11996, file 0-34-2, vol. 3, Minutes of the Ninth Meeting of the Defence Research Board, Ottawa, 20 December 1948.

11 Lawrence Aronson, "Planning Canada's Economic Mobilization for War: The Origins and Operation of the Industrial Defence Board, 1945–1951," *American Journal of Canadian Studies* 15, no. 1 (1985): 38–58.

12 NAC, RG 24, vol. 11996, file 1-0-34-2, vol. 2, Minutes of the Seventh Meeting of the Defence Research Board, 20–1 June 1948.

13 Ibid.

14 Ibid.

15 UT OMS, B93/0041/0033, file DRB Addresses 1946–1948, Address to the Montreal United Service Institute, 27 October 1948.

16 Solandt interview, 24 February 1986, part 2; NAC, RG 24, vol. 11996, file 1-0-34-2, vol. 2, Minutes of the Eighth Meeting of the Defence Research Board, Held at Valcartier 26–7 September 1948.

17 NAC, RG 24, vol. 11996, file: 1-0-34-2, vol. 2, Minutes of the Eighth Meeting of the Defence Research Board, Held at Valcartier 26–7 September 1948.

18 UT OMS, B93-0041/0033, file DRB Addresses 1946–1948, Address to the Industrial and Scientific Research Conference, CMA, 27 May 1948.

19 Ibid.

20 UT OMS, B93/0041/033, file DRB Addresses, 1946–1948, Address to Canadian Manufacturers' Association, Toronto, 28 May 1948.

21 Ibid.

22 Ibid.

23 These included the Canadian Ordnance Association, "which is a voluntary association of industrialists who are interested in contributing toward industrial preparedness in Canada." They worked with other government agencies such as Canadian Arsenals Limited (operating and maintaining a nucleus of munitions factories), Canadian Commercial Corporation (purchases for the Armed Forces and other agencies in government such as the Department of Trade and Commerce, Mines and Resources). UT OMS, B93/0041/033, file DRB Addresses, 1946–1948, Address to Canadian Manufacturers' Association, Toronto, 28 May 1948.

24 See NAC, RG 24, vol. 11996, file 1-0-34-2, vol. 2, Report of the Special Meeting of the Defence Research Board, 26 September 1948, to Consider the Report of the Chairman as Recorded under Minute 3.6 of the Sixth Meeting of the Board; NAC, RG 24, vol. 11996, file 1-0-34-2, vol. 4, Annual Report of the Chairman, Defence Research Board, September 1950.

25 UT OMS, B93/0041/033, file DRB Addresses, 1946–1948, Address to Canadian Manufacturers' Association, Toronto, 28 May 1948; NAC, RG 24, vol. 11996, file 1-0-34-2, vol. 4, Annual Report of the Chairman, Defence Research Board, September 1950.

26 Doctor Johnston had been critical of naval wartime research on degaussing of Canadian ships to protect against magnetic mines. Avery, *The Science of War*, 85.

27 Doctor Shrum was head of the Department of Physics at UBC. NAC, RG 24, vol. 10339, file DRB 1954, Statement of the Defence Research Board, n.d.

28 Doctor Field had worked on sea mines during the war and directed Canada's wartime anti-submarine research program in conjunction with the United States and Great Britain. In 1941, while a passenger on the HMS *Rodney* en route to Canada from England, he was present when the warship engaged the German battleship *Bismarck*. They were accompanied by three destroyers and participated in the famous sinking of the ship. NAC, RG 24, vol. 10339, file DRB 1953, Statement by OM Solandt, Chairman, Defence Research Board, 29 September 1953.

29 Solandt interview, 24 February 1986, part 2.

30 Pennie interview, part 1.

31 NAC, RG 24, vol. 11996, file 0-34-2, vol. 3, Report of the 13th Meeting of the Selection Committee, 20 March 1949.

32 Pennie interview, part 1.

33 NAC, RG 24, vol. 11996, file 1-0-34-2, vol. 2, Minutes of the Eighth Meeting of the Defence Research Board, Held at Valcartier, 26–7 September 1948.

34 NAC, RG 24, vol. 11996, file 1-0-34-2, vol. 4, Annual Report of the Chairman, Defence Research Board, September 1950.

35 NAC, RG 24, vol. 11996, file 1-0-34-2, vol. 3, Minutes of the Sixth Meeting of the Standing Committee on Extra-Mural Research of the Defence Research Board, 19 March 1949.

36 NAC, RG 24, vol. 11996, file 1-0-34-2, vol. 3, Minutes of the Tenth Meeting of the Standing Committee on Extra-Mural Research of the DRB, 19 March 1950.

37 NAC, RG 24, vol.10339, file DRB 1953, Directorate of Public Relations, Army HC, Ottawa, untitled, n.d.

38 Ibid.

39 Stanley Brice Frost, *McGill University: For the Advancement of Learning.* Vol. 2, *1985–1971* (Montreal and Kingston: McGill-Queen's University Press, 1984), 339.

40 Solandt interview, 27 and 29 November 1986, part 1.

41 Goodspeed, *History of the Defence Research Board*, 204.

42 R.J. Whitehead, *Memoirs of a Boffin*, http://www.whitehead-family.ca/drrennie/radar.html.

43 Goodspeed, *History of the Defence Research Board*, 200–1.

44 NAC, RG 24, vol. 10339, file DRB 1953, Statement by O.M. Solandt, Chairman, Defence Research Board, 29 September 1953.

45 NAC, RG 24, vol. 10339, file DRB 1955, Defence Research Board, 12 July 1955.

46 Whitehead, *Memoirs of a Boffin*.

47 These included Leading Aircraftman R. Flood, member of RCAF Air Defence Command, Flight Lieutenant A.S. Matthews, member of RCAF Air Defence Command. NAC, RG 24, vol. 10339, file DRB 1953, Directorate of Public Relations, Army HC, Ottawa, untitled document, n.d.

48 Chief NRC members on the project were Mister Ballard and his assistant, Doctor D.W. McKinley. NAC, RG 24, vol. 10339, file DRB 1953, Statement by O.M. Solandt, Chairman, Defence Research Board, 29 September 1953.

49 Goodspeed, *History of the Defence Research Board*, 201. Members of the RCA research team who worked with Whitehead included H. Laks, E.A. Pinnell, and W. Patterson, RCA Victory maintenance team leader. NAC, RG 24, vol. 10339, file DRB 1953, Directorate of Public Relations, Army HC, Ottawa, untitled document, n.d.

50 "On behalf of DRB, I was soon able to contract with an advanced development group at the RCA Victor Company in Montreal to develop an engineering model. The group was headed by Dr Ross Warren, who was a delight to work with. He was particularly strong on the theoretical aspects and we

wrote a major, even definitive report on the subject together. Stan Pinnell and Herb Lax of Ross Warren's group looked after the Engineering aspects most effectively." Whitehead, *Memoirs of a Boffin*.

51 NAC, RG 24, vol. 11997, file 1-0-24-2, vol. 6, Minutes of the Twenty-Seventh Meeting of the Defence Research Board, Held at Shirley's Bay, ON, 16 October 1953.

52 Ibid.

53 NAC, RG 2, vol. 2748, file CDC Conclusions, vol. 8, Minutes of the Ninety-Fifth Meeting of the CDC, 6 October 1953.

54 SPIDERWEB entailed procuring, installing, and operating radio equipment based on the McGill Fence in a system to enable RCAF to determine operational usefulness of simple Doppler early warning radar and to enable DRB to advance developments further in anticipation of a possible service requirement. NAC, RG 24, vol. 11997, file 1-0-24-2, vol. 6, Annual Report of the Chairman, Defence Research Board, October 1953.

55 In 1953 the USAF argued that unmanned systems were critical to North America's continental air defence system. So in July 1953, DRB provided more suitable equipment for possible "unattended operation in the remote areas of Northern Canada." This is project LONEWOLF. "The development phase of SPIDERWEB was transferred to Project LONEWOLF." New contracts with RCA Victor Company Limited to do RND work for the detection portion of the system were generated. The DRB's Radio Propagation Laboratory reworked its agenda to do intensive work in propagation and communications for this system. NAC, RG 24, vol. 11997, file 1-0-24-2, vol. 6, Annual Report of the Chairman, Defence Research Board, October 1953.

56 NAC, RG 2, vol. 2748, file CDC Conclusions, vol. 7, Minutes of Hundred-and-First Meeting of the CDC, 12 November 1954.

57 UT OMS, B93/0041/049, file Arlikor vs USA Depositions, "Meeting at Ritz-Carlton Hotel, 1 June 1951."

58 UT OMS, B93/0041/049, file Arlikor vs USA Depositions, Solandt to Watter, date likely July 1954 in response to a Treasury Board request for clarification of Hebb's grant on 26 July 1954.

59 Ibid.

60 Collins, *In the Sleep Room*.

61 Ibid., 118. Cameron also interviewed Solandt for her work.

62 Solandt interview, 10 May 1986, part 5.

63 Solandt interview, 8 May 1986, part 3.

64 Collins, *In the Sleep Room*, 135.

65 UT OMS, B93/0041/049, file Arlikor vs USA Depositions, Solandt to Turner, 14 October 1988.

66 Solandt interview, 9 September 1985, part 5.

67 Solandt interview, 24 February 1986, part 1.

68 NAC, RG 2, vol. 2748, file CDC Conclusions, vol. 8, Minutes of the Forty-Third Meeting of the CDC, 15 April 1948.

69 NAC, RG 2, vol. 2748, file CDC Conclusions, vol. 8, Minutes of the Forty-Fourth Meeting of the CDC, 2 June 1948.

70 Goodspeed, *History of the Defence Research Board*, 103–4.

71 Pennie interview, part 1.

72 Solandt admitted to Wansbrough he had "shamelessly urged [Davies] to stay. He has now been with me as my Deputy for nearly six weeks and I find him to be all that can be desired in a deputy." Wansborough couldn't blame Solandt for snatching Davies and took it as a "plagiaristic compliment." Solandt had high regard for Davies, who "had done a magnificent job and further undermined my rather tenuous delusion of indispensability." NAC, RG 24, vol. 4243, file Solandt/Wansborough-Jones, Solandt to Wansborough-Jones, 20 May 1949.

73 See Goodspeed, *History of the Defence Research Board*; NAC, RG 24, vol. 11996, file 1-0-34-2, vol. 2, Report of the Special Meeting of the DRB, 26 September 1948.

74 NAC, RG 24, volume 4243, file Solandt/Wansborough-Jones, Solandt to Wansborough-Jones, 6 September 1949.

75 NAC, RG 24, vol. 11996, file 0-34-2, vol. 3, Annual Report of the Chairman of the Defence Research Board, September 1949.

76 Ibid.

77 NAC, RG 24, vol. 11997, file 1-0-24-2, vol. 6, Annual Report of the Chairman, October 1952.

78 Tom Bower, *The Paperclip Conspiracy: The Battle for the Spoils of and Secrets of Nazi Germany* (London: Paladin Grafton Books, 1987); Samuel A. Goudsmit, *Alsos* (Los Angeles: Tomash, 1986); Clarence G. Lasby, *Project Paperclip: German Scientists and the Cold War* (New York: Atheneum, 1971); Jeremy Bernstein, *Hitler's Uranium Club: The Secret Recordings at Farm Hall* (London: Springer, 2001); David Cassidy, "German Scientists and the Nazi Atomic Bomb," *Dimensions* 10, no. 2 (1996): 15–22; Cassidy, "Controlling German Science I: U.S. and Allied Forces in Germany, 1945–1947," *Historical Survey for the Physical and Biological Sciences* 24, no. 1 (1994): 197–235; Cassidy, "Controlling German Science II: Bizonal Occupation and the Struggle over West German Science, 1946–1949," *Historical Survey of the Physical and Biological Sciences* 26, no. 2 (1996): 197–240. For comparative purposes, see John Gimble, "Project Paperclip: German Scientists, American Policy and the Cold War," *Diplomatic History* 14, no. 3 (1990): 343–65; Gimbel, *Science,*

Technology and Reparations: Exploitation and Plunder in Postwar Germany (Stanford: Stanford University, 1990).

79 In June 1948, G.D. Mallory, director of the Industrial Division (Foreign Trade Service) of the Department of Trade and Commerce, asked for the DRB's input on German scientists. This was specifically in reference to an application by "a Paul Steffen reputed to be an expert on guided missiles." Shrum noted that universities were keen on German scientists and that there was an institute in Montreal helping these people get to Canada. The board agreed to go on record as "being in favour of bringing highly qualified German scientists to Canada but suggested that they be employed initially in universities and in industry." They would not be employed at DRB establishments. NAC, RG 24, vol. 11996, file 1-0-34-2, vol. 2, Minutes of the Seventh Meeting of the Defence Research Board, 20–1 June 1948.

80 NAC, RG 24, vol. 11996, file 1-0-34-2, vol. 5, Minutes of the Twentieth Meeting of the Defence Research Board, 29 September 1951. Engel's name is not mentioned in the rest of the minutes, nor is he mentioned in the history of the Canadian Armament Research and Development Establishment (CARD), where guided missile work was done. See Alain Gelly and H.P. Tardif, *Defence Research Establishment Valcartier, 1945–1995: 50 Years of History and Scientific Progress* (Ottawa: Minister of Public Works, Canada, 1995).

81 The Egyptian government had no such qualms, and by 1952, Engel made an international scene as an ex-Nazi scientist designing weapons systems for Egypt and trained on the Jewish state of Israel. Michael Neufeld, "Rolf Engel vs the German Army: A Nazi Career in Rocketry and Repression," *History and Technology* 13 (1996): 53-72.

82 NAC, RG 2, vol. 2748, file CDC Conclusions, vol. 5, The Seventy-Second Meeting of the Cabinet Defence Committee, 20 May 1951.

83 UT OMS, B93-0041/0033, file Addresses, 1946–1949, Address to the Twenty-First Congregation of the University of British Columbia, 29 October 1947.

84 Solandt interview, 24 February 1986, part 2.

85 NAC, RG 24, vol. 11996, file 1-0-34-2, vol. 2, Minutes of the 20–1 June 1948 Meeting of the DRB.

86 Law interview.

87 NAC, RG 24, vol. 11996, file 0-34-2, vol. 3, Minutes of the Meeting of the DRB, 20 March 1950.

88 NAC, RG 24, vol. 1196, file 1-0-34-2, vol. 2, Minutes of the Sixth Meeting of the DRB, 20–1 June 1948.

89 Solandt did not want these addresses confused with his voluminous public lectures, which do exist at both the University of Toronto archive and the NAC. Solandt interview, 24 February 1986, part 3.
90 Fordyce interview, part 1.
91 NAC, RG 24, vol. 11996, file DRB Minutes 1946, Minutes of the Fourth Meeting of the DRB, 16 September 1947.
92 Fordyce interview, part 1.
93 NAC, RG 24, vol. 11996, file DRB Minutes 1946, Minutes of the Fifth Meeting of the DRB, 15 December 1947.
94 Ibid.
95 NAC, RG 24, vol. 11996, file 1-0-34-2, vol. 2, DRB Minutes 1946, Minutes of the Fifth Meeting of the DRB, 15 December 1947.
96 NAC, RG 24, vol. 11996, file 1-0-34-2, vol. 2, Minutes of the Seventh Meeting of the DRB, 20–1 June 1948.
97 RG 24, vol. 11996, file 1-0-34-2, vol. 3, Minutes of the Ninth Meeting of the Defence Research Board, Ottawa, 20 December 1948.
98 NAC, RG 24, vol. 11996, file 1-0-34-2, vol. 3, Minutes of the Tenth meeting of the DRB, 29 March 1949.
99 NAC, RG 24, vol. 11996, file 1-0-34-2, vol. 3, Annual Report of the Chairman of the Defence Research Board, September 1949.
100 NAC, RG 24, vol. 11996, file 1-0-34-2, vol. 3, Minutes of the Thirteenth Meeting of the DRB, 1 December 1949.
101 NAC, RG 24, vol. 11996, file 1-0-34-2, vol. 4, A Five-Year Plan for Defence Research and Development, 1950–1954.
102 Signals research.
103 NAC, RG 24, vol. 11996, file 1-0-34-2, vol. 4, A Five-Year Plan for Defence Research and Development, 1950–1954.
104 NAC, RG 24, vol. 11996, file 1-0-34-2, vol. 4, DRB Minutes, 1950–1951, Minutes of the Sixteenth Meeting of the DRB, 16 and 17 October 1950.
105 NAC, RG 24, vol. 11996, file 1-0-34-2, vol. 5, DRB Minutes, 1951–1952, Minutes of the Twentieth Meeting of the DRB, 29 September 1951.
106 NAC, RG 24, vol. 11996, file 1-0-34-2, vol. 5, DRB Minutes 1951–1952, Annual Report of the Chairman, Defence Research Board, September 1951.
107 NAC, RG 24, vol. 11996, file 1-0-34-2, vol. 6, Defence Research and Development: Program and Estimates, 6 March 1952.
108 NAC, RG 24, vol. 11997, file 1-0-24-2, vol. 6, Annual Report of the Chairman, October 1952.
109 NAC, RG 24, vol. 11997, file 1-0-24-2, vol. 6, Minutes of the Twenty-Fourth Meeting of the DRB, 18 October 1952.
110 Pennie interview, part 1.

12 Rockets, Germs, and UFOs

1 UT OMS, B93/0041/0033, file DRB Addresses 1952–1955, Address to the Canadian Club of Ottawa, 7 January 1953.

2 This is done in part because Goodspeed's history is guarded about each of these subjects, given their sensitive nature.

3 See UT OMS, B93/0041/0033, file DRB Addresses 1946–1955.

4 The DRB section would recruit for all groups, conduct inter-service OR, and supply a secretariat of coordinating and advisory committees. The COSC agreed and the DRB built up a strong and critical Operational Research Group that farmed out talent and resources to the services to assist in myriad projects. Goodspeed, *History of the Defence Research Board*, 160.

5 NAC, RG 24, vol. 11996, file 1-0-34-2, vol. 2, Report of the Special Meeting of the Defence Research Board, 26 September 1948 to Consider the Report of the Chairman as Recorded under Minute 3.6 of the Sixth Meeting of the Board.

6 UT OMS, B93/0041/033, file DRB Addresses, 1946–1948, Address to the Montreal United Services Institute, 27 October 1948.

7 NAC, RG 24, vol. 11996, file 1-0-34-2, vol. 3, Minutes of the Eleventh Meeting of the DRB, Halifax, 12 June 1949.

8 NAC, RG 24, vol. 11996, file 1-0-34-2, vol. 3, Annual Report of the Chairman of the Defence Research Board, September 1949.

9 NAC, RG 24, vol. 11996, file 1-0-34-2, vol. 3, Minutes of the Twelfth Meeting of the DRB, 17 September 1949.

10 NAC, RG 24, vol. 11996, file 1-0-34-2, vol. 3, Minutes of the Thirteenth Meeting of the DRB, 1 December 1949.

11 Wood, *Strange Battleground*, 188; NAC, RG 24, vol. 11996, file 1-0-34-2, vol. 5, Annual Report of the Chairman, Defence Research Board, September 1951.

12 NAC, RG 24, vol. 11996, file 1-0-34-2, vol. 3, General Report of the Chairman, Defence Research Board, September 1949.

13 NAC, RG 24, vol. 11996, file 1-0-34-2, vol. 5, Annual Report of the Chairman, Defence Research Board, September 1951.

14 NAC, RG 24, vol. 11997, file 1-0-24-2, vol. 6, Annual Report of the Chairman, Defence Research Board, October 1952.

15 Law interview.

16 George Lindsey, "In Memoriam: Omond Solandt," *Canadian Defence Quarterly* 22, no. 6 (July 1993): 44.

17 For distinct wartime origins for most of these establishments, see John Bryden, *Deadly Allies* (Toronto: McClelland and Stewart, 1989); Goodspeed, *History of the Defence Research Board*, 1–100.

18 John R. Longard, *Knots, Volts and Decibels: An Informal History of the Naval Research Establishment, 1940–1967* (Dartmouth: Defence Research Establishment Atlantic, 1993); John Vardalas, "From Datar to the FP-6000 Computer: Technological Challenge in a Canadian Industrial Context," *EEE Annals of the History of Computing* 16, no. 2 (1994), http://ieee.ca/millennium/fp6000/fp6000_datar.html.

19 Goodspeed, *History of the Defence Research Board*, 216–22.

20 Jockel, *No Boundaries Upstairs*, 64, 66, 67; Richter, *Avoiding Armageddon*, 41, 55–6, 68; Goodspeed, *History of the Defence Research Board*, 160–75.

21 Goodspeed, *History of the Defence Research Board*, 233–41.

22 NAUK, AIR 20/7531, Defence Research Board, Department of National Defence, Canada, *A Twilight Computer and Planisphere for High Altitudes Navigation*; NAC, RG 24, vol. 10339, file DRB 1954, *Armed Forces News*, Statement of the Defence Research Board, 31 August 1954; Goodspeed, *History of the Defence Research Board*, 175–89.

23 Pennie interview.

24 Solandt interview, 24 February 1986.

25 NAC, RG 24, vol. 11996, file 1-0-34-2, vol. 2, Report of the Special Meeting of the Defence Research Board, 26 September 1948, Quebec City, to Discuss the Programme and Future Development of the Canadian Armament Research and Development Establishment.

26 The sabot principle involved the use of a high-density, high-velocity, sub-calibre projectile in the carrier, which is discarded. "The Chairman mentioned that the sub-calibre projectile in this ammunition is made of tungsten carbide which, if adopted and required in large quantities, would raise the problem of additional supplies of this scarce material." NAC, RG 24, vol. 11996, file: 1-0-34-2, vol. 2, Report of the Special Meeting of the Defence Research Board, 26 September 1948, Quebec City, to Discuss the Programme and Future Development of the Canadian Armament Research and Development Establishment.

27 NAC, RG 24, vol. 11996, file 1-0-34-2, vol. 4, Annual Report of the Chairman, Defence Research Board, September 1950.

28 NAC, RG 24, vol. 11996, file 1-0-34-2, vol. 2, Report of the Special Meeting of the Defence Research Board, 26 September 1948.

29 NAC, RG 24, vol. 11996, file 0-34-2, vol. 3, Minutes of the Thirteenth Meeting of the Defence Research Board, Held at U of T, 1 December 1949.

30 NAC, RG 24, vol. 11996, file 1-0-34-2, vol. 4, Minutes of the Fifteenth Meeting of the DRB, in Trenton, 9 and 10 June 1950.

31 NAC, RG 24, vol. 11996, file 1-0-34-2, vol. 5, Minutes of the Twentieth Meeting of the Defence Research Board, 29 September 1951.

32 NAC, RG 24, vol. 11996, file 1-0-34-2, vol. 5, Minutes of Twenty-Second Meeting of the Defence Research Board, 14 March 1952.

33 NAC, RG 24, vol. 11996, file 1-0-34-2, vol. 5, Annual Report of the Chairman, Defence Research Board, October 1953.

34 Solandt interview, 24 February 1986, part 1.

35 NAC, RG 24, vol. 11996, file 1-0-34-2, vol. 1, DRB Minutes 1946, Second DRB Meeting 14 May 1947.

36 Gelly and Tardiff, *Defence Research Establishment Valcartier*, 79.

37 NAC, RG 2, vol. 2748, file CDC Conclusions, vol. 8, Minutes of the Sixty-Sixth Meeting of the CDC, 3 October 1950; NAC, RG 24, vol. 11996, file 1-0-34-2, vol. 4, Annual Report of the Chairman, Defence Research Board, September 1950; and Gelly and Tardiff, *Defence Research Establishment Valcartier*, 81–3.

38 Gelly and Tardiff, *Defence Research Establishment Valcartier*, 81–3.

39 For a timeline of the developments of the A.V. Roe Arrow, see the Avro Arrow Museum Website, http://www.avromuseum.com/.

40 Solandt interview, 24 February 1986, part 1.

41 While Solandt had an interest in the value of space and space science during his chairmanship, he felt that CARDE's interest in designing a solid-fuel rocket to launch a payload into orbit was "too far in the future" to be useful during his tenure. Ibid., part 2.

42 The British came to this assessment in 1950. Agar and Balmer, "British Scientists and the Cold War," 219–24.

43 G.B. Carter, *Chemical and Biological Defence at Porton Down, 1916–2000* (London: Her Majesty's Stationery Office, 2000), 68. In 1945, Canadian Major V.P. Ignatieff was appointed to Combined Intelligence Objective Sub Committee (CIOS) Group Leaders, and led CIOS Group 6 in investigations of German chemical warfare facilities, including the discovery of the nerve gas Tabun. However, no connection has yet been found on their discoveries and Canada's post-war chemical warfare work. V.P. Ignatieff, "The Capture of the Main Chemical Warfare Experimental Facilities of the German Army: A Personal Memoir," *Canadian Defence Quarterly* 12, no. 1 (Summer 1982): 43–5.

44 UT OMS, B93/0041/0033, file DRB Addresses, 1946–1948, Address to the Montreal United Service Institute, 27 October 1948.

45 Ibid.

46 Stacey, *Arms, Men and Government*, 509.

47 For discussion of Canada's interwar and wartime work on chemical warfare research, see Avery, *The Science of War*, 122–76; Bryden, *Deadly Allies*.

48 Bryden, *Deadly Allies*; Cecil Law interview.

49 See chapter 14.
50 NAC, RG 24, vol. 11996, file 1-0-34-2, vol. 2, Minutes of the Seventh Meeting
   of the Defence Research Board, 20–1 June 1948.
51 NAC, RG 24, vol. 11996, file 1-0-34-2, vol. 2, Minutes of the Eighth Meeting
   of the Defence Research Board, Held at Valcartier, 26–7 September 1948.
52 Ibid.
53 NAC, RG 24, vol. 11996, file 1-0-34-2, vol. 3, Minutes of the Ninth Meeting
   of the Defence Research Board, Ottawa, 20 December 1948.
54 Ibid.
55 Ibid.
56 The inaugural course ran as follows: CW, Ottawa, 7–19 April 1949; BW,
   Ottawa, 20–3 April 1949; AW, Chalk River, 25 April–4 May 1949; BW/CW,
   Suffield, 9–21 May. NAC, RG 24, vol. 11996, file 1-0-34-2, vol. 3, Minutes of
   the Tenth Meeting of the DRB, 21 March 1949.
57 NAC, RG 24, vol. 11996, file 1-0-34-2, vol. 3. General Report of the Chairman,
   Defence Research Board, September 1949.
58 NAC, RG 24, vol. 11996, file 1-0-34-2, vol. 2, Minutes of the Eighth
   Meeting of the Defence Research Board, Held at Valcartier, 26–7 September
   1948.
59 NAC, RG 24, vol. 11996, file 1-0-34-2, vol. 3, Minutes of the Thirteenth
   Meeting of the Defence Research Board, Held at U of T, 1 December 1949.
60 NAC, RG 24, vol. 11996, file 1-0-34-2, vol. 3, General Report of the Chairman,
   Defence Research Board, September 1949.
61 Carter, *Chemical and Biologic69al Defence at Porton Down*, 75.
62 See Sheldon Harris, *Factories of Death* (London: Routledge, 1994); Global
   Security Organization, "Weapons of Mass Destruction," Country Profile,
   Japan, http://www.globalsecurity.org/wmd/world/japan/cw.htm; "4 [*sic*]
   Arrested in Scandal over Japan's Removal of WWII Chemical Weapons in
   China," *International Herald Tribune*, 23 April 2008, http://www.iht.com/
   articles/ap/2008/04/23/news/Japan-China-Weapons-Disposal.php.
63 Cecil Law interview.
64 NAC, RG 24, vol. 11996, file 0-34-2, vol. 3, Minutes of the Fourteenth Meeting
   of the Defence Research Board, Ottawa, 20 March 1950.
65 NAC, RG 24, vol. 11996, file 1-0-34-2, vol. 4, Best to Solandt on BW Review
   Committee meeting at RMC, n.d. (late November, early December 1950?).
66 NAC, RG 24, vol. 11996, file 1-0-34-2, vol. 4, Minutes of the Seventeenth
   Meeting of the DRB, Ottawa and Fort Churchill, Manitoba, 6 and
   9 December 1950.
67 NAC, RG 24, vol. 11996, file 1-0-34-2, vol. 5, Annual Report of the Chairman,
   Defence Research Board, September 1951.

68 NAC, RG 24, vol. 11996, file 1-0-34-2, vol. 5, Minutes of the Twentieth
   Meeting of the Defence Research Board, 29 September 1951.
69 Ibid.
70 See NAC, RG 2, vol. 2748, file CDC Conclusions, vol. 8, Minutes of the
   Eighty-Seventh Meeting of the CDC, 26 August 1952; and NAC, RG 2,
   vol. 2748, file CDC Conclusions, vol. 8, Minutes of the Eighty-Ninth
   Meeting of the CDC, 9 October 1952.
71 See NAC, RG 2, vol. 2748, file CDC Conclusions, vol. 8, Annual Report of
   the Chairman, Defence Research Board, October 1953.
72 NAC, RG 25, vol. 4220, file 723-935-267, file US Chemical Corps, Edgewood,
   Bacteriological Warfare Projects, Major General E.F. Bulline, US Chemical
   officer, to Solandt, 7 August 1951.
73 NAC, RG 25, vol. 4220, file 723-935-267, US Chemical Corps, Edgewood,
   Bacteriological Warfare Projects. Lieutenant-Colonel H.E. Staples, Canadian
   Army technical representative to United States to DRB Headquarters,
   3 February 1954.
74 NAC, RG 24, vol. 11997, file 1-0-34-2, vol. 3, Annual Report of the Chairman,
   Defence Research Board, October 1952.
75 The board noted that Solandt's speaking engagements had their
   support, since they "served a very important function in keeping up
   interest in the work of the Board." NAC, RG 24, vol. 11996, file 1-0-34-2,
   vol. 2, Minutes of the Seventh Meeting of the Defence Research Board,
   20–1 June 1948.
76 Foulkes first announced his fear on this front in late 1947. NAC, RG 24,
   vol. 11996, file 1-0-34-2, vol. 1, DRB Minutes 1946, Minutes of the Fifth
   Meeting of the Defence Research Board, 17 December 1947.
77 UT OMS, B93/0041/0033, File DRB Addresses, Lecture 24 April 194.
78 Ibid.
79 J.L. Black, "Kanada-Votchina Amerikanskogo Imperializmo: Canada and
   the Canadian Communists in the Soviet 'Coming War' Paradigm, 1946–1951,"
   Canadian Department of Foreign Affairs Website: http://www.dfait-
   maeci.gc.ca/hist/coldwar_section06-en.asp.
80 "Charles A. Pope: Publicity Chief for Defence Board," *Ottawa Journal*,
   14 February 1952.
81 Cyril Bassett, "Science: Our Armed Forces' $35 Million Fourth Arm,"
   Financial Post 46, 16 February 1952.
82 Ibid.
83 NAC, RG 24, vol. 10341, file Press Clippings, March 1947–November 1952,
   Ron Kenyon, "Bacterial Bunk! Endicott's Charge of Germ Warfare Is Proved
   Absurd," *Toronto Telegram*, 21 April 1952.

84 NAC, RG 24, vol. 10341, file Press Clippings, March 1947–November 1952,
   Ron Kenyon, "Bacterial Bunk! Endicott's Charge of Germ Warfare Is Proved
   Absurd," *Toronto Telegram*, 21 April 1952.
85 Ibid.
86 NAC, RG 24, vol. 10341, file Press Clippings, March 1947–November 1952,
   "Saucers 'Not Nonsense' Canada's Scientists Warn," *Toronto Daily Star*, 17
   April 1952.
87 NAC, Defence Research Board, Minutes of the First Meeting of the
   Committee Set Up to Deal with 'Flying Saucer' Sightings, 24 April 1952.
88 Solandt's reputation has been somewhat warped within the community of
   those who believe in UFOs. Some UFO enthusiast websites include alleged
   transcripts of phone conversations with Solandt and scanned letters from
   him not long before his death in 1993. Solandt did admit that engineer
   Wilbert Smith, who worked for the DRB on an electromagnetic propulsion
   project that failed, often talked about UFOs. In what appears to be a genuine
   letter to UFO researcher Grant Cameron on the otherwise hyperbolic and
   bizarre UFO site *Presidential UFO*, Solandt makes it clear that DRB always
   kept an open mind about UFOs, but Smith's research turned out to be
   bunk and no evidence was ever found. I intend to confront this issue more
   fully in a professional academic article, but it required mention here, given
   the number of newspaper accounts written during the 1950s. See http://
   www.presidentialufo.com/omond_solandt.htm.
89 Solandt told the press in 1952 that "for the time being they are being left with
   a staff officer with whom we leave those things we don't understand." NAC,
   RG 24, vol. 10341, file Press Clippings, March 1947–November 1952, "Defence
   Research Board Chairman Solandt Reports ... Science Mystified, Busily
   Prying into Riddle of the Flying Saucer," 16 May 1952, publication unknown.
90 UT OMS, B93/0041/0033, DRB Addresses, 1952–1955, Address to the
   Canadian Club of Ottawa, 7 January 1953.

13 Canada's Defence Research Diplomat

 1 NAC, RG 24, vol. 11996, file 01-34-2, vol. 3, Minutes of the Eleventh Meeting
   of the DRB, Halifax, 12 June 1949.
 2 Avery, *The Science of War*, 249.
 3 Ibid.
 4 Ibid., 52–9.
 5 Solandt interview, 9 September 1985, part 4.
 6 NAC, RG 24, vol. 11996, file 1-0-34-2, vol. 5, Minutes of the Twenty-Third
   Meeting of the Defence Research Board, held in Kingston, 19 June 1953.

7 Solandt interview, 24 February 1986, part 1.
8 DHH Raymont Fonds, 73/1223, box 59, file 130 Chiefs of Staff Committee Minutes 1946, 20 August 1946; Goodspeed, *History of the Defence Research Board*, 82.
9 NAC, RG 24, vol. 4243, file Solandt/Wansbrough-Jones, Wansbrough-Jones to Solandt, 9 April 1951. Others, however, did not share this view. DRB comptroller Alec Fordyce disagreed. He assessed Carrie as a socialite who did little work but bragged a lot because he was protected by McNaughton's friendship. Fordyce interview, part 2.
10 NAUK, DEFE 9/36. See letters between Solandt and Tizard 1947–50. Solandt also advised Tizard of a possibly unfortunate statement he made regarding Canada's value in defence research, that it was best to go to Canada first when he wanted information on the United States. Out of context, this read like a breach of security that would harm U.S./Canada relations. NAUK, DEFE 9/ 36, Tizard to Sir William Morgan, British Joint Services Mission, 14 March 1946.
11 NAUK, DEFE 9/36, Tizard to Solandt, 11 February 1948.
12 NAUK, DEFE 9/36, Matters for Discussion by Sir Henry Tizard in Canada: Note by the Scientific Advisors, n.d. (summer 1947?).
13 NAUK, DEFE 9/21, Sir Henry Tizard's Visit to Canada (Brief by the Controller of the Navy), n.d. (late May 1951?).
14 Digital Automatic Tracking and Resolving computer system for managing naval battle data. See Vardalas, "From Datar to the FP-6000 Computer."
15 He also suggested that Tizard talk to the services about this, since the services would feel flattered and be able to "let their hair down" to a scientist without allegiance to the DRB. NAUK, DEFE 9/21, Sir Henry Tizard's Visit to Canada (Brief by the Controller of the Navy), n.d. (late May 1951?).
16 This included towed sonar, an ice barge as a 1,000-ton displacement as anti-pressure mine sweep (though not seen to be valuable at the moment), anti-mine nets, and the hydrofoil *Massawippi*.
17 NAUK, DEFE 9/21, Controller's Visit to USA and Canada, April/May 1951.
18 Reports were also written by a representative of Lieutenant-General Crawford, controller of Supplies (Munitions), generally of high praise. NAUK, DEFE 9/21, Wansbrough-Jones report, April/May 1951.
19 Solandt attended Royal Navy Exercise Trident 27 April–1 May 1949 and the War Office exercise "Britannia" 23–7 May 1949, a civil defence exercise. Solandt was keenest on "Britannia." In February 1949, Solandt visited "Harness" with Fred Wilkins. See NAC, RG 24, vol. 4243, file Solandt/ Wansbrough-Jones, Wansbrough Jones to Solandt, 24 January 1949; Solandt to Wansbrough-Jones, 8 February 1948.

20 British policy regarding Pakistan and India included restrictions on discussing equipment concerns with them, so some felt there was little point discussing research work. Others in attendance were Sir Henry Tizard (chair) Sir Ben Lockspier (Department of Scientific and Industrial Research); Vice-Admiral Sir Charles Daniel (controller of the Navy); Rear Admiral R.A.B. Edwards (assistant chief of Naval Staff); Doctor J.A. Carroll (Admiralty); Mister F. Brundett (Admiralty); Doctor O.H. Wansbrough-Jones (War Office); Air Vice Marshal C.B.R. Polly (Air Ministry); Doctor R. Cockburn (Air Ministry) Lieutenant-General Sir Frederick Wrisberg (Ministry of Supply); Air Marshall Sir Alec Coryton (Ministry of Supply); Mister H.M. Garner (Ministry of Supply), and the secretariat of Mister I. Montgomery, Lieutenant-Commander H. Winter, RN, and Mister N.S. Forward. NAUK, DEFE 10/10, DRPC Minutes, 3 May 1949.

21 This included Sir Henry Tizard (chair); Sir Frederick Brudrett (Ministry of Defence); Rear Admiral E.M. Evnas-Lombe (Admiralty); Mister W.J.R. Cook (Admiralty); Lieutenant-General Sir John Whiteley (War Office); Mister H.A. Sargeant (War Office); Air Chief Marshal Sir Ralph Cochrane (Air Ministry); Lieutenant-General Sir Kenneth Crawford (Ministry of Supply); Air Vice Marshal C.P. Brown (Ministry of Supply); and Sir John Cockcroft (Ministry of Supply). Also included with Solandt were Mister R.A. Hughes (Ministry of Agriculture); Mister R.R. Powell (Ministry of Defence); Air Chief Marshal Sir Alec Coryton (Ministry of Supply); Sir William Slater (Agricultural Research Council); General Sir William Morgan); Doctor Galloway (Foot and Mouth Research Institute); and Mister J.M. Wilson (Ministry of Supply). NAUK, DEFE 10/10, DRPC Minutes, 4 September 1951.

22 Ibid.

23 Ibid.

24 NAUK, DEFE 10/10, DRPC Minutes, 4 September 1951.

25 NAUK, DEFE 10/10, DRPC Minutes, 8 July 1952. Sir Henry Tizard (chair); Sir Ben Lockspeiser (DSIR); Sir Frederick Brudrett (Ministry of Defence); Admiral Sir Michael Denny (Admiralty); Rear Admiral G. Barnard (Admiralty); Mister W.J.R. Cook (Admiralty); Lieutenant-General Sir John Whiteley (War Office); Doctor A. Cockburn (Air Ministry); Air Marshal Sir John Boothman (Ministry of Supply); Mister H.A. Sargeant (War Office); Air Chief Marshal Sir Ralph Cochrane (Air Ministry); Lieutenant-General Sir Kenneth Crawford (Ministry of Supply); and Mister S.S.C. Mitchell (Ministry of Supply). Also attending were Doctor W.G. Whitman and Sir Alwyn Crow.

26 There was an amendment attached to these minutes that Mister Whitman's estimates were 2.5 times too small.

27 NAUK, DEFE 10/10, DRPC Minutes, 8 July 1952.

28 Ibid.

29 He estimated this at $10 million.

30 NAUK, DEFE 10/10, DRPC Meeting Minutes, 8 July 1952.

31 This was an automatic blind bombing system. NAUK, DEFE 10/10 DRPC, DRPC Minutes, 8 July 1952.

32 NAC, RG 24, vol. 4243, file Solandt/Wansbrough-Jones Wansbrough-Jones to Solandt, 6 January 1947.

33 Agar and Balmer, "British Scientists and the Cold War."

34 Briefly, the organization removed the word *commonwealth* from their organization to appease King's concerns, referring to themselves as the "ACDS." NAC, RG 24, vol. 4243, file Solandt/Wansbrough-Jones, Solandt to Wansbrough-Jones, 28 February 1947; Wansbrough-Jones to Solandt, 3 September 1948.

35 NAC, RG 24, vol. 4243, file Solandt/Wansbrough-Jones, Wansbrough-Jones to Solandt, 22 April 1947.

36 NAUK, DEFE 9/36, Matters for Discussion by Sir Henry Tizard in Canada: Note by the Scientific Advisors, n.d. (summer 1947?).

37 NAUK, DEFE 10/10, Minutes of the CACDS, 4 November 1947.

38 Solandt debated with Tizard on the differing needs of radar air defence in Canada. Solandt said Canada's northern border concerns were in greater need for lateral range, as her warning stations could be advanced several hundreds of miles ahead of her population centres. Tizard argued that Canada's problems with air defence were simpler than the U.K.'s, since "accuracy of warning need not be so high as in the United Kingdom." They discussed doing interceptor trials in Canada. Solandt said that "from the Canadian point of view, with the possibility of attack over the Polar region it was of importance to know whether bombers with the very long range that would be required would in fact have the speed which had been assumed throughout this discussion. Although the experiments which Sir Henry Tizard has suggested were technically not difficult to carry out, practical difficulties would be considerable." NAUK, DEFE 10/10, Minutes of the CACDS, 6 November 1947.

39 NAUK, DEFE 10/10, Minutes of the CACDS, 7 November 1947.

40 The Canadian note on this issue outlines work done in the Commonwealth in the areas. Further work needed to be done on long-term efforts on prevention regarding research on the psychological basis for warfare, methods of psychological warfare, as well as problems of development and maintenance for the armed forces. Solandt was struck by the interest of members in this note, saying Canada had a strong academic background

in these fields. Most work had been on personnel selection more than psychological bases for warfare, "a subject which he felt it was important not to neglect." Ibid.

41 Solandt said Canada, like the rest of the Commonwealth, had the problem of deciding how seriously to take the threat of air attack and how best to meet it. Population and cold winters made Canadian experience unique. "He thought that the most important unknown factor in Civil Defence was the reaction of the population in various countries to atomic attack or other comparable disasters." Doctor Morton said some comparable data collected on Canadian reactions to natural disasters would be useful, and Tizard said the United Kingdom would be grateful for any data Canada could provide. NAUK, DEFE 10/11, Minutes of the CACDS Meeting, 1949.

42 Mister E.C. Williams, one of Blackett's disciples in OR, led the British section on OR discussions. There were four main divisions of OR and all were closely related, but "the statistical method was the main one used," though there were others of great value. Two that he had not mentioned that were critical were "direct enquiry and experiment." To be valuable, experiments or exercises had to be realistic. To be truly effective, OR must employ the joint thinking of the service man, the scientists, and the engineer. Solandt introduced the Canadian paper and mentioned that statistical method was less appropriate to Army OR than the other services. "In Canada there was one Operational Research Group within the Defence Research Board and this also supplied the staff for operational research sections which were in the process of being established in each Service. This arrangement should ensure good co-ordination of methods between the Services." NAUK, DEFE 10/10, Minutes of the CACDS Meeting, 4 July 1950.

43 "[Solandt's] experience showed that good results could be obtained within one country by achieving co-ordination between various branches of geophysics – those that study meteorological, magnetic, ionosphere and cosmic ray effects – without which meteorology was in danger of becoming isolated." Ibid.

44 NAUK, DEFE 10/10, Minutes of the CACDS Meeting, 5 July 1950.

45 Ibid.

46 Doctor B.K. Blount wanted to canvass the Commonwealth for its efforts and stressed the need to pool resources to get a good view of scientific capabilities of outside countries, especially those with stringent security measures on scientific and technical matters. Solandt said Canada's effort was modest but growing. He stressed the work done on Canada's arctic and the importance of using overt methods. NAUK, DEFE 10/10, Minutes of the CACDS Meeting, 7 July 1950.

47 Ibid.
48 Solandt said that major changes had occurred in this area as the result of the discovery of new oil fields in Alberta. Still, they were a great distance away from industrial sectors of Canada, adding cost to the fuel, but the long-term result was the growing import of Alberta. "The development of this supply of natural petroleum had removed the economic pressure in Canada of developing substitutes for oil, but in the long term Canada's reserves of such mineral resources as bituminous sands were of considerable interest. It was not generally appreciated that Canada had hitherto always had to import power on a large scale: there were prospects that her position might now come into balance." NAUK, DEFE 10/10, Minutes of the CACDS Meeting, 11 July 1950.
49 Ibid.
50 Ibid.
51 Ibid.
52 Ibid.
53 NAUK, DEFE 9/36, Solandt to Tizard, 24 July 1950.
54 NAC, RG 24, vol. 4243, file Solandt/Wansbrough-Jones, Solandt to Wansbrough-Jones, 22 August 1946; Wansbrough-Jones to Solandt, 28 August 1946.
55 NAC, RG 24, vol. 4243, file Solandt/Wansbrough-Jones, Wansbrough-Jones to Solandt, 20 February 1947.
56 Solandt felt that it was none of his business and yet provided a short survey of the problems he had been advised about on this project by the Canadian Army, a project he had originally backed and now had concerns about. See NAC, RG 24, vol. 4243, file Solandt/Wansbrough-Jones, Solandt to Wansbrough-Jones, 21 May 1951.
57 No first name was used. His current identity is unclear.
58 "It is possible though that we can go this far. If you state that the work on these materials was being done independently of the authorised BW programme, then perhaps it need not be automatically revealed to the United States. If, further, you can state on behalf of your Government that it is desired that work on this job should be carried out in Canada, and that all the necessary precautions to prevent any possibility whatever of any harm arising would be taken, then we might be able to settle these obvious difficulties." NAC, RG 24, vol. 4243, file Solandt/Wansbrough-Jones, Wansbrough-Jones to Solandt, 20 June 1951.
59 Balmer, *Britain and Biological Warfare*, 128–54.
60 The illegal nature of the research is mentioned in NAUK, DEFE 10/10, DRPC Minutes, 4 September 1951.

61 NAC, RG 24, vol. 4243, file Solandt/Wansbrough-Jones, Wansbrough-Jones to Solandt, 2 February 1956.

62 NAC, RG 24, vol. 4243, file Solandt/Bush, Solandt to Bush, 11 March 1946.

63 Ibid.

64 NAC, RG 24, vol. 4243, file Solandt/Bush, Bush to Solandt, 18 December 1946.

65 Allan Needle, *Science, Cold War, and the American State: Lloyd V. Berkner and the Balance of Professional Ideals* (Washington: Harwood Academic Publishers, 2000).

66 Solandt mentioned that General Henry had suggested this could be initiated by discussions with General Tom C Rives, chief, Electronics Subdivision, Air Materials Command. Rives was on the Committee on Electronics, JRDB, and backed the discussions. NAC, RG 24, vol. 4243, file Solandt/ Bush, Solandt Bush to Solandt, 6 May 1947.

67 OMS UT, B93/0041/033, file DRB Address, 1946–1948, Notes on Conference 3 October 1947, The Pentagon.

68 NAC, RG 24, vol. 4243, file Solandt/Bush, Solandt to Bush, 2 January 1948.

69 NAC, RG 24, vol. 4243, file Solandt/Bush, Bush to Solandt, 20 January 1948.

70 Solandt interview, 8 September 1985, part 4.

71 NAC, RG 24, vol. 4243, file Solandt-Bush, Bush to Solandt, 22 July 1948; Solandt interview, 8 September 1985, part 4.

72 RG 24, vol. 4243, file Solandt/Wansbrough-Jones, Solandt to Wansbrough-Jones, 10 June 1952.

73 NAC, RG 24, vol. 11996, file 1-0-34-2, vol. 2, Minutes of the Sixth Meeting of the DRB, 19 March 1948.

74 NAC, RG 24, vol. 11996, file 1-0-34-2, vol. 2, Minutes of the Seventh Meeting of the Defence Research Board, Held at Valcartier, 26–7 September 1948.

75 NAC, RG 24, vol. 11996, file 1-0-34-2, vol. 1, DRB Minutes 1947, Minutes of the Fifth Meeting of the DRB, 15 December 1947.

76 On 23 December 1947, Solandt reported to the COSC that the United States and the United Kingdom agreed to conduct biological warfare trails at sea, and reconnaissance for right place to be done shortly. DRB representatives had been present at the discussions, and four or five DRB employees would be part of reconnaissance effort and trials. While this was within DRB responsibility, U.K. COS wanted COSC approval, which was given. DHH Raymont Fonds, 73/1223, box 59, file 130, Chiefs of Staff Committee Minutes 1947, 23 December 1947.

77 NAC, RG 24, vol. 11996, file 1-0-34-2, vol. 2, Minutes of the Sixth Meeting of the DRB, 19 March 1948.

78 For instance, several of the 250 monkeys flown in from India for the tests escaped in October 1948, much to the public's amusement. Balmner, *Britain and Biological Warfare*, 104–10.
79 DHH Raymont Fonds, 73/1223, box 61, file 1307, Chiefs of Staff Committee Minutes 1949, Minutes of the COSC Meeting, 2 February 1949.
80 Balmer, *Britain and Biological Warfare*, 104–10.
81 NAC, RG 24, vol. 11996, file 0-1-34-2, vol. 3, Minutes of the Tenth Meeting of the DRB, Ottawa, 21 March 1949.

14 Atomic Realist

1 See Maloney, *Learning to Love the Bomb*, 19–51; Stephen Twigge and Len Scott, *Planning Armageddon: Britain, the United States, and the Command of Western Nuclear Forces 1945–1964* (London: Harwood Academic Publishers, 2000), 147–94; and Gaddis, *The US and the Origins of the Cold War*, 164–98.
2 UT OMS, B93/0041/0033, file DRB Addresses, 1954–55, Address to University of Toronto Convocation, 16 June 1954.
3 Omond Solandt, "The Nuclear Age," Sixth Francis Sheperd Memorial Lecture, *Medical Service Journal [Canada]* 4, no. 10 (November 1958): 681.
4 NAC, RG 24, vol. 11996, file 1-0-34-2, vol. 1, DRB Minutes 1946, Minutes of the Fifth Meeting of the DRB, 15 December 1947.
5 Between 1946 and 1956, Solandt gave ten public addresses on nuclear weapons, and dozens on modern warfare that included discussions of atomic weapons. See UT OMS, 0093/0044/0031, file DRB addresses, 1946–1956; Jason S. Ridler, "From Nagasaki to Toronto: Omond Solandt and the Defence Research Board's Early Vision of Atomic Warfare, 1945–1947," *Canadian Military History*, forthcoming.
6 NAC, RG 24, vol. 10339, file DRB 1953, untitled news release, 11 May 1953.
7 NAC, RG 24, vol. 11996, file 1-0-34-2, vol. 1, DRB Minutes 1946, Minutes of the Fifth Meeting of the DRB, 15 December 1947.
8 NAC, RG 2, vol. 2748, file CDC Conclusions, vol. 8, Minutes of the Forty-Third Meeting of the CDC, 15 April 1948.
9 NAC, RG 24, vol. 11996, file 1-0-34-2, vol. 3, Minutes of the Ninth Meeting of the Defence Research Board, Ottawa, 20 December 1948.
10 NAC, RG 24, vol. 11996, file 1-0-34-2, vol. 3, Minutes of the Eighth Meeting of the Defence Research Board, Held at Valcartier, 26–7 September 1948.
11 NAC, RG 24, vol. 11996, file 1-0-34-2, vol. 2, Minutes of the Seventh Meeting of the DRB, 20–1 June 1948.
12 Solandt felt research here was not going as fast as he desired as the result of a shortage of civil engineers, and there was little prospect of improvement

with the Building Research Division of NRC understaffed. NAC, RG 24, vol. 11996, file 1-0-34-2, vol. 4, Minutes of the Sixteenth Meeting of the Defence Research Board, held at Ottawa and Chalk River, 16–17 October 1950.

13 NAC, RG 24, vol. 11996, file 1-0-34-2, vol. 5, Annual Report of the Chairman, Defence Research Board, September 1951.

14 DHH Raymont Fonds, 733/1223, box 61, series 3, file 1307, Minutes of the Chiefs of Staff Committee (Regular and Special) 1952–1953, Minutes for 9 November 1953. See Steve Harris, "Civil Defence," *Canadian Encyclopedia*, 2013, http://www.thecanadianencyclopedia.com/en/article/civil-defence/.

15 DHH Raymont Fonds, 733/1223, box 61, series 3, file 1307, Minutes of the Chiefs of Staff Committee (Regular and Special) 1952–1953, Minutes of the Meeting of the COSC, 12 March 1954.

16 Ibid.

17 DHH Raymont Fonds, 733/1223, box 61, series 3, file 1308, Minutes of the Chiefs of Staff Committee, Minutes for 8 February 1955.

18 DHH Raymont Fonds, 73/1223, box 59, file 130, Chiefs of Staff Committee Minutes 1946, Minutes of the Meeting of the COSC, 28 March 1946.

19 DHH Raymont Fonds, 73/1223, box 59, file 130, Chiefs of Staff Committee Minutes 1946, COSC Minutes, 28 March 1946.

20 Avery, *The Science of War*, 176–202.

21 The Chalk River laboratory was amalgamated into the Crown corporation Atomic Energy Canada Limited in 1952, under Mackenzie. For the history of Canada's atomic energy programs, see Robert Bothwell, *Nucleus: A History of Atomic Energy of Canada, Ltd.* (Toronto: University of Toronto Press, 1988).

22 NAC, RG 24, vol. 11996, ile: 0-1-34-2, vol. 3, Minutes of the Ninth Meeting of the Defence Research Board, Ottawa, 20 December 1948.

23 DHH Raymont Fonds, 73/1223, box 61, file 1307, Chiefs of Staff Committee Minutes 1949, Minutes of the COSC Meeting, 8 March 1949.

24 He announced that DND and the Department of Trade and Commerce wanted him to attend a meeting of the Atomic Energy Control Board. Its Atomic Energy Panel "will be the forum for policy discussions concerned with military problems," and that the Chiefs of Staff were to attend meetings dealing with national defence. However, it was agreed that "the actual research problems associated with the military aspects of atomic energy will be directed by the Defence Research Board." NAC, RG 24, vol. 11996, file 0-1-34-2, vol. 3, Minutes of the Tenth Meeting of the DRB, Ottawa, 21 March 1949.

25 NAC, RG 24. vol. 11996, file 0-1-34-2, vol. 3, Minutes of the Tenth Meeting of the DRB, Ottawa, 21 March 1949.

26 NAC, RG 24, vol. 11996, file 0-1-34-2, vol. 3, Minutes of the Eleventh Meeting of the DRB, Halifax, 12 June 1949.

27 While there was no information available concerning bomb production or increase in destructive qualities, Solandt noted, information concerning uranium deposits and certain phases of production was available. From this would come valuable deductions for the Services and DRB. DHH Raymont Fonds, 73/1223, box 61, file 1307, Chiefs of Staff Committee Minutes 1949, Minutes of the COSC Meeting, 8 March 1949.

28 DHH Raymont Fonds, 73/1223, box 61, file 1307, Chiefs of Staff Committee Minutes 1949, 8 March 1949.

29 Atomic, biological, and chemical warfare.

30 NAC, RG 24, vol. 11996, file 01-34-2, vol. 3, General Report of the Chairman, Defence Research Board, September 1949.

31 No mention of this group is included in the chief history of Canada's nuclear energy program. Bothwell, *Nucleus*.

32 NAC, RG 24, vol. 11996, file 0-1-34-2, vol. 3, Minutes of the Fourteenth Meeting of the Defence Research Board, Ottawa, 20 March 1950.

33 NAC, RG 24, vol. 11996, file 1-0-34-2, vol. 4, Annual Report of the Chairman, Defence Research Board, September 1950.

34 NAC, RG 24, vol. 11996, file 1-0-34-2, vol. 3, Minutes of the Eighth Meeting of the Defence Research Board, Held at Valcartier, 26–7 September 1948.

35 See chapter 12.

36 Solandt, quoted in Maloney, *Learning to Love the Bomb*, 80.

37 Solandt interview, 10 May 1986, part 6; Manuel Zack, Lawrence Martin, and Alvin A. Lee, *Harry Thode: Scientist and Builder at McMaster University* (Hamilton, ON: McMaster University Press, 2003), 60–1.

38 NAC, RG 24, vol. 11996, file 1-0-34-2, vol. 4, Minutes of the Fourteenth Meeting of the Standing Committee on Extra Mural Research of the Defence Research Board, 9 March 1951.

39 NAC, RG 24, vol. 11996, file 1-0-34-2, vol. 5, Annual Report of the Chairman, Defence Research Board, September 1951.

40 Rhodes, *Dark Sun*, 129–30, 229.

41 DHH Raymont Fonds, 73/1223, box 61, file 1307, Chiefs of Staff Committee Minutes 1949, 13 September 1949.

42 DHH Raymont Fonds, 73/1223, box 61, file 1307, Chiefs of Staff Committee Minutes 1949, 3 October 1949.

43 DHH Raymont Fonds, 73/1223, box 61, file 1307, Chiefs of Staff Committee Minutes 1949, 3 October 1949.

44 According to Solandt, the United States had dispersed the locations of atomic storage establishments into the central part of the country, away

from the coasts, to decrease strategic vulnerability. To ensure against atomic attack, installations were underground. The initial cost of these installations was high and they still had many technical concerns. As such, manufacturing and storage should remain in the United States. Some Americans had suggested a storage facility in Canada, as a means of overcoming U.S. objection to having atomic bombs stored in Europe. Solandt felt the United Kingdom might block this move, as they wanted bombs on hand, too. DHH Raymont Fonds, 73/1223, box 61, file 1307, Chiefs of Staff Committee Minutes 1949, Minutes of the COSC Meeting, 3 October 1949.

45 DHH Raymont Fonds, 73/1223, box 61, file 1307, Chiefs of Staff Committee Minutes 1949, Minutes of the COSC Meeting, 3 October 1949.

46 Ibid.

47 Solandt interview, 24 February 1986, part 1; DHH Raymont Fonds, 73/1223, box 61, file 1307, Chiefs of Staff Committee Minutes 1949, 15 November 1949.

48 Solandt interview, 24 February 1986, part 2.

49 DHH Raymont Fonds, 733/1223, box 61, series 3, file 8 (1307S), Minutes of the Chiefs of Staff Committee (Regular and Special) 1952–1953, Minutes of the COSC Meeting, 3 April 1952.

50 NAC RG 24 Vol. 11996 File: 1-0-34-2 vol. 5, Annual Report of the Chairman, Defence Research Board, September 1951.

51 Ibid.

52 NAC RG 24 Vol. 11996 File: 1-0-34-2 vol. 5, Minutes of the Twenty-Fourth Meeting of the Defence Research Board 18 October 1952.

53 Ibid.

54 NAC RG 24 Vol. 11996 File: 1-0-34-2 vol. 5, Annual Report of the Chairman, Defence Research Board, October 1953.

55 In 1948, he attended the RAF Exercise "Pandora," which was to test "technical doctrines and technical equipment in simulated conflict between the U.K. and a mythical enemy." "Pandora" had two phases. The first was based on neither combatant using an atomic bomb, the second with both sides having it. "All Services in the United Kingdom, United States and Canada were represented by senior officers, and following the exercise a general discussion was held of the results." Solandt's general impression was that the United Kindgom was pessimistic about scientific and technical development in the next few years, "since it was necessary for them to put so much emphasis on industrial recovery." The United States was more optimistic about current scientific developments. NAC, RG 24, vol. 11996, file 1-0-34-2, vol. 2, Minutes of the Seventh Meeting of the DRB, 20–1 June 1948.

56 Margaret Gowing and Lorna Arnold, *Independence and Deterrence: Britain and Atomic Energy, 1945–1952* (London: Macmillan, 1974), 326.

57 J.L. Symonds, *A History of British Atomic Tests in Australia* (Canberra: Australian Government Publishing Services for the Department of Resources and Energy, 1985), 53.

58 Symonds does not, however, provide evidence to support this argument. Ibid., 60.

59 Solandt was making this claim as late as 1954. UT OMS, B93/0041/0033, file DRB Addresses, 54–55. Address to University of Toronto Convocation, 16 June 1954.

60 NAC, RG 2, vol. 2748, file CDC Conclusions, vol. 8, Minutes of the Ninetieth Meeting of the CDC, 14 November 1952; Symonds, *History of British Atomic Tests in Australia* (North Melbourne: Ruskin, 1985).

61 Solandt interview, 24 February 1986, part 2.

62 Ibid.

63 Maloney, *Learning to Love the Bomb*, 21–2.

64 Maloney, *War without Battles*, 69–73.

65 DHH Raymont Fonds, 733/1223, box 61, series 3, file 1307, Minutes of the Chiefs of Staff Committee (Regular and Special), 1952–1953, Minutes of the COSC Meeting, 16 February 1953.

66 NAC, RG 24, vol. 11996, file 1-0-34-2, vol. 5, Minutes of the Twenty-Fifth Meeting of the Defence Research Board, 17 February 1953.

67 DHH Raymont Fonds, 733/1223, box 61, series 3, file 1307, Minutes of the Chiefs of Staff Committee (Regular and Special), 1952–1953, Minutes of the Meeting of the COSC, 23 October [*sic*] 1953.

68 DHH Raymont Fonds, 733/1223, box 61, series 3, file 1307, Minutes of the Chiefs of Staff Committee (Regular and Special), 1952–1953, Minutes of the Meeting of the COSC, 7 October 1953.

69 NAC, RG 24, vol. 11996, file 1-0-34-2, vol. 5, Annual Report of the Chairman, Defence Research Board, October 1953.

70 Ibid.

71 NAC, RG 24, vol. 11997, file 1-0-24-2, vol. 6, Minutes of the Twenty-Seventh Meeting of the Defence Research Board, held at Shirley's Bay, ON, 16 October 1953.

72 Ibid.

73 Solandt interview, 24 February 1986, part 2.

74 Wayne Reynolds, "Atomic War, Empire Strategic Dispersal and the Origins of the Snowy Mountain Scheme," *War and Society* 14, no. 1 (May 1996): 121–44; Reynolds, "Planning the Defence of World War III: The Post-war British Empire and the Role of Australia and South Africa, 1943–1957,"

Historia 43, no. 1 (May 1998): 72–90; Reynolds, "Defence Science Research, Higher Education and the Australian Quest for the Atomic Bomb, 1945–1960," *History of Education* 26, no. 2 (1997): 225–42; Reynolds, "Rethinking the Joint Project: Australia's Bid for Nuclear Weapons, 1945–1960," *Historical Journal* 41, no. 3 (1998): 853–73.

75 Solandt interview, 24 February 1986, part 2.

76 Ibid.

77 NAC, RG 24, vol. 11996, file 1-0-34-2, vol. 5, Minutes of the Twenty-Seventh Meeting of the Defence Research Board, Held at Shirley's Bay, ON, 16 October 1953.

78 Foulkes initiated the creation of a Joint Special Weapons Policy Committee on 25 January 1954, to "determine and co-ordinate requirements of the services concerning atomic matters and to disseminate information as it becomes available." Solandt pushed for consistency regarding the new committee, and the existing Joint Special Weapons Committee, which had done excellent work, and the COSC agreed that the old committee could be a subcommittee of the policy committee. The COSC agreed that the chairmanship of both committees would rotate annually between the services but both committees should be chaired by the same service. In February, Solandt suggested Maas be made advisor to the committee and attend a meeting of the JSWP committee, as it would deal with a wide range of scientific and technical topics and should have other suitable advisors. Solandt also suggested that DRB's Committee on Defence Aspects of Atomic Warfare and the Chairman's Panel on Radiation, Protection and Treatment would be suitable advisors to the committee. DHH Raymont Fonds, 733/1223, box 61, series 3, file 1307, Minutes of the Chiefs of Staff Committee (Regular and Special) 1952–1953, Minutes of the Meeting of the COSC, 25 January 1954.

79 NAC, RG 24, vol. 11997, file: 1-0-34-2, vol. 7, Minutes of the Twenty-Ninth Meeting of the Defence Research Board, 11 June 1954.

80 Ibid

81 This briefing covered technical aspects, civil defence, OR studies, and intelligence information. Ibid.

82 Ibid.

83 This emphasized the threat of atomic retaliation to threats from aggressor nations as a means to conserve dollars on manpower by investing them in technological advantage. It was also referred to as "massive retaliation." See Gaddis, *Strategies of Containment*, 127–63.

84 NAC, RG 24, vol. 11997, file 1-0-34-2, vol. 7, Minutes of the Twenty-Ninth Meeting of the Defence Research Board, 11 June 1954.

85 DHH Raymont Fonds, 733/1223, box 61, series 3, file 1307, Minutes of the Chiefs of Staff Committee (Regular and Special) 1952–1953, Minutes of the Meeting of the COSC, 30 July 1954.

86 NAC, RG 24, vol. 11997, file 1-0-34-2, vol. 7, Minutes of the Thirtieth Meeting of the Defence Research Board, 9 October 1954.

87 Ibid.

88 Ibid.

89 Ibid.

90 DHH Raymont Fonds, 733/1223, box 61, series 3, file 1307, Minutes of the Chiefs of Staff Committee (Regular and Special) 1952–1953, Minutes of the Meeting of the COSC, 3 November 1954.

91 NAC, RG 2, vol. 2748, file CDC Conclusions, vol. 7, Minutes of the Hundred and First Meeting of the CDC, 12 November 1954.

92 Ibid.

93 The best analysis of these policies is Maloney, *Learning to Love the Bomb*, 99–123.

15 Conclusion

1 Solandt interview, 24 February 1986, part 3.

2 Ibid.

3 Chute interview, part 2.

4 UT OMS, B93/0041/012, file Job Offers 1948–1968, Solandt to Fitzgerald, 8 March 1952.

5 Solandt interview, 24 February 1986, part 3.

6 UT OMS, B93/0041/0035, file CNR Correspondence, Donald Gordon to Solandt, 14 September 1955.

7 Solandt interview by Grenville, 24 February 1986, part 3. Later in life, Paul Hellyer, who began his work in the Department of Defence shortly before Solandt left, asked Solandt why he left the DRB in 1956. Solandt's response was that he had been there long enough. Hellyer interview.

8 Paul Hellyer served as Campney's aide in the 1950s and often attended COSC meetings where Solandt was present. When Hellyer heard this story about Campney's conduct, he confirmed that this sounded just like Campney's approach: blunt and direct. Helleyer interview.

9 Solandt interview, 24 February 1986, part 3.

10 See chapter 11; Goodpseed, *History of the Defence Research Board of Canada*, 249.

11 Solandt interview, 24 February 1986, part 3.

12 Fordyce interview, part 1.

13 On the DRB and Canada's space programme, see Andrew Godefroy, "Defence and Discovery: Science, National Security, and the Origins of the Canadian Rocket and Space Program" (PhD diss. Royal Military College of Canada, 2004).

14 Law interview; Fordyce interview.

15 Quoted in Maloney, *Learning to Love the Bomb*, 110.

16 Lindsey interview, part 2.

17 UT OMS, B93/0041/0033, file 4, Douglas Letterman, "Secrecy Curtain Demolished: Canada to Need Nuclear Weapons, Solandt Says," *Hamilton Spectator*, 7 March 1956.

18 Ibid.

19 J.L.C. Carrier, "A Versatile 'Helix' of the National System of Innovation," Canadian Force College Paper NSSC 5, 23 March 2003, http://wps.cfc. forces.gc.ca/papers/nssc/nssc5/carrier.htm.

20 Douglas Bland, *The Administration of Defence Policy in Canada* (Kingston: Ronald P. Frye, 1987), 142–5.

21 Office of the Auditor General, Canada, *1994 Report of the Auditor General* (1994), chap. 24, "National Defence and Defence Management Systems," http://www.oag-bvg.gc.ca/internet/English/aud_ch_oag_1994_24_e_5929.html.

22 Gordon Watson, "Why the Bureaucrats Secretly Carved up the DRB: It Worked too Well," *Science Forum* 47 (October 1975): 22–5.

23 Quoted in Bland, *Administration of Defence Policy in Canada*, 56.

24 Omond Solandt, "The Defence Research Board's Untimely End: What It Means for Military Science," *Science Forum* 47 (October 1975): 19–21.

25 Solandt CV.

26 P.J. Sanford, "The Origins and Growth of the Canadian Operational Research Society," *Canadian Operational Research Society Journal* 1, no. 1 (December 1963): 1–12.

27 "Trustees Solandt, Haywood, End Service as Board Members," *MITRE Matters* 11, no. 10 (October 1969): 1.

28 In a letter to Sir Edward Playfair at the Ministry of Defence (U.K.), Solly Zuckerman, one of Solandt's colleagues from the war and scientific advisor to the Ministry of Defence, mentioned that General Norstad was actively considering Solandt for the job as scientific advisor to SHAPE. Mysteriously, Zuckerman remarked, "When I see you on Tuesday there is a point about this matter which I should like to tell you by word of mouth, rather than in a letter. You will understand my reticence when we meet." In the next paragraph Zuckerman speaks of Solandt's old rival, Patrick Johnson, though the two points do not appear linked. In the end, the job went to Doctor Harold Agner, a nuclear scientist who had worked with

Enrico Fermi on the world's first nuclear chain reaction at the University of Chicago in 1942 and was one of the chief scientists on the Manhattan project. UEA SZ/CSA/113/1/63, Zuckerman to Playfair, 27 May 1961.

29 Solandt CV.

30 Quoted in Robert Havener, "The Legacy of Omond M. Solandt: Organizing and Managing the Practical Application of Science at the International Level," in Lindsey, Law, and Grenville, *Perspectives in Science*, 173.

31 Goodspeed, *A History of the Defence Research Board*.

32 Solandt always felt that the DRB's influence in industry and academia was poorly reflected in Goodspeed's history. Solandt interview, 24 February 1986, part 2.

33 Havener, "Legacy of Omond M. Solandt," 176.

34 C.P. Snow, *The Two Cultures* (1959; Cambridge: Cambridge University Press, 1996); Snow, *Science and Government* (Cambridge: Harvard University Press, 1960).

35 Omond Solandt, "Foreword," in Lindsey, Law, and Grenville, *Perspectives in Science*, xiv.

Bibliography

Primary Sources

Personal Papers

Directorate of History and Heritage (Ottawa)
    DHH 733/1223, Charles Foulkes Papers
The Imperial War Museum (London, U.K.)
    Ivor Evans Papers, box P426, file A2/A2
    Basil Schonland Papers, 86/63/01
National Archive (Kew, U.K.)
    DEFE 9, Papers of Sir Henry Tizard, Chairman of the Defence Research
        Policy Committee, 1945–1952
    Ministry of Defence Records
National Archives of Canada (Ottawa)
    MG 32, B5, vol. 221, Brooke Claxton Papers
Queen's University Archive (Kingston)
    Queen's Historical Collection D, Alumni files 3599 Solandt, Donald M.;
        Young, Edith
    Queen's Student Register, vol. 5, call #1161, file 2885
    Adam Shortt Correspondences, call #2047, box 5, files 29–30
University of East Anglia Archive (Cambridge, U.K.)
    Solly Zuckerman fonds, Oxford Extra Mural Unit (OEMU), 21/3/41
University of Toronto Archive (Toronto)
    Omond McKillop Solandt Fonds
    0041/31, Ffile: "International Symposium on Military Operational
        Research."
    B1991-0015/011, file Atomic Bomb, Japan, 1945

B88/0016/001, file Diplomas and Certificates
B93/0041/007, file Activities File, 1953–1988
B93-0041/007, file Canadian Aeronautics
B93/0041/012, file Associations and Committees, 1967–1985
B93/0041/031, file Operational Research, 1933–1989
B93/0041/033, file Defence Research Board of Canada, 1946–1956
B93/0041/049, file Mrs David Orlikow vs United States Government,
    1951–1988
B93/0044/033, file Defence Research Board Addresses 1946–1952
Charles Best Papers
Frederick Banting Papers
The Wellcome Trust Archive (London, U.K.)
    Papers of Dr William Poachin
Wilfrid Laurier Archive (Waterloo)
    Ronnie Shepard Papers and Collection

Private Papers

Michael Bliss, correspondences and transcribed interviews with Omond Solandt
David Grenville, personal research papers
Cecil Law, documents from the Solandt Symposium Archives

Government Record Groups

Directorate of History and Heritage (Ottawa)
    73/1223, boxes 59–63, Chiefs of Staff Committee Minutes, 1945–1956
National Archive of Canada Record Groups (Ottawa)
    RG 2, volume 2748, file CDC Conclusions
    RG 24, volume 4243, Chairman of the Defence Research Board
        Correspondences 1945–1956
    RG 24, volume 10339, DRB Press Releases
    RG 24, volumes 11996–7, Defence Research Board Minutes, volumes 1–7,
        1946–1954
National Archive, United Kingdom, Medical Research Council Records
    (Kew, U.K.)
    FD 1/5892, Medical Research Council Blood Supply Depots, 1940
    FD 1/6384, Medical Research Council Body Protection Committee Minutes
        1940– 1941
    FD 1/6391, Research on Steel Helmets and Body Armour: Armoured Fighting
        Vehicles Sub Committee

FD 1/6392, Medical Research Committee and Medical Research Council, files, Working Conditions in Armoured Vehicles, 1940–1941

FD 1/6393, Medical Research Committee, Working Conditions in Armoured Vehicles, 1941–1942

FD 1/6394, Medical Research Committee and Medical Research Council, Working Conditions in Armoured Vehicles, 1942–1943

FD 1/6398, Medical Research Committee and Medical Research Council, Working Conditions in Armoured Vehicles, 1944–1952

FD 1/6424, Clothing for Tank Crews, 1941–1942

FD 1/7507, Sub Committee on Armoured Fighting Vehicles: Minutes

National Archive, United Kingdom, Ministry of Defence Records (Kew, U.K.)

DEFE 9/21, Visits of Vice Admiral Sir M. Denny and Sir Henry Tizard to USA and Canada, 1951

DEFE 9/36, Commonwealth Defence: Canada

DEFE 10/10, Commonwealth Advisory Committee on Defence Science: Minutes of Meetings and Memoranda, 1947–1949

DEFE 10/11, Commonwealth Advisory Committee on Defence Science: Minutes of Meetings and Memoranda, 1950

National Archive, United Kingdom, War Office Records (Kew, U.K.)

WO 188/229, Records of the Ordnance Office and Its Successors at the War Office, Army Operational Research Group, memo no. 201, Progress of Tank Design

WO 222/73, Medical Historians' Papers: First and Second World Wars, 1942

WO 233/22, War Office: Directorate of Air: Papers, Army Operational Research Group: Minutes of Control Meetings

WO 291/22, Ministry of Supply and War Office: Military Operational Research Unit, Successors and Related Bodies: Reports and Papers

WO 291/595, Ministry of Supply and War Office: Military Operational Research Unit, Successors and Related Bodies: Reports and Papers, Army Operational Research Group: Memoranda, 1944

Ontario Public Archive (Toronto)

RD 1–116 [b268616], Ontario Air Service

Interviews

Interviews conducted by David Grenville*
R. Adie, Cambridge, UK, 30 November 1985

---

*Permission required. Access granted by David Grenville.

Herb Bailie, Toronto, Ontario, 6 September 1986
Helen Chute (Reid), Loretto, Ontario, 12 September 1985
Laurie Chute, Loretto, Ontario, 12 September 1985
D.M. Coolican, Ottawa, Ontario, 17 April 1986
C. Crawley, Cambridge, U.K., 29 November 1985, 30 November 1985
W.S. Feldberg, London, U.K., 18 November 1985
A.M. Fordyce, Ottawa, 20 April 1986, part 1; 20 April 1986, part 2
Barbara Griffin, Toronto, Ontario, 26 February 1986
R. Haist, Toronto, Ontario, 25 November 1986
G.H. Hattersley-Smith, London, U.K., 21 November 1985
G.D. Kaye, Ottawa, Ontario, 18 September 1985
P. Lapp, Toronto, Ontario, 7 May 1986
A. Ellis Lewis, London, U.K., 30 November 1985
Lloyd Lillico, Toronto, Ontario, 28 November 1986
George Lindsey, Ottawa, Ontario, 23 August 1985, 17 September 1985,
    3 August 1986
Maggie Mollison, Whitchurch, U.K., 16 October 1985
Patrick Mollison, London, U.K., 19 November 1985
Archie Pennie, Ottawa, Ontario, 21 April 1986
Eliot Rodger, Ottawa, Ontario, 23 April 1986
G.R. Rowley, Ottawa, 19 April 1986
H.A. Sargeaunt, Sway, U.K., 4 December 1985
Ronnie Shephard, Shrivenham, U.K., 3 December 1985
Omond McKillop Solandt, Bolton, Ontario, 8–10 September 1985,
    13 November 1985, 24–8 February 1986, 1 March 1986, 6–10 May 1986,
    9–11 August 1986, 17 September 1986, 27–9 November 1986, 19 April 1987,
    14 March 1990
E.A. Treadwell, Combe Martin, UK, 8 December 1985
Interviews conducted by Jason S. Ridler (transcripts available from Jason S. Ridler)
Brian Austin, Leeds, UK, 5 May 2006
David Grenville interview Sutton, 15 December 2007.
Paul Hellyer, Toronto, 26 April 2007
Cecil Law, Kingston, 21 September 2004, 27 May 2007

Secondary Sources

Official Histories and Government Reports

Board, Suzanne. *A Brief History of the Defence Research Establishment Ottawa, 1941–2001*. Ottawa: Defence Research Establishment Ottawa, 2002.

Dunn, C.L. *Official Medical History of the War of 1939–1945: The Emergency Medical Services*. Vol. 1, *England and Wales*. London: Her Majesty's Stationery Office, 1952.

The Effects of Atomic Bombs at Hiroshima and Nagasaki: Report of the British Mission to Japan. London: His Majesty's Stationery Office, 1946.

Green, F.H.K., and Gordon Covell. *Official Medical History of the War of 1939–1945: Medical Research*. London: Her Majesty's Stationery Office, 1953.

Holness, D. *The Office of the Counsellor Defence Research and Development: London (CDRD[L]): A Brief History 1940–1993*. Ottawa: Directorate of International Research and Development, CDAD/DND, 1994.

Kennedy, J. de N. *History of the Department of Munitions and Supply: Canada in the Second World War*. Vol. 1, *Production Branches and Crown Companies*. Ottawa: King's Printer and Controller of Stationery, 1950.

– Volume 2, *Controls, Services and Finance Branches, and Units Associated with the Department*. Ottawa: King's Printer and Controller of Stationery, 1950.

Mayne, J. W. "History of Canadian Army Operational Research Establishment." In *Defence Research and Analysis Establishment Report 15*, volumes 1 and 2 (April and June 1970). Ottawa: Department of National Defence, Canada.

– "Operational Research in the Canadian Armed Forces during the Second World War." Volumes 1 and 2. *Operational Research and Analysis Establishment*, report no. R68 (June 1978). Ottawa: Department of National Defence, Canada.

Medical Research Council. *Medical Research in War: Report of the Medical Research Council for the Years 1939–1945*. London: Her Majesty's Stationery Office, 1947.

Postan, M.M., D. Hay, and J.D. Scott. *Design and Development of Weapons: Studies in Government and Industrial Organization*. London: Her Majesty's Stationery Office and Longmans, Green, 1964.

Proceedings of the XVth International Physiological Congress, Leningrad—Moscow August 9th to 16th 1935. Sechenov Journal of Physiology of the USSR. Moscow: State Biological and Medical Press, 1938.

Shepard, Ronald W. *Readings on Early Military Operational Research (with Particular Reference to Army OR)*. Department of Management Sciences Working Paper OR/WP/8. Shrivenham: Royal Military College of Science, Operational Research Branch, March 1984.

Stacey, C.P. *Arms, Men and Government: The War Policies of Canada, 1939–1945*. Ottawa: Queen's Printer, 1970.

The United States Strategic Bombing Survey Summary Report (European War). 30 September 1945.

Wood, H.F. *Strange Battleground: The Official History of the Canadian Army in Korea*. Ottawa: Queen's Printer, 1966.

Unpublished Papers, Dissertations, and Theses

Godefroy, Andrew. "Defence and Discovery: Science, National Security, and the Origins of the Canadian Rocket and Space Program." PhD diss., Royal Military College of Canada, 2004.

Hancock, Jay. "Determined Victor: Canada's Role in the Prosecution of Class 'A' Japanese War Criminals." MA thesis, Royal Military College of Canada War Studies Program, Kingston, ON, 2002.

Hennessy, Michael A. "The Rise and Fall of a Canadian Maritime Policy, 1939–1965: A Study of Industry, Navalism and the State." PhD diss., University of New Brunswick, 1995.

Maloney, Sean. "Canadian Shield: Canada's National Security Strategy and Nuclear Weapons, 1951–1971." PhD diss., Temple University, 1998.

Ridler, Jason S. "Emergence: Towards a Historiography of Defence Research in Canada during the Second World War." *International Journal of Canadian Studies*, forthcoming.

– "From Deeds to Words: An Initial Historiography of Operational Research in Great Britain during the Second World War." Unpublished.

– "The Road to Ryerson: Reverend Donald McKillop's Solandt's Journey to Canadian Publishing, 1889–1936." Unpublished.

– "The Unintended Road: Omond Solandt's Journey to Military Affairs, 1927–1940." Unpublished.

Russell, Steven Paul Isinger. "The AVRO Canada CF-105 Arrow Programme: Decisions and Detriments." MA thesis, University of Saskatchewan, September 1997.

Solandt, O.M. "The Duration of the Recovery Period, and the Respiratory Quotient of the Excess Metabolism, following Strenuous Muscular Exercise." MA thesis, University of Toronto, 1932.

Published Books and Articles

Acheson, Dean. *Present at the Creation: My Years in the State Department*. New York: W.W. Norton, 1969.

Agar, John, and Brian Balmer. "British Scientists and the Cold War: The Defence Research Policy Committee and Information Networks, 1947–1963." *Historical Studies in the Physical and Biological Sciences* 28, no. 2 (1998): 209–52.

Aldrich, Richard. "British Intelligence and the Anglo-American 'Special Relationship' during the Cold War." *Review of International Studies* 24 (1998): 331–51.

Andrew, Christopher. "Intelligence and International Relations in the Early Cold War." *Review of International Studies* 24 (1998): 321–30.

Andrew, Christopher, and Vasili Mitrokhin. *The Sword and Shield: The Mitrokhin Archive and the Secret History of the KGB*. New York: Basic Books, 1999.

Aronson, Lawrence. "Planning Canada's Economic Mobilization for War: The Origins and Operation of the Industrial Defence Board, 1945–1951." *American Journal of Canadian Studies* 15, no. 1 (1985): 38–58.

Austin, Brian. *Schonland: Soldier Scientist*. London: London Institute of Physics, 2003.

Avery, Donald. "Allied Scientific Co-operation and Soviet Espionage in Canada, 1941–1945." *Intelligence and National Security* 8, no. 3 (July 1993): 100–28.

– "Atomic Scientific Co-operation and Rivalry among Allies: The Anglo-Canadian Montreal and the Manhattan Project, 1943–1943." *War in History* 2, no. 3 (1995): 274–305.

– *The Science of War: Canadian Scientists and Allied Military Technology during the Second World War*. Toronto: University of Toronto Press, 1998.

Balmer, Brian. *Britain and Biological Warfare, Expert Advice and Science Policy, 1935–65*. London: Palgrave, 2001.

Baxter, James Phinney, III. *Scientist against Time*. Boston: Little, Brown, 1952.

Bennet, M.R. "Obituary: Sir Bernard Katz (1911–2003)." *Journal of Neurocytology* 32 (2003): 431–6.

Bercuson, David. *True Patriot: The Life of Brooke Claxton, 1898–1960*. Toronto: University of Toronto Press, 1993.

Bernal, J.D. *The Social Function of Science*. London, 1939.

Bernstein, Jeremy. *Hitler's Uranium Club: The Secret Recordings at Farm Hall*. London: Springer, 2001.

Bertrand, Guy J.C. "Canadian Military Research and Development: Success—Failings—Prospects." *Canadian Defence Quartlery* 1, no. 1 (Summer 1971): 19–25.

Best, Charles, and Norman Taylor. *The Human Body and Its Functions*. New York: Henry Holt, 1932.

Best, H.M. *Margaret and Charley: The Personal Story of Dr Charles Best, Co-Discoverer of Insulin*. Toronto: Dundurn Group, 2003.

Bijker, Wiebe, Thomas Hughes, and Trevor Pinch, eds. *The Social Construction of Technological Systems*. Boston: MIT Press, 1989.

Black, J.L. "Kanada-Votchina Amerikanskogo Imperializmo: Canada and the Canadian Communists in the Soviet 'Coming War' Paradigm, 1946—1951." Canadian Department of Foreign Affairs, http://www.dfait-maeci.gc.ca/hist/coldwar_section06-en.asp.

Blackett, P.M.S. "Scientists at the Operational Level" (1941), reprinted in 1948 in "Operational Research," *Advancement of Science* 17 (1948), collected in *Studies of War: Nuclear and Conventional*. Edinburgh: Oliver & Boyd, 1962.

– *Studies in War: Nuclear and Conventional.* Edinburgh: Oliver and Boyd, 1962.

Bland, Douglas. *The Administration of Defence Policy in Canada.* Kingston: Ronald P. Frye, 1987.

Bliss, Michael. *Banting: A Biography.* Toronto: McClelland and Stewart, 1984.

– *William Osler: A Life in Medicine.* Toronto: University of Toronto Press, 1999.

Bothwell, Robert. *Eldorado: Canada's National Uranium Company.* Toronto: University of Toronto Press, 1984.

– *Nucleus: The History of the Atomic Energy of Canada Limited.* Toronto: University of Toronto Press, 1988.

Bothwell, Robert, and William Kilbourne. *C.D. Howe: A Biography.* Toronto: McClelland and Stewart, 1979.

Botti, Timothy J. *The Long Wait: The Forging of the Anglo-American Nuclear Alliance, 1945–1958.* New York: Greenwood, 1987.

Bower, Tom. *The Paperclip Conspiracy: The Battle for the Spoils and Secrets of Nazi Germany.* London: Paladin Grafton Books, 1987.

Bronowski, Jacob. *Science and Human Values.* London: Hutchinson of London, 1961.

Bryden, John. *Deadly Allies.* Toronto: McClelland and Stewart, 1989.

Bud, Robert, and Philip Gummett, eds. *Cold War, Hot Science: Applied Research in Britian's Defence Laboratories 1945–1990.* London: Science Museum, 1999.

Buderi, Robert. *The Invention That Changed the World: How a Small Group of Radar Pioneers Won the Second World War and Launched a Technological Revolution.* New York: Touchstone, 1998.

Bush, Vannevar. *Pieces of the Action.* New York: Morrow, 1970.

Bywaters, E.G.L., and D. Beall. "Crush Injuries with Impairment of Renal Function." *British Medical Journal* 1 (1941): 427–32.

Cain, Frank. "Missiles and Mistrust: U.S. Intelligence Response to British and Australian Missile Research." *Intelligence and National Security* 3 no. 4 (October 1988): 5–22.

Campbell, Sandra. "Nationalism, Morality, and Gender: Lorne Pierce and the Canadian Literary Cannon, 1920–1960." *Papers of the Bibliographical Society of Canada* 32, no. 2 (Fall 1994): 135–60.

Carter, G.B. *Chemical and Biological Defence at Porton Down, 1916–2000.* London: Stationery Office, 2000.

Carter, G. B., and Graham Pearson. "North Atlantic Chemical and Biological Research Collaboration." *Journal of Strategic Studies* 19, no. 1 (March 1996): 74–103.

Cassidy, David. "Biological Warfare and Biological Defence in the United Kingdom, 1940–1979." *Royal United Service Institute* 137, no. 6 (December 1992): 67–74.

– "Controlling German Science, II: Bizonal Occupation and the Struggle over West German Science Policy, 1946–1949." *Historical Studies in the Physical and Biological Sciences* 26, no. 2 (1996): 197–240.

– "German Scientists and the Nazi Atomic Bomb," *Dimensions* 10, no. 2 (1996): 15–22.

Clark, Ronald W. "Science and Technology, 1919–1945." In *A Guide to the Sources of British Military History*, edited by Robert Higham, 542–65. Berkeley: University of California Press, 1971.

– *Tizard*. London: Methuen, 1965.

Collins, Anne. *In the Sleep Room: The Story of CIA Brainwashing Experiments in Canada*. Toronto: Key Porter Books, 1988.

Conant, James B. *Harvard to Hiroshima and the Making of the Nuclear Age*. Stanford: Stanford University Press, 1993.

Copp, Terry. "Counter-Mortar Operations in 21 Army Group." *Canadian Military History* 3, no. 2 (1994): 45–52.

– , ed. *Montgomery's Scientists: Operational Research in Northwest Europe, the Work of No. 2 Operational Research Section with 21st Army Group, June 1944–July 1945*. Waterloo, ON: Laurier Centre for Military Strategic and Disarmament Studies, 2000.

– "Operational Research in 21 Army Group." *Canadian Military History* 3, no. 1 (1994): 71–84.

Cooke, Owen. *The Canadian Military Experience 1867–1995: A Bibliography*. 3rd ed. Ottawa: Directorate of History and Heritage, Department of National Defence, 1997.

Crook, Paul. "Science and War: Radical Scientists and the Tizard-Cherwell Area Bombing Debate in Britain." *War & Society* 12, no. 2 (October 1997): 69–101.

Crowther, J.G., and R. Whiddington. *Science at War*. London: His Majesty's Stationery Office, 1947.

Cummings, Bruce. *The Origins of the Korean War*. Vol. 1, *Liberation and the Emergence of the Separate Regimes, 1945–1947*. Princeton: Princeton University Press, 1981.

– *The Origins of the Korean War*. Vol. 2, *Roaring of the Cataract, 1947–1950*. Princeton: Princeton University Press, 1990.

Dale, Henry. "Science in War and Peace." *Nature* 160, no. 4061 (30 August 1947): 281.

Deibel, Terry L., and John Lewis Gaddis. *Containment: Concept and Policy in Two Volumes*. Vol. 2. Washington: National Defense University Press, 1986.

Dockrill, Michael. *British Defence since 1945*. Oxford: Basil Blackwell, 1988.

– *The Cold War, 1943–1963*. Atlantic Highlands: Humanities, 1988.

Dockrill, Saki. *Britain's Policy for West German Rearmament, 1950–1955.*
   Cambridge: Cambridge University Press, 1991.
Douglas, W.A.B. *The Creation of a National Air Force: The Official History of the
   Canadian Air Force.* Vol. 2. Toronto: University of Toronto Press / Department
   of National Defence / Canadian Government Publishing. Centre, Supply
   and Services Canada, 1986.
– *No Higher Purpose: The Operational History of the Royal Canadian Navy in the
   Second World War, 1939–1945.* St Catharines, ON: Vanwell Publishing, 2002.
Drury, A., C. Luttwak-Mann, and O.M. Solandt. "The Inactivation of Adenosine
   by Blood with Especial Reference to Cat's Blood." *Quarterly Journal of
   Experimental Physiology* 27 (1938): 215–36.
Duffin, Jacalyn. *History of Medicine.* Toronto: University of Toronto Press, 2004.
Eayrs, James. *In Defence of Canada.* Vol. 3, *Peacemaking and Deterrence.* Toronto:
   University of Toronto Press, 1973.
– *In Defence of Canada.* Vol. 4, *Growing Up Allied.* Toronto: University of Toronto
   Press, 1980.
Edgerton, David. *The Warfare State: Britain, 1920–1970.* Cambridge: Cambridge
   University Press, 2006.
Eggleston, Wilfrid. *Canada's Nuclear Story.* Toronto: Clarke, Irwin, 1965.
– *Scientists at War.* London: Oxford University Press, 1950.
Etzold, Thomas H., and John Lewis Gaddis. *Containment: Documents on American
   Policy and Strategy, 1945–1950.* New York: Columbia University Press, 1978.
Feldberg, W., and O.M. Solandt, "The Effects of Drugs, Sugars and Allied
   Substances on the Isolated Small Intestine of the Rabbit." *Journal of
   Physiology* 101, no. 2 (1942): 137–17.
– "The Stimulating Effect of Glucose and Pyruvate on the Rabbit's Guts."
   Proceedings of the Physiological Society (4 May 1940): 97–100.
Finan, James, and W. Hurley. "McNaughton and Canadian Operational
   Research at Vimy." *Journal of Operation Research* 48 (1997): 10–14.
Fordyce, A.M. "How It All Started: The Goforth Paper." *Canadian Defence
   Quarterly* 1, no. 4 (Spring 1972): 15–16.
Franklin, K.J. "A Short History of the International Congresses of Physiologists."
   Annals of Science 3, no. 3 (15 July 1938): 241–334.
Fraser, D.A. "Present at the Creation: Recollections of the Very First Anti-
   Submarine Operations Using Leigh-Light." *Canadian Defence Quarterly* 9,
   no. 4 (Spring 1980): 43–6.
Frost, Stanley Brice. *McGill University: For the Advancement of Learning.*
   Montreal and Kingston: McGill-Queen's University Press, 1984.
Gaas, Saul, and Arjang A. Assad. *An Annotated Timeline of Operational Research:
   An Informal History.* Boston: Kluwer Academic, 2005.

Gaddis, John Lewis. "How Relevant Was U.S. Strategy in Winning the Cold War?" U.S. Army War College Strategic Studies Institute, 13 February 1992.
– *The Landscape of History: Historians Map the Past.* Oxford: Oxford University Press, 2002.
– *The Long Peace: Inquiries into the History of the Cold War.* New York: Oxford University Press, 1987.
– *Strategies of Containment: A Critical Appraisal of Postwar American National Security Policy.* Oxford: Oxford University Press, 1982.
– *Surprise, Security, and the American Experience.* Cambridge, MA: Harvard University Press, 2004.
– *The United States and the End of the Cold War: Implications, Reconsiderations, Provocations.* New York. Oxford University Press, 1992.
– *The United States and the Origins of the Cold War.* New York: Columbia University Press, 1972.
– *We Know Now: Rethinking the Cold War.* Oxford: Oxford University Press, 1997.
Gaddis, John Lewis, and Ernest May, eds. *Cold War Statesmen Confront the Bomb: Nuclear Diplomacy since 1945.* Oxford: Oxford University Press, 1999.
Geade, Robert L., and Jarold M. Merklinger. *Seas, Ships, and Sensors: An Informal History of the Defence Research Establishment Atlantic, 1968–1995.* Dartmouth: Defence Research Establishment Atlantic, 2002.
Gelly, Alain and H.P. Tardif. *Defence Research Establishment Valcartier, 1945–1995: 50 Years of History and Scientific Progress.* Ottawa: Minister of Public Works, Canada, 1995.
Gentile, Gian P. "Shaping the Past Battlefield 'For the Future': The United States Strategic Bombing Survey's Evaluation of the American Air War against Japan." *Journal of Military History* 64 (October 2000): 1085
Gilbert, Martin. *In Search of Churchill.* Toronto: Harper Collins, 1994.
Gimble, John. "Project Paperclip: German Scientists, American Policy and the Cold War." *Diplomatic History* 14, no. 3 (1990): 343–65.
– *Science, Technology and Reparations: Exploitation and Plunder in Postwar Germany.* Stanford: Stanford University Press, 1990.
Gold, Lorne. *The Canadian Habbakuk Project: A Project of the National Research Council of Canada.* Cambridge, UK: International Glaciological Society, 1993.
Gordon, Charles. *Postscript to Adventure: The Autobiography of Ralph Connors.* New York: Farrar Rinehart, 1938.
Goodpseed, D.J. *A History of the Defence Research Board of Canada.* Ottawa: Queen's Printer and Controller of Stationery, 1958.
Goudsmit, Samuel A. *Alsos.* Los Angeles: Tomash, 1986.
Gowing, Margaret. *Britain and Atomic Energy, 1939–1945.* London: MacMillan, 1965.

– *Independence and Deterrence: Britain and Atomic Energy, 1945–1952*. Vol. 1, *Policy Making*. London: MacMillan, 1974.

Granatstein, J.L. *Canada 1957–1967: The Years of Uncertainty and Innovation*. Toronto: McClelland and Stewart, 1986.

– *The Generals*. Calgary: University of Calgary Press, 2005.

– *The Ottawa Men: The Civil Service Mandarins, 1935–1957*. Toronto: Oxford University Press, 1982.

– *Who Killed Canadian History?* Toronto: Harper Collins, 1998.

Granatstein, J.L., and David Bercuson. "Robert Leckie." In *Dictionary of Canadian Military History*, 113. Toronto: Oxford Universty Press, 1992.

Granatstein, J.L., and Peter Neary, eds. *The Good Fight: Canadians and World War II*. Toronto: Copp Clark, 1995.

Gridgeman, N.T. *Biological Science at the NRC: The Early Years to 1952*. Waterloo, ON: Wilfrid Laurier University Press, 1979.

Grodzinski, John. "'Kangaroos at War': The History of the 1st Canadian Armoured Personnel Carrier Regiment." *Canadian Military History* 4, no. 2 (1995): 43–50.

Guillemin, Jeanne. "Life or Death?: The Life Sciences and Weapons Research." *Social Studies of Science* 32, no. 4 (2002): 624–30.

Gummett, Philip. *Scientists in Whitehall*. Manchester, UK: Manchester University Press, 1980.

Hacker, Barton. "Military Institutions, Weapons, and Social Change: Toward a New History of Military Technology." *Technology and Culture* 35 (1994): 768–834.

Hackmann, Willem D. "Sonar Research and Naval Warfare 1914–1954: A Case Study of Twentieth-Century Establishment Science." *Historical Studies in the Physical and Biological Sciences* 16, no. 1 (1986): 83–110.

Hammond, Paul Y. "NSC-68: Prologue to Rearmament." In *Strategy, Politics, and Defense Budgets*, edited by Warner R. Schilling, Paul Y. Hammond, and Glenn H. Snyder, 267–378 (New York: Columbia University Press, 1962)

Harris, Sheldon. *Factories of Death*. London: Routledge, 1994.

Hartcup, Guy, and T.E. Allbione. *Cockcroft and the Atom*. Bristol: Adam Hilger, 1984.

Haycock, Ronald. *Clio and Mars in Canada: Teaching Military History*. Athabasca, AB: Athabasca University Press, 1995.

Hennessy, Michael. "The Industrial Front: The Scale and Scope of Canadian Industrial Mobilization during the Second World War." In *Forging a Nation: Perspectives on the Canadian Military Experience*, edited by Bernd Horn, 135–54. St Catharines, ON: Vanwell, 2002.

Herken, Gregg. *Cardinal Choices: Presidential Science Advisors from the Atomic Bomb to SDI*. Stanford: Stanford University Press, 2000.

Hersey, John. *Hiroshima*. London: Vintage, 1946.

Hill, A.V. *The Ethical Dilemma of Science and Other Writings*. New York: Rockefeller Institute Press, 1960.

Hitsman, J.M. "David Thompson and Defence Research." *Canadian Historical Review* 40, no. 4 (December 1959): 315–18.

Hogan, Michael, ed. *America in the World: The Historiography of American Foreign Relations since 1941*. Cambridge: Cambridge University Press, 1995.

– *A Cross of Iron: Harry S. Truman and the Origins of the National Security State*. Cambridge: Cambridge University Press, 1998.

– , ed. *The End of the Cold War: Its Meanings and Implications*. Cambridge: Cambridge University Press, 1992.

– *The Marshall Plan: America, Britain, and the Reconstruction of Western Europe, 1947–1952*. New York: Cambridge University Press, 1987.

– *Paths to Power: The Historiography of American Foreign Relations to 1941*. Cambridge: Cambridge University Press, 2000.

Hogan, Michael, and Thomas G. Paterson, eds. *Explaining the History of American Foreign Relations*. Cambridge: Cambridge University Press, 1991.

Holloway, David. *Stalin and the Bomb: The Soviet Union and Atomic Energy 1939–1956*. New Haven, CT: Yale University Press, 1994.

Horn, Bernd, and Stephen Harris, eds. *Warrior Chiefs: Perspectives on Senior Canadian Military Leaders*, Toronto: Dundurn, 2001.

Howard, Michael. *The Lessons of History*. New Haven, CT: Yale University Press, 1991.

Hughes, Thomas. *American Genesis: A Century of Invention and Technological Enthusiasm, 1870–1970*. Chicago: University of Chicago Press, 1989.

– "Emerging Themes in the History of Technology." *Technology and Culture* 20 (1979): 697–711. Originally published in *Geschichte und Gesellschaft* 4 (1978): 258–71.

– *Rescuing Prometheus: Four Monumental Projects That Changed the Modern World*. New York: Vintage Books, 1998.

Hyland, H.H., W.J. Gardner, F.C. Heal, W.A. Ollie, and O.M. Solandt, "Acute Anterior Poliomyelitis (A Review of Sixty-Six Adult Cases Which Occurred in the 1937 Ontario Epidemic)," *Canadian Medical Association Journal* 39 (1938): 1–12.

Ignatieff, V.P. "The Capture of the Main Chemical Warfare Experimental Facilities of the German Army: A Personal Memoir." *Canadian Defence Quarterly* 12, no. 1 (Summer 1982): 43–5.

Irving, Larry, O.M. Solandt, Donald Y. Solandt, and K.C. Fisher. "Respiratory Characteristics of the Blood of the Seal." *Journal of Cellular and Comparative Physiology* 6, no. 3 (August 1935): 393–403.

– "The Respiratory Metabolism of the Seal and Its Adjustment to Diving." *Journal of Cellular and Comparative Physiology* 7, no. 1 (October 1935): 137–51.

Jarrell, R.A., and N.R. Ball. "The Study of the History of Canadian Science and Technology." In *Science, Technology, and Canadian History*, edited by R. A. Jarrell and N. R. Ball, 1–8. Waterloo, ON: Wilfrid Laurier University Press, 1978.

Jockel, Joseph. *No Boundaries Upstairs*. Vancouver: University of British Columbia Press, 1985.

Kaiser, Walter. "A Case Study in the Relationship of History of Technology and of General History: British Radar Technology and Neville Chamberlain's Appeasement Policy." *ICON: Journal of the International Committee for the History of Technology* 2 (1996): 29–52.

Kanigal, Robert. *The One Best Way: Frederick Winslow Taylor and the Enigma of Efficiency*. London: Penguin, 1997.

Kaplan, K. *The Short March: The Communist Takeover in Czechoslovakia, 1945–1948*. London: C. Hurt, 1987.

Keegan, John. *The Second World War*. London: Viking, 1989.

Kennan, George F. *At Century's Ending: Reflections, 1982–1995*. New York: W.W. Norton, 1996.

– *Memoirs: 1925–1950*. New York: Pantheon Books, 1967.

Kennan, George F, and John Lukacs. *George F. Kennan and the Origins of the Containment, 1944–1946*. Columbia: University of Missouri Press, 1997.

Kerwick, R.A. "Sir Alan Nigel Drury 3 November 1889–2 August 1980." *Biographical Memoirs of the Fellows of the Royal Society* 27 (1981): 173–98.

Killian, James. *Sputnik, Science and Eisenhower: A Memoir of the First Special Assistant to the President for Science and Technology*. Boston: MIT Press, 1978.

Kindsvatter, Andrew. *American Soldiers: Ground Combat in the World Wars, Korea, and Vietnam*. Lawrence: University of Kansas Press, 2003.

Kirby, Maurice. *Operational Research in War and Peace: The British Experience from the 1930s to 1970s*. London: Imperial War College Press.

Klienman, Daniel Lee. *Politics on the Endless Frontier: Post-War Research Policy in the United States*. Durham, NC: Duke University Press, 1995.

Kolko, Gabriel. *Politics of War: The World and United States Foreign Policy, 1943–1945*. New York: Random House, 1968.

Krige, John. "Scientists as Policymakers: British Physicists' 'Advice' to Their Government on Membership of CERN (1951–1952)." In *Solomon's House Revisited: The Organization and Institutionalization of Science*, edited by Tore Frangsmyr, 270–91. Stockholm: Science History Publications, 1990.

Kuhn, Thomas. *The Copernican Revolution*. Chicago: University of Chicago Press, 1957.

– *The Essential Tension*. Chicago: University of Chicago Press, 1977.

– *The Structure of Scientific Revolutions*. Chicago: University of Chicago Press, 1962.

Kuniholm, Bruce. *The Origins of the Cold War in the Near East: Great Power Conflict and Diplomacy in Iran, Turkey and Greece*. Princeton: Princeton University Press, 1980.

Lasby, Clarence G. *Project Paperclip: German Scientists and the Cold War*. New York: Atheneum, 1971.

Law, C.E. "Dr Omond McKillop Solandt: Operational Research Pioneer." *INFOR* 31, no. 2 (May 1993).

Lee, James, and David Bellamy. "Dr R.J. Sutherland: A Retrospective." *Canadian Defence Quarterly* (Summer 1987): 41–6.

LeFerber, Walter. America, Russia, and the Cold War, 1945–1984. New York: Knopf, 1997.

Leffler, Melvyn. *A Preponderance of Power: National Security, the Truman Administration, and the Cold War*. Stanford: Stanford University Press, 1992.

Li, Alison. "Expansion and Consolidation: The Associate Committee and the Division of Medical Research of the NRC, 1938–1959." *Scientia Canadensis* 15, no. 2 (1991): 89–103.

Lindsey, George. "Canada-US Defence Relations in the Cold War." In *Fifty Years of Canada–United States Defence Cooperation: The Road from Obdensburg*, edited by Joseph Jockel and Joel Sokolsky (Lewiston: Edwin Mellen, 1992).

– "In Memoriam: Omond Solandt." *Canadian Defence Quarterly* vol. 22, no. 6 (July 1993): 44.

– , ed. *No Day Long Enough: Canadian Science in the Second World War*. Toronto: Canadian Institute for Strategic Studies, 1997.

– "Some Personal Recollections of Army Operational Research in World War II." *Canadian Military History* 4, no. 2 (1995): 69–74.

Lindsey, G.R., and D.M. Grenville, eds. *Perspectives in Science and Technology: The Legacy of Omond Solandt*. Proceedings of symposium, Donald Gordon Centre, Queen's University, Kingston, Ontario, 8–10 May 1994. Kingston: Queen's Quarterly, 1994.

Longard, John R. *Knots, Volts and Decibels: An Informal History of the Naval Research Establishment, 1940–1967*. Dartmouth: Defence Research Establishment Atlantic, 1993.

Lund, Wilfred G.D. "Vice-Admiral Grant: Father of the Post-War Royal Canadian Navy." In *Warrior Chiefs: Perspectives on Senior Canadian Military Leaders*, edited by Bernd Horn and Steve Harris, 193–7 (Toronto: Dundurn, 2001).

The Mackenzie King Diaries. Libraries and Archives Canada. http://www.bac-lac.gc.ca/eng/discover/politics-government/prime-ministers/william-lyon-mackenzie-king/Pages/search.aspx.

Marcum, James A. "The Development of Heparin in Toronto." *Journal of the History of Medicine*" 52 (July 1997): 310–37.

Maloney, Sean. "The Canadian Army and Tactical Nuclear Warfare Doctrine." *Canadian Defence Quartlery* 23, no. 2 (December 1993): 23–9.

– "General Charles Foulkes: A Primer on How to Be CDS." In *Warrior Chiefs: Perspectives on Senior Canadian Military Leaders*, edited by Bernd Horn and Steve Harris, 219–36. Toronto: Dundurn, 2001.

– *Learning to Love the Bomb: Canada's Nuclear Weapons during the Cold War.* Dulles, VA: Potomac Books, 2007.

– "The Mobile Striking Force and Continental Defence, 1948–1955." *Canadian Military History* 2, no. 2 (1993): 75–89.

– *War without Battles: Canada's NATO Brigade in Germany, 1951–1993.* Toronto: McGraw-Hill-Ryerson, 1997.

Mastny, Vojtech. *The Cold War and Soviet Insecurity: The Stalin Years.* Oxford: Oxford University Press, 1996.

– *Russia's Road to the Cold War: Diplomacy, Warfare, and the Politics of Communism, 1941–1945.* New York: Columbia University Press, 1979.

May, Ernest, ed. *American Cold War Strategy: Interpreting NSC 68.* Boston: Bedford Books, 1993.

– *"Lessons" of the Past: The Use and Misuse of History in American Foreign Policy.* New York: Oxford University Press, 1973.

Mayers, David. "George F. Kennan and the Soviet Union, 1933–1938: Perceptions of a Young Diplomat." *International History Review* 5, no. 4 (November 1984), 525–49.

Mayon-White, R., and O.M. Solandt, "A Case of Limb Compression Ending Fatally in Uraemia." *British Medical Journal* 1 (22 March 1941), 434–5; reprinted in *Crush Injuries and Kidney Function: Reprint of Three Articles from the "British Medical Journal," March 22, 1941, with Reports of Cases Collected by the M.R.C. and Editorials*, 25–8. London: Office of the British Medical Association, 1941.

McIsaac, David. *Strategic Bombing in World War Two: The Story of the United States Strategic Bombing Survey.* New York: Garland, 1976.

McNeil, William. *The Pursuit of Power.* Chicago: University of Chicago Press, 1984.

Middleton, W.E. Knowles. *Mechanical Engineering at the National Research Council of Canada, 1929–1951.* Waterloo, ON: Wilfrid Laurier University Press, 1984.

– *Physics at the National Research Council of Canada, 1929–1952.* Waterloo, ON: Wilfrid Laurier University Press, 1979.

– *Radar Development in Canada: The Radio Branch of the National Research Council of Canada 1939–1946.* Waterloo, ON: Wilfrid Laurier University Press, 1981.

Miscamble, Wilson. "The Foreign Policy of the Truman Administration: A Post–Cold War Appraisal." *Presidential Studies Quarterly* 24, no. 3 (Summer 1994): 479–94.

Moore, Richard. "A JIGSAW Puzzle for Operational Researchers: British Global War Studies, 1954–1962." *Journal of Strategic Studies* 20 no. 2 (June 1997): 75–91.

Morse, Philip. *The Canadian Patriotic Fund: A Record of Its Activities from 1914 to 1918.* (Ottawa?): n.p., 1920.

Morton, Desmond. *A Military History of Canada.* 4th ed. Toronto: McClelland and Stewart, 1999.

Morton, N.W. "A Brief History of the Development of Canadian Military Operational Research." *Operations Research: The Journal of the Operational Research Society of America* 4, no. 2 (April 1956): 187–92.

Mott, Neville. *A Life in Science.* London: Taylor and Francis, 1986.

National Research Council. *The War History of the Division of Biology.* Ottawa: n.p., 1946.

National Research Council. *The War History of the Radio Branch.* Ottawa: n.p., 1948.

Needle, Allan. *Science, Cold War, and the American State: Lloyd V. Berkner and the Balance of Professional Ideals.* Washington: Harwood Academic Publishers, 2000.

Nesmith, Tom. "Commentary: Reflections on the State of the History of Canadian Science and Technology." *History of Science and Technology in Canada Bulletin* 3, no. 4 (August 1978): 5–6.

Neufeld, Michael. "Rolf Engel vs the German Army: A Nazi Career in Rocketry and Repression." *History and Technology* 13 (1996): 53–72.

Nitze, Paul, with Anne Smith and Stephen L. Rearden. *From Hiroshima to Glasnost: At the Centre of Decision.* New York: Grove Weidenbfeld, 1989.

"Operational Research in War and Peace." *Nature* 160, no. 4072 (15 November 1947): 660.

Overy, Richard. *Why the Allies Won.* London: Plimco, 1995.

Pearson, Lester B. *Mike: The Memoirs of the Rt Hon. Lester B. Pearson.* Vol. 1. Toronto: University of Toronto Press, 1972.

Pennie, Archie. "The Defence Research Board: A Quarter Century of Achievement." *Canadian Defence Quarterly* 1, no. 4 (Spring 1972): 1–16.

– "On What Happened to DRB." *Canadian Defence Quartlery* 4, no. 2 (Autumn 1974): 50–2.

Pennycuick, K. "Army Operational Research in the Far East, 1952–54." *Operational Research Quarterly* (1954): 121–9.

Pile, Frederic. *Ack-Ack: Britain's Defence against Air Attack during the Second World War.* London: George G. Harrap, 1949.

Phillipson, D.J.C. *Associate Committees of the NRCC, 1916–1974.* Ottawa: National Research Council of Canada, 1985.

– *International Scientific Liaison and the NRCC, 1916–1974.* Ottawa: National Research Council of Canada, 1985.

– "The National Research Council of Canada: Its Historiography, Its Chronology, Its Bibliography." *Scientia Canadensis* 15, no. 2 (1991): 177–200.

Pierce, Lorne, ed. *The Chronicle of a Century, 1829–1929: The Record of One Hundred Years of Progress in the Publishing Concerns of the Methodist, Presbyterian and Congregational Churches in Canada.* Toronto: Ryerson, 1929.

Popper, Karl. *The Logic of Scientific Discovery.* New York: Basic Books, 1959.

Reynolds, Wayne. "Atomic War, Empire Strategic Dispersal and the Origins of the Snowy Mountain Scheme." *War and Society* 14, no. 1 (May 1996): 121–44.

– "Defence Science Research, Higher Education and the Australian Quest for the Atomic Bomb, 1945–60." *History of Education* 26, no. 2 (1997): 225–42.

– "Planning the Defences of World War III: The Post-War British Empire and the Role of Australia and South Africa, 1943–1957." *Historia* 43, no. 1 (May 1998): 72–90.

– "Rethinking the Joint Project: Australia's Bid for Nuclear Weapons, 1945–1960." *Historical Journal* 41, no. 3 (1998): 853–73.

Rhodes, Richard. *Dark Sun: The Making of the Hydrogen Bomb.* New York: Touchstone, 1995.

– *The Making of the Atomic Bomb.* New York: Touchstone, 1988.

Richter, Andrew. *Avoiding Armageddon: Canadian Military Strategy and Nuclear Weapons, 1950–1963.* Vancouver: University of British Columbia Press, 2002.

Ridler, Jason S. "From Nagasaki to Toronto: Omond Solandt and the Defence Research Board's Early Vision of Atomic Warfare, 1945–1947." *Canadian Military History* 18, no. 2 (2009), 26–40.

Roland, Alex. *The Military-Industrial Complex.* City: American Historical Association and the Society for the History of Technologyd, 2001.

Romano, Terrie M. "The Associate Committees on Medical Research of the National Research Council and the Second World War." *Scientia Canadensis* 15, no. 2 (1991): 71–88.

– "Technology and War: A Bibliographic Essay." In *Military Enterprise and Technological Change: Perspectives on the American Experience,* edited by Meritt Roe Smith, 347–79. Cambridge, MA: MIT Press, 1985.

Rosser, F.T., and M.W. Thistle. *War Stories from Biology: A "Popular" Account of the Work of the Division of Applied Biology, of the National Research Council, during the Second World War.* Ottawa: n.p., 1955.

Rowe, A.P. *One Story of Radar.* Cambridge: Cambridge University Press, 1948.

Rutty, Christopher. "The Middle-Class Plague: Epidemic Polio and the Canadian State, 1936–1937." *Canadian Medical Bulletin* 13 (1996): 277–314.

Segal, Howard P. "Technology, History and Culture: An Appreciation of Melvin Kranzberg." *Virginia Quarterly Review* 74 (1998): 641–53.

Sheppard, Ronnie. *A Bibliography of (Mainly) Army Operational Research, 1939–1948*, Department of Management Sciences Working Paper OR/WP/21. Shrivenham: Royal Military College of Science.

– "The Influence of Solandt on the Development of Operational Research in Britain." In Law, Lindsey, and Grenville, *Perspectives in Science*, 30–47.

Shermer, Michael. *The Borderlands of Science: Where Sense Meets Nonesense.* Oxford: Oxford University Press, 2001.

Sibley, Katherine A.S. "Soviet Industrial Espionage against American Military Technology and U.S. Responses, 1930–1945." *Intelligence and National Security* 14, no. 2 (Summer 1999): 94–123.

Simpson, John. *The Independent Nuclear State: The Untied States, Britain, and the Military Atom.* London: MacMillan, 1986.

Smith, David. "Sir Owen Wansbrough-Jones." *Journal of Operation* 43, no. 2 (1983): 105–6.

Smith, George. "The Franks Flying Suit in Canadian Aviation Medical History, 1939–1945." *Canadian Military History* 8, no. 2 (1999): 35–41.

Snow, C.P. *Science and Government.* Cambridge: Harvard University Press, 1960.

– *The Two Cultures.* 1959; Cambridge: Cambridge University Press, 1996.

Solandt, O. M. "The Challenge of Science." In *Canada's Tomorrows*, edited by G.P. Gilmour, 300–2 (Toronto: MacMillan, 1954).

– "Civilian Benefits from Defence Research." *Proceedings of the Royal Canadian Institute*, series IIIA 14 (session 1948–9): 73–4.

– "The Defence Research Board's Untimely End: What It Means for Military Science." *Science Forum* 47 (October 1975): 19–21.

– "Defence Research in Canada." *Engineering Journal* (August 1951): 765–8.

– "Defence Research in Canada." *Precambrian* (November 1950): 8–11, 46.

– "The Emphasis on Forces-in-Being." *Bulletin of the Canadian Industrial Preparedness Association*, 14 November 1955, 2–3, 12.

– "The Engineer in Industry and the Armed Forces." *Canadian Chemistry and Processes Industries* (February 1950): n.p.

– "The Interpretation of Respiratory Quotients," *University of Toronto Medical Journal* 19, no. 6 (1932): 214–18.

– "Massive Collapse of Lung." *American Journal of Medical Science* 3, no. 7 (September 1937): 428–43.

– "The Nuclear Age." *Medical Service Journal Canada* 4, no. 10 (November 1958): 679–94.

– "Observation, Experiment, and Measurement in Operational Research." *Journal of Operational Research* 3 no. 1 (February 1955): 1–14.

- "Recent Advances in the Treatment of Cyanide Poisoning." *Canadian Public Health Journal* (December 1934): 592–8.
- "Research, Defence, and Industry." *Industry Canada,* July 1948, 129–32.
- "Science as It Relates to Defence." John Waddell Lecture, Royal Military College of Canada, Kingston, 1977.
- "Scientific Research in the Modern World." The Samuel Robertson Memorial Lecture, Prince of Wales College, Charlottetown, PEI, 28 March 1957.
- "The Sex Hormones of the Pituitary Gland." *Medical Journal,* University of Toronto (1936?).
- "Some Observations upon Sodium Alginate." *Quarterly Journal of Experimental Physiology and Cognate Medical Sciences* 31, no. 1 (1941): 25–31.
- "Technical Fields." In *Wanted … More Experts!* 15–18. Ottawa: Government of Canada, Department of Labour, 1958.
- "The Work of a London Emergency Blood Supply Depot." *Canadian Medical Association Journal* 44 (1941): 189–91.

Solandt, O.M., and C.R. Cowan. "The Duration of the Recovery Period Following Strenuous Muscular Exercise, Measured to a Base Line of Steady, Mild Exercise." *Journal of Physiology* 89, no. 4 (1937): 462–6.

Solandt, O.M., Alan Drury, and C. Lutwak-Mann. "The Inactivation of Adenosine by Blood, with Special Reference to Cat's Blood." *Quarterly Journal of Experimental Physiology* 27 (1938): 215–36.

Solandt, O.M., and G.C. Ferguson. "The Effects of Strenuous Exercise of Short Duration upon the Sugar Content of the Blood." *Transactions of the Royal Society of Canada,* section 5 (1932): 173–9.

Solandt, O.M., John F. Palmer, and Charles Herbert Best. "The Distribution of Chlorine." *Biochemical Journal* 29, no. 10 (1935): 2278–84.

Solandt, O.M., and Jesse Ridout. "The Duration of the Recovery Period Following Strenuous Muscular Exercise." *Proceedings of the Royal Society,* B, 113 (1933): 327–44.

Solandt, O.M., Donald Y. Solandt, and R.W. Gerard. "Methemoglobin and Methylene Blue as Cyanide Antagonists." *Proceedings of the Society for Experimental Biology and Medicine* 31 (1934): 539–41.

Solandt, O.M., D.Y. Solandt, and Larry Irving. "Nerve Impulses from Branchial Pressure Receptors in the Dogfish." *Journal of Physiology* 84, no. 2 (1935): 187–90.

Schachtman, Tom. *Laboratory Warriors: How Allied Science and Technology Tipped the Balance in World War II.* New York: Perennial, 2003.

Spinardi, Graham. "Aldermaston and British Nuclear Weapons Development: Testing the 'Zuckerman Thesis.'" *Social Studies of Science* 24, no. 4 (1997): 547–82.

Stacey, C.P. *Canada in the Age of Conflict.* Vol. 2. Toronto: MacMillan, 1981.

– *A Date with History: Memoirs of a Canadian Historian.* Ottawa: Deneau, 1983.

Staudenmaier, John. "Rationality, Agency, Contingency: Recent Trends in the History of Technology." *Reviews in American History* 30 (2002): 168–81.

– *Technology's Storytellers: Reweaving the Human Fabric.* Cambridge, MA, and London: Society for the History of Technology and the MIT Press, 1985.

– "What SHOT Hath Wrought and What SHOT Hath Not: Reflections on Twenty-Five Years of the History of Technology." *Technology and Culture* 25 (1984): 713–17.

Swettenham, John. *McNaughton.* 3 vols. Toronto: Ryerson, 1968.

Symonds, J.L. *A History of British Atomic Tests in Australia.* North Melbourne: Ruskin, 1985.

Tardiff, H.P. *DREV, 1945–1995: Fifty Years of Progress.* Ottawa: Department of National Defence, 1995.

Thistle, Mel. "British, the United States, and the Development of NATO Strategy, 1950–1964." *Journal of Strategic Studies* 19, no. 2 (June 1996): 260–81.

– , ed. *The Mackenzie-McNaughton Wartime Letters.* Toronto: University of Toronto Press, 1975.

Tizard, Henry. "Science and the Services." *Royal United Service Institute Journal* 91, no. 563 (August 1946): 346.

Twigge, Stephen, and Alan MacMillan. "Learning to Love the Bomb: The Command and Control of British Nuclear Forces, 1953–1964." *Journal of Strategic Studies* 22 no. 1 (March 1999): 29–53.

Twigge, Stephen, and Len Scott. *Planning Armageddon: Britain, the United States and the Command of Western Nuclear Forces 1945–1964.* Amsterdam: Harwood Publishers, 2000.

Vance, Jonathan. *Death So Noble: Memory, Meaning and the First World War* Vancouver: University of British Columbia Press, 1997.

– *High Flight: Aviation and the Canadian Imagination.* Toronto: Penguin Canada, 2002.

Watson, Brent Byron. *Far Eastern Tour: The Canadian Infantry in Korea, 1950–1953.* Montreal and Kingston: McGill-Queen's University Press, 2007.

Watson, Gordon. "Why the Bureaucrats Secretly Carved Up the DRB: It Worked Too Well." *Science Forum* 47 (October 1975): 22–5.

Watson-Watt, Robert. *Three Steps to Victory: A Personal Account by Radar's Greatest Pioneer.* London: Odhams, 1957.

Watt, D. Cameron. *What about the People? Abstraction and Reality in History and the Social Sciences.* Inaugural lecture, London School of Economics and Political Science, 1983.

Weisberg, Robert. *Creativity: Genius and Other Myths.* New York: W.H. Greeman, 1986.

Whitaker, Reginald, and Gary Marcuse. *Cold War Canada: The Making of a National Insecurity State, 1945–1957.* Toronto: University of Toronto Press, 1994.

White, Richard. *The Skule Story: The University of Toronto Faculty of Applied Science and Engineering.* Toronto: University of Toronto Press, 2002.

Wilford, Hugh. "The Information Research Department: Britain's Secret Cold War Weapon Revealed." *Review of International Studies* 24 (1998): 353–69.

Wilkie, Tom. *British Science and Politics since 1945.* Oxford: Blackwell, 1991.

Willis, Kirk. "War, Diplomacy, and Nuclear Strategy in Britain." *Journal of British Studies* 38, no. 1 (1999): 125–31.

Zacahary, G. Pascal. *Endless Frontier: Vannevar Bush, Engineer of the American Century.* Boston: MIT Press, 1999.

Zack, Manuel, Lawrence Martin, and Alvin A. Lee. *Harry Thode: Scientist and Builder at McMaster University.* Hamilton, ON: McMaster University Press, 2003.

Zimmerman, David. "A First Step in Comparative History: The Royal Australian and Canadian Navies and High Technology in the Second World War." *Scientia Canadensis* 17, nos 1 and 2 (1993): 255–67.

– *The Great Naval Battle of Ottawa.* Toronto: University of Toronto Press, 1989.

– "The Organization of Science for War: The Management of Canadian Radar Development, 1939–1945," *Scientia Canadensis* 10, no. 2 (Autumn/Winter 1986): 93–108.

– "Radar and Research Enterprises Limited: A Study of Wartime Industrial Failure." *Ontario History* 80, no. 2 (June 1988): 121–42.

– "The Royal Canadian Navy and the National Research Council, 1939–1945." *Canadian Historical Review* 69, no. 2 (1988): 203–21.

Zubok, Vladislav, and Constantine Pleshakov. *Inside the Kremlin's Cold War: From Stalin to Khrushchev.* Cambridge, MA: Harvard University Press, 1996.

Zuckerman, Solly. *From Apes to Warlords.* London: Harper and Row, 1978.

– *Scientists and War: The Impact of Science on Military and Civil Affairs.* London: Hamish-Hamilton, 1966.

Index

Printed in the USA
CPSIA information can be obtained
at www.ICGtesting.com
LVHW040554281024
794557LV00007B/130/J

9 781442 647473